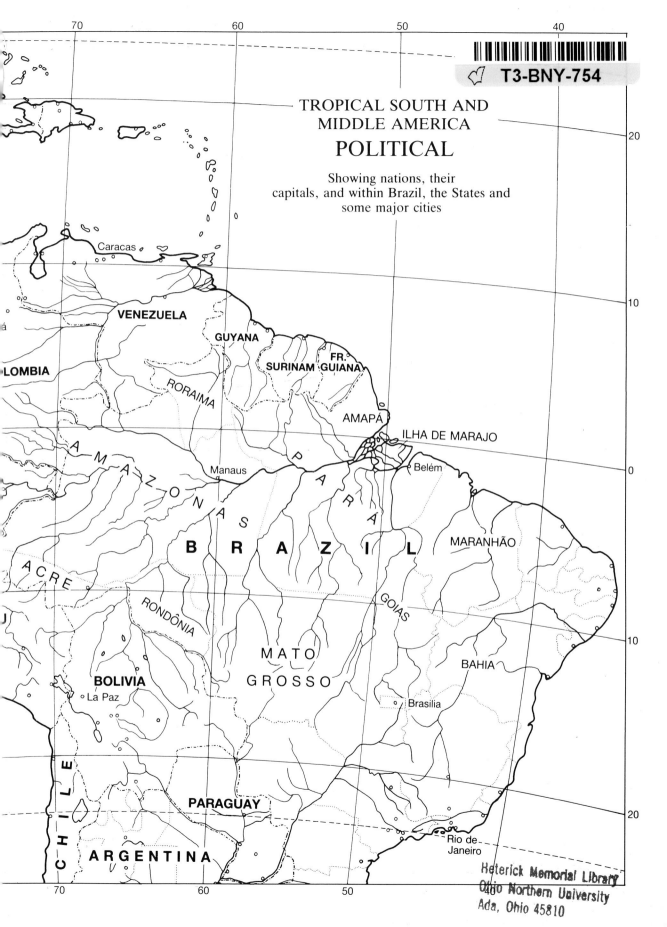

OXFORD MONOGRAPHS ON BIOGEOGRAPHY

Editors: W. GEORGE, A. HALLAM, AND T. C. WHITMORE

OXFORD MONOGRAPHS ON BIOGEOGRAPHY

Editors
W. George, Department of Zoology, University of Oxford.
A. Hallam, Department of Geological Sciences, University of Birmingham.
T. C. Whitmore, Department of Plant Sciences, University of Oxford.

In an area of rapid change, this series of Oxford monographs will reflect the impact on biogeographical studies of advanced techniques of data analysis. The subject is being revolutionized by radio-isotope dating and pollen analysis, plate tectonics and population models, biochemical genetics and fossil ecology, cladistics and karyology, and spatial classification analyses. For both specialist and non-specialist, the Oxford Monographs on Biogeography will provide dynamic syntheses of the new developments.

1. T. C. Whitmore (ed.): *Wallace's line and plate tectonics*
2. Christopher J. Humphries and Lynne R. Parenti: *Cladistic biogeography*
3. T. C. Whitmore and G. T. Prance (ed.): *Biogeography and Quaternary history in tropical America*

Biogeography and Quaternary History in Tropical America

Edited by

T. C. WHITMORE

University Senior Research Officer
Department of Plant Sciences, Oxford

and

G. T. PRANCE

Senior Vice-President for Science
New York Botanical Garden

CLARENDON PRESS · OXFORD
1987

Oxford University Press, Walton Street, Oxford OX2 6DP

Oxford New York Toronto
Delhi Bombay Calcutta Madras Karachi
Petaling Jaya Singapore Hong Kong Tokyo
Nairobi Dar es Salaam Cape Town
Melbourne Auckland
and associated companies in
Beirut Berlin Ibadan Nicosia

Oxford is a trade mark of Oxford University Press

Published in the United States
by Oxford University Press, New York

British Library Cataloguing in Publication Data
Biogeography and quaternary history in
tropical America. – (Oxford monographs in
biogeography)
1. Paleobiogeography – Latin America
2. Paleobiogeography – Tropics
3. Paleontology – Pleistocene
I. Whitmore, T. C.   II. Prance, Ghillean T.
574.98   QE721.2.P24
ISBN 0-19-854546-0

Library of Congress Cataloging-in-Publication Data
Biogeography and Quaternary history in tropical America.
(Oxford monographs on biogeography)
1. Biogeography – Latin America.   2. Paleobiogeography
– Latin America.   3. Paleoecology – Latin America.
4. Geology, Stratigraphic – Quaternary.   5. Geology –
Latin America.   I. Whitmore, T. C. (Timothy Charles)
II. Prance, Ghillean T., 1937- .   III. Series.
QH106.5.B54   1986   581.98   86-26844
ISBN 0-19-854546-0

Set by Hope Services, Abingdon, Oxon
Printed in Great Britain by
R. J. Acford, Chichester, Sussex

PREFACE

Since the early 1970s a revolutionary view of tropical biogeography (based on extensive studies in South America) has gained much attention. Geoscientific evidence, vegetation analysis, and animal distributions have all contributed to a theory of tropical evolution in palaeoecological refugia. This theory has been adopted enthusiastically and often uncritically by some and rejected *in toto* by others. In fact both the geoscientific and biological evidence are fragmentary and, in some cases, equivocal. In this book an attempt is made at a critical appraisal of all the evidence.

The geoscientific evidence, from geology, geomorphology, palaeoclimatology, and pedology is reviewed (Chapters 1 and 2). Vegetation is described and mapped, using the results of the recent aerial survey of the Brazilian part of the Amazon basin, and likely past changes in vegetation in times of drier climate are suggested (Chapter 2). A model of places likely to have remained clothed by humid tropical forest is then constructed.

The distribution patterns are analysed for the three best known groups, plants, birds, and butterflies, each by the leading specialist, (Chapters 3, 4, and 5). Much new information is incorporated in these accounts. A brief review is given of the published studies which cover the biogeography of many other animal groups.

Man has been in South America for about 14 000 years. The chief importance of the model of climatic and vegetation change for archaeology, anthropology, and also for biogeography is to provide an interpretative framework and to focus on what now needs study (Chapter 6). In addition, on a practical level the refuge model is being utilized for land-use planning, including conservation planning, in the humid neotropics.

This book is addressed to students seeking a perspective of these exciting new developments in neotropical biogeography, to ecologists, foresters and planners seeking the latest knowledge on Amazonian vegetation and to research workers looking for a critical appraisal of the lines of evidence which have led to the proposal of the palaeoecological refuge model in the neotropics.

Oxford,   T. C. W.
New York   G. T. P.

ACKNOWLEDGEMENTS

As always, a work so complex as this is the product of many contributions besides those of the editors and authors.

We thank the Director, Natural Climatic Center, Ashville, North Carolina, USA for permission to use the photograph on which Fig. 1.4 is based.

C. P. Burnham and A. van Wambeke kindly read a draft of the subchapter on soils.

The many persons and institutions who have helped provide the data for Chapter 4 are fully listed in Brown (1979). Special thanks are due to P. M. Sheppard FRS (who initially stimulated the butterfly studies), J. R. G. Turner, P. E. Vanzolini, W. W. Benson, R. F. D'Almeida, G. Lamas, M. D. Gifford (deceased), I. Lepsch, L. D. Miller, A. N. Ab'Sáber, L. E. Gilbert, O. H. H. Mielke, and M. G. Emsley for important contributions. Chapters four and seven were prepared while one of us (K.S.B.) was a visiting professor in the group of P. P. Feeny, Department of Entomology, Cornell University, Ithaca, New York, who is warmly thanked for facilities and supplies.

The distribution maps of birds (Chapter 5) are partly based on unpublished locality records obtained during study of the collections of the following museums, whose authorities are thanked: American Museum of Natural History (AMNH), Field Museum of Natural History, Chicago (FM), Carnegie Museum of Natural History, Pittsburgh (CM), Zoological Museum A. Koenig, Bonn. We are most grateful to Dr J. W. Fitzpatrick (Chicago) who gave much of his time preparing the excellent sketches of birds illustrating the distribution maps and to the following persons for critical comments on the text: Drs F. Vuilleumier, E. Eisenmann (deceased), M. A. Traylor, J. Fitzpatrick. In addition Drs H. Sick, D. Snow, M. A. Traylor, J. O'Neill, and J. Eley kindly supplied information on the distribution of certain species.

The editors would like to thank all those at the Oxford Forestry Institute, Oxford who assisted with the book as a whole, especially Janette Inglis who drew the base distribution map, Jenny Max who redrafted many figures, Jon Lovett who helped with indexing, and the three indefatigable typists Tina Hodgkinson, Coral Taylor, and Pam Taylor, aided by Katy Whitmore.

The whole book was read in draft by Dr W. George, Dr A. Sugden, Dr J. R. G. Turner, and Dr P. E. Vanzolini to whom we are all indebted for encouragement and constructive criticism.

The book was originally commissioned by the senior editor (T.C.W.) and he would like here to express his thanks to Dr Prance for adding his assistance at a time when it looked otherwise as though it would never be completed, and for compiling the index.

CONTENTS

CONTRIBUTORS

K. S. BROWN, Jr,
Instituto de Biologia,
Universidade Estadual de Campinas,
São Paulo,
Brazil

J. HAFFER,
Tommesweg 60,
4300 Essen-1,
West Germany

B. J. MEGGERS,
National Museum of Natural History,
Smithsonian Institution,
Washington, D.C,
USA

G. T. PRANCE,
New York Botanical Garden,
The Bronx,
New York,
USA

INTRODUCTION

The humid tropics, especially the parts clothed in tropical rain forest, are well known for their extreme richness in species of plants and animals and lowland tropical rain forests are amongst the world's most complex ecosystems. It has traditionally been assumed that this great richness and complexity has arisen because of the very favourable growing conditions and that these have remained the same through the past, untouched by the climatic vicissitudes associated with ice ages which have had a strong effect on all higher latitudes.

The notion of climatic stability in the humid tropical lowlands has, however, recently given way to the realization that there have been alternating phases of extensive perhumid climates and high sea level, as are found today, and other phases of lower rainfall, more extensive areas of seasonal climate and lower sea level. These alternating phases are correlated respectively with the inter-glacial and glacial periods of higher latitudes. In the equatorial region they were first discovered in high mountains where vegetation belts including the tree-line and ice caps on the highest peaks have been high and low as global climate has been warm and cool. Data for the tropical lowlands still remain scattered. For the neotropics a number of different geoscientific sources now clearly show that there have been repeated phases during the last two million years when dry climates have been much more extensive than at the present day.

During the late nineteen-sixties it was independently demonstrated for birds (Haffer 1967a, 1969) and for certain lizards (Vanzolini and Williams 1970) that there are in the lowlands of tropical South America centres rich in species or characters separated by regions with a poorer or more mixed fauna. These centres were interpreted to coincide with refugia to which the rain forest was restricted at past times of drier, more seasonal climate. This revolutionary hypothesis, striking at the foundations of the then generally accepted views on the humid tropical biome, rapidly attracted attention. Numerous biologists examined other groups, found heterogeneity in the ranges of component members and interpreted them on the same lines. By 1980 four international symposia had been held to discuss tropical refuge phenomena, mainly in the neotropical lowlands, and there were numerous other independent studies.

Our knowledge about species distributions in the neotropics remains very incomplete. The Amazonian hylaean* rain forest is huge. Early biological investigation was along the rivers and the coast and near the scattered main settlements. Over the past few decades Amazonia has yielded up a few more of its secrets but for many plants and animals information on ranges remains woefully scanty and patchy.

The 'refuge model', as it has become known has, in some cases for which the data base remains sparse, been espoused enthusiastically but uncritically. In this book we set out a critical examination of the biogeography of tropical South America, especially the humid lowlands. The geoscientific evidence for the hypothesis of isolated humid refugia during glacial maxima is examined. The three groups of organisms, plants, birds, and butterflies, for which the greatest body of data has been gathered are reviewed and evidence for their centres of species richness and endemism is compared with that from geoscientific sources.

Man has been in South America for at least 14 000 years. In the humid tropical lowlands his artefacts date only from about 5000 years BP. Subsidiary to the main theme an examination is made of the extent to which archaeological, ethnographic, and linguistic distributions may reflect past differences in climate and vegetation.

The geoscientific sources are the geology of Quaternary deposits, the geomorphology of land forms, palaeoclimatological reconstructions based on numerical models and certain features of the soil. There are a few pollen cores from lowland South America. These show rain forest alternating with savanna forest with much grass, i.e. alternating perhumid and seasonal climates, but are too few to help build a picture of how vegetation belts may have moved.

The present vegetation of the Brazilian Amazon is described. There is a range of vegetation types spanning the series of climates from perhumid and aseasonal to strongly seasonal. Soil can be an important factor. The continual, extensive cloud cover has

*Hylaea is a term introduced by the German explorer A. von Humboldt for the tropical rain forests of the Amazon lowlands plus the eastern and southern portion of the Guiana region south of the Orinoco.

prevented conventional aerial survey of the hylaean rain forest but during the 1970s aerial survey using radar (which penetrates cloud) was made in Brazil and some other Amazonian countries. A vegetation map is presented based on this new information. Previous opinions on the extent of the various forest formati~ns have been radically transformed, and some kinds of forest discovered whose existence had been unsuspected. Terra firme rain forest occupies only about half of Brazilian Amazonia. Within it the patches of rain forest on white sand are more extensive than had been believed. The rain forest is surrounded in part by a belt of transition forest which mostly occurs in humid seasonal climates. Both to the north and south cerrado and Amazonian savanna occur mainly in seasonal climates but also with strong edaphic control. Transition forest is interspersed with gallery forest along rivers and by enclaves of cerrado and Amazonian savanna. It is conjectured that in past times of drier climate transition forest became much more extensive with even greater interdigitation, yet this forest formation is poorly known and is today the one most rapidly disappearing at the hand of man.

The various different lines of geoscientific evidence enable a map to be drawn of the places most likely to have remained covered in part by humid tropical forest during periods of dry climate. The ranges of the groups of plants, butterflies, and birds studied in detail enable maps to be drawn showing centres of species endemism. These do not coincide with present day physiographic or climatic barriers. There is a remarkably close correlation between all these maps and between them and the geoscientific model. It is emphasized that maps are not very meaningful for organisms whose ranges are poorly known.

The rates of species formation, their dispersion and their response to climatic change are different from one group to another. Insect species, for example, are believed to be mainly of pre-Pleistocene and early Pleistocene and bird species of late Tertiary to late Pleistocene origin. There have been many alternating periods of relatively wet and dry climate since the Tertiary. Different cycles will all have affected ranges. The last, Wisconsin–Würm, glacial maximum at 18 000 BP will in most cases have been the main determinant of present-day distributions, but its effects are superimposed on ranges already influenced by one or several earlier glaciations. Thus the pattern perceived today is the summation of a whole series of past patterns.

It is concluded that, allowing for this complex of past changes, it is not by chance that numerous unrelated organisms show much the same centres of species endemism which in many cases overlap with the forest refugia deduced independently from geoscientific evidence.

Response to climatic fluctuation must not be considered the sole determinant of patterns perceived in the distributions of neotropical plants and animals. It is not the sole reason for diversification. On a longer timescale than the Pleistocene isolations in response to geotectonic events are also likely to have played a role. Fragmentation of species ranges due to climatic-vegetational fluctuations and tectonic events are both important in the field which has been named 'vicariance biogeography' by some workers. In addition considerations of population ecology show that centres of endemism within the complex rain forest ecosystem can arise by differentiation even in the absence of environmental pressure. These different models of diversification must all be considered.

The relevance of the refuge model rests upon the importance of geographical isolation for genetic differentiation and hence allopatric speciation, a view adhered to by most geneticists, though not of all (Endler 1977; White 1978). The latest critical review of the rather weak evidence for non-allopatric speciation is by Futuyma and Mayer (1980). In addition the refuge model indicates that conditions of geographical isolation leading to extensive differentiation at the subspecies and species levels probably continued in the tropics from the Tertiary into the Pleistocene. This contrasts with the traditional view that all or most tropical species and their adaptations and distribution patterns are old, i.e. Tertiary in age.

The refuge model helps identify places, topics and organisms profitable now to study. As with other general hypotheses this focusing is one of its salient aspects and it is an important contribution to archaeology and anthropology. For now new criteria are clearly identified against which to appraise human history in the region. On a practical level the major role of refuge theory has been to help lay a rational basis for land use planning and conservation. In the neotropics, as elsewhere around the equator, conversion of virgin tropical rain forest is currently being planned and achieved at a great rate. It is valuable that biologists are able to identify areas in tropical America, namely the endemic centres and their peripheries, most vital for the conservation and perpetuation of the genetic stocks of the region.

1

QUATERNARY HISTORY OF TROPICAL AMERICA

SUMMARY

Studies of the Quaternary sediments and of their pollen content as well as analyses of the surface land forms in South America have revealed dramatic and wide-ranging reversals in the climate of the tropical latitudes which probably caused vast vegetational fluctuations in the tropical lowlands during the past million years. Forests expanded during humid periods and survived in restricted 'refuge' areas during dry climatic periods; the opposite was true for savanna regions.

Geological, mineralogical, and micropalaeontological analyses of seabottom cores taken in the Caribbean Sea and off northeastern South America indicate climatic aridity during the glacial phases in adjacent onshore regions. Certain soil formations, dissected and gullied land surfaces as well as crusts (stone lines) found in portions of Amazonia and in other humid lowland regions of tropical South America, indicate the previous existence of sparse vegetation in these areas pointing to at least one former period of drier climate than today. Palynological investigations of Quaternary sediment cores from several study sites in tropical South America permit analysing the changing vegetation at these localities through time. In contrast to the highlands of the northern Andes (for which region a rich palynological documentation of the entire Quaternary period exists) pollen data from the neotropical lowlands are still insufficient as far as the Pleistocene is concerned; most of the available lowland sections are Holocene (i.e. post-Pleistocene) in age. The vegetational fluctuations recognized for the neotropical lowlands on the basis of palynological data of the last 10 000 years together with the results from a few Pleistocene pollen cores taken in southern Amazonia and in the coastal lowlands of the Guianas support the notion of vast changes in the distribution of forest and non-forest vegetation during the Pleistocene. Model studies of the July climate of 18 000 years BP are also in agreement with this interpretation.

To be sure, the neotropical forest and non-forest vegetation is old having originated during the Cretaceous and in Tertiary times. In the areas of their uninterrupted occurrence forests and savannas continuously offered ecologically stable and equable conditions to the species of tropical animals. It is the overall geographical distribution of these vegetation types that has varied dramatically in response to the changing world climates of the Pleistocene. The geoscientific studies reviewed in this chapter have completely destroyed the earlier notion of the 'stable tropics' which supposedly offered constant environmental conditions over large continental areas during the Quaternary period.

INTRODUCTION

For several decades many scientists assumed that the climatic fluctuations of the last two million years (Quaternary period) did not affect the tropical lowlands to any major degree. The rain forest and savanna flora and fauna of South America, Africa, and southeast Asia supposedly survived in an equable climate relatively unchanged since Tertiary times. Tropical organisms as well as their adaptations and distribution patterns were thought to be at least Tertiary in age and, in any case, much older than those of the North and South Temperate zone floras and faunas which have been deeply influenced by the vicissitudes of the Pleistocene climatic and vegetational changes.

Intensified fieldwork during the last twenty years proved the concept of environmental stability in the tropics during the Quaternary to be ill-founded and erroneous. Although the temperature fluctuations in the equatorial lowlands were rather small, alternating dry and humid climatic periods probably led to vast changes in the distribution of forest and savanna vegetation. Forests expanded during humid periods and survived in restricted 'refuge' areas during dry periods; the opposite was true for savanna regions.

The following brief introductory review of general

information on the surface relief and climate in tropical America may serve as an ecogeographical basis for the discussion of the Quaternary history. Some background knowledge of present conditions is necessary to appreciate the effects of the climatic fluctuations and associated vegetational changes which occurred during the recent geological past.

Relief

Most of tropical South America is below 500 m elevation. Flat or gently rolling lowlands stretch from the eastern base of the Andes across the entire continent to the Atlantic coast (Fig. 1.1). The vast Amazon valley lies at less than 200 m above sea-level and is more or less separated from the equally low-lying llanos plains of eastern Colombia and southern

Venezuela by the Guiana Shield which extends from the interior Guianas and southern Venezuela westward into southeastern Colombia. Elevations are above 200 m and reach 1000–3000 m in the central portion where spectacular table mountains (tepuis) rise above the partially forested lowlands. Rivers draining the Guiana Shield are, among many others, the Rio Negro and the Rio Branco (with their headwaters) which flow south into the Amazon River; the Rio Orinoco flows north and then east into the Atlantic Ocean. The Rio Negro and Rio Orinoco drainages are connected by the Caño Casiquiare, a fact which caused considerable discussion among early geographers.

The Brazilian tableland to the south of Amazonia has elevations from 600 to 1300 m reaching near 3000 m in several peaks of the mountain ranges near the Atlantic coast (Serra do Mar). Large southern

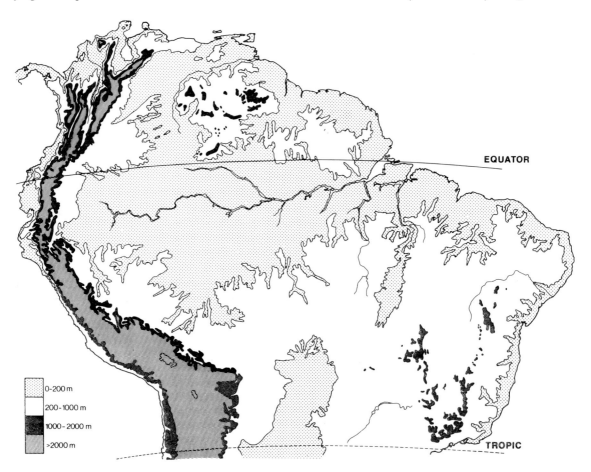

EQUATOR

TROPIC

0-200 m
200-1000 m
1000-2000 m
>2000 m

Fig. 1.1. General relief map of tropical South America. (After Simpson and Haffer 1978.)

tributaries of the Amazon River draining the area of the Brazilian Shield are the Rio Tocantins, Araguaia, Xingú, Tapajós, and Madeira.

The narrow mountain ranges of the Andes follow the entire western margin of the South American continent reaching maximum width and height in the central portion of Bolivia and Peru. The mountains decrease in elevation northward to about 2000 m in the narrow ranges of northern Peru where the Andes swing from a northwesterly to a north-northeasterly direction. The cordilleras again widen rapidly in northern Ecuador and southern Colombia where, in a northward direction, the Ecuadorian Andes split in two separate ranges (the Cordillera Occidental and Cordillera Central), and a third (the Cordillera Oriental) is added in the east. The Eastern Cordillera is a separate geological unit and only weakly linked to the main Andes in southeastern Colombia by a narrow mountain chain barely reaching 1500 m elevation. The Eastern Cordillera continues into Venezuela following the Caribbean coast to northern Trinidad although a moderate discontinuity, the Lara Depression, exists in northcentral Venezuela, and a large one in the northeast. A side-chain, the Sierra de Perijá, follows the Colombia/Venezuela border west of Lake Maracaibo to near the Caribbean coast. In this region of northeastern Colombia, the isolated massif of the Sierra Nevada de Santa Marta rises abruptly from sea level to 5775 m elevation.

A narrow band of trans-Andean lowlands follows the Pacific coast from Chile to northwestern Colombia and here merges broadly with the Caribbean lowlands of Colombia and western Venezuela north of the Andes.

The isthmus of Middle America is formed by a backbone of mountains and rather narrow lowlands along the Pacific and Caribbean coasts.

Climate

Three humid tropical climates (Af, Am, Aw) and one arid tropical climate (Bs) in the Köppen classification determine weather patterns in tropical America (Figs. 1.2–1.4). The characteristics of these climates describe the observed regional variations in a broad schematic manner. Because relief, soil, and vegetation are interdependent there are rapid changes of climate and vegetation where pockets of humid conditions or large rain-shadows exist.

The *Af climate* is the permanently humid tropical climate. The average monthly precipitation never drops below 60 mm. Western Amazonia from the Andes to the lower Rio Negro and also the Pacific lowlands of Colombia have a uniformly warm and continuously humid climate with two inconspicuous seasons of slightly reduced rainfall. However, variations from year to year are very noticeable. Total annual precipitation is mostly above 3000 mm and annual temperature variations are usually less than 3.5 °C. The Andean foothill regions in southeastern Colombia and in eastern Peru represent narrow climatic extensions of humid upper Amazonia receiving locally over 6000 mm of rain per year. In these regions the humid air is forced to rise, and it loses its moisture in the form of mist and frequent rain.

Another small area with an Af climate comprises the coastal lowlands of the Guianas.

The *Am climate* is transitional between the perhumid Af climate and the seasonal Aw climate. Although in regions with an Am climate the influence of the equatorial air mass is felt during most of the year, the drier northern and southern tropical air masses move in during part of the year (i.e. during the low position of the sun), causing a relatively dry season with at least one month with average rainfall below 60 mm. However, the dry season does not have a pronounced effect on the vegetation because there is sufficient moisture in the mornings and high precipitation during the rest of the year. This effect increases with decreasing total annual rainfall and/or the severity of the dry season itself. Annual precipitation under the Am climate is mostly 2000–3000 mm but may be less in areas where the dry season is not pronounced.

North of the equator, the northern trade winds blow from the northeast and, south of the equator, the southern trade winds blow from the southeast. Both trade wind belts move northward and southward with the position of the sun. These steady easterly winds pick up moisture from the Atlantic Ocean and transfer it onto the South American continent and Middle America. Rains fall where the movement of the tropical air masses is slowed, e.g. on the windward side of a barrier such as a hilly region or a mountain range. Therefore, the windward (eastern) slopes of the mountains from Mexico through northern South America to Brazil have a humid climate. In general, however, the trade winds have a stabilizing effect on the climate over level onshore areas and induce a regional dry season despite their high moisture content (e.g. Eidt 1969).

The area with an Am climate north of upper Amazonia is narrow due to a rapid transition from a humid to a dry climate in the llanos region of Colombia–Venezuela but is very wide south of the Amazon River where annual rainfall reaches a maximum of just over 2500 mm between the upper Rio

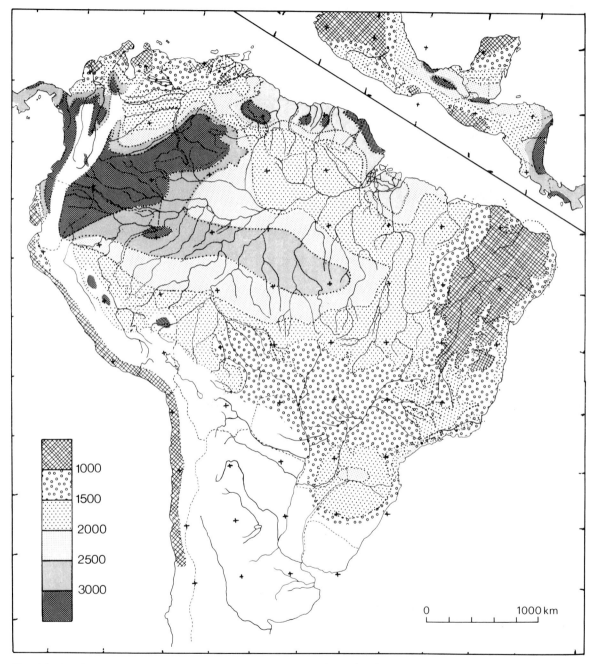

Fig. 1.2. Total annual rainfall (mm) in lowland tropical America. (Adapted from Ratisbona 1976, Reinke 1962, Trewartha 1961, Paes de Camargo *et al.* 1977, and Escoto 1964).

Madeira and the Rio Tapajós (Lábrea 2746 mm, Manicoré 2613 mm, Novo Aripuanã 2604 mm; Brown 1979; 3000 mm was shown for this region by Simpson and Haffer 1978, Fig. 3, in error). The Am climate also characterizes extensive portions of western Brazil south of the Amazon River and parts of eastern Peru.

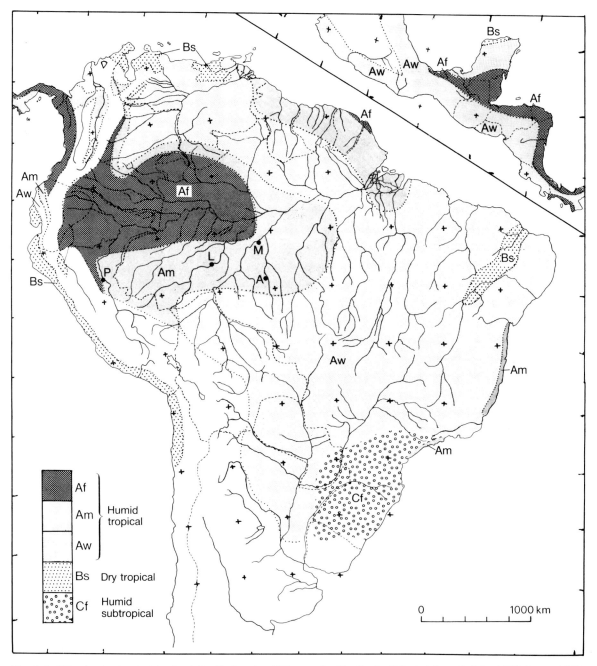

Fig. 1.3. Climates in tropical lowland South America (Köppen classification). (Adapted from Reinke 1962. Lauer 1952, and Trewartha 1954.) A humid tropical climate; B dry tropical climate; C humid subtropical climate. See text for details. A Novo Aripuaná, L Lábrea, M Manicoré, P Pucallpa.

Rainfall decreases below 2000 mm per year and a pronounced winter dry season occurs in the middle Rio Ucayali Valley of eastern Peru and in the hilly border region between Brazil and Peru. Annual rainfall is locally even less than 1500 mm (e.g. Pucallpa 1473 mm). However, rainfall data are insufficient in this region to map a more extensive area with less than 1500 mm of rain per year. The narrow elongated area along the Brazilian/Peruvian border receiving less than 2000 mm of rain separates the more humid forested lowlands of south-central Amazonia from the superhumid Andean foothill region of eastern Peru (Fig. 1.2).

The Am climate is further found in the Atlantic coastal lowlands of the Guianas south to the mouth of the Amazon River. In the latter region the season

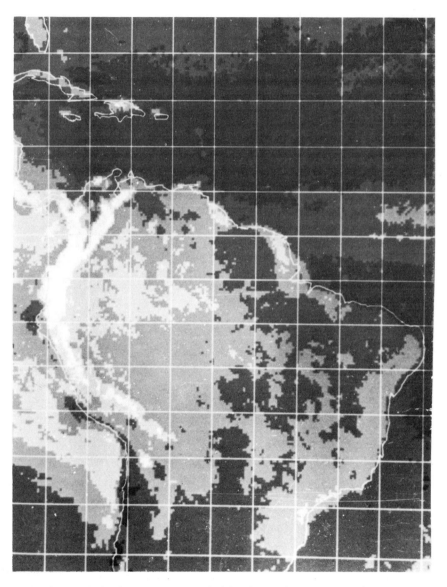

Fig. 1.4. Average cloud cover during four years over tropical South America as measured from a satellite. Cloud cover was measured at 1400 h by an eight-point scale of brightness shown here as increasing whiteness. Note the similarity with climate type shown in Fig. 1.3.

with reduced rainfall is shifted from July–August toward the spring (October–November). The winds rise over the coastal slopes of the Guianas and lose part of their moisture thus causing a rainshadow effect inland on the leeward side of the Guiana mountains within the relatively dry transverse zone of lower Amazonia (see below).

A humid climate extends from northwestern Colombia along the Caribbean slope of Middle America north to southern Mexico. The Pacific slope is mostly dry but a humid climate occurs along the Pacific slopes of the Sierra Madre (especially between Chiapas and western El Salvador) as well as in Pacific southern Costa Rica and northwestern Panama.

The seasonal *Aw climate* is characterized by a reduced annual precipitation and a definite dry season. The limit between the Am and Aw climates corresponds approximately to a line separating regions with more or less than nine humid months during the year, respectively (Lauer 1952; Reinke 1962). The Aw climate is found in northern and central South America surrounding humid western Amazonia (Fig. 1.3). Both these regions are connected in lower Amazonia by a climatic Aw 'bridge' which, probably rather discontinuously, crosses the Amazon between Óbidos-Santarém and the mouth of the Rio Xingu. A wet season occurs during the period of high sun position and a dry season is caused by the stabilizing trade winds during the period of low sun position, i.e. December–March (northern hemisphere) or June–September (southern hemisphere). The dry season may be alleviated in some years due to a varying influence of the humid equatorial air mass. Tropical temperatures are still rather constant with annual variations not exceeding 5 °C. Precipitation is usually less than 2000 mm per year.

An important aspect of the humid tropical climates generally is the conspicuous variability from year to year. Stations with an annual average of 1400 mm of rain have recorded only 800 mm in some years and as much as 2000 mm in others. The season with reduced rainfall can vary from 0 to 6 months in areas where 2 or 3 months are expected. Frosts can occur unpredictably in some areas near the Andes, and reach lowland southwestern Brazil with cooling effects reaching eastwards in Amazonia as far as Manaus at the mouth of the Rio Negro.

The dry tropical *B climate* is found in the Caribbean lowlands of northern Venezuela to northeastern Colombia and in northeastern Brazil. Rainfall is less than 1000 mm per year in most areas. The narrow Pacific coastal lowlands from southwestern Ecuador southward are also arid (getting to less than 100 mm per year south of central Peru, essentially rainless in southern Peru) as are many interior valleys of the Andes where the humidity is caught by the surrounding mountains. The cool winds which blow down into these valleys from the highland have an additional desiccating effect, as they are warmed up in the lower tropical elevations.

The subtropical *Cf climate* is permanently humid with hot summers and cold winters with frosts. There is no conspicuous dry season. A Cf climate occurs in southern Brazil, eastern Paraguay, and Uruguay.

Summaries of climatic data for tropical America can be found in Schwerdtfeger (see Ratisbona 1976), Eidt (1969), Miller and Feddes (1971), FAO/UNESCO (1971, 1976), and Salati and Marques (1984).

THE LOWLANDS

The physical features of South America – its surface relief, coastal delineation, and river systems – had developed toward the end of the Tertiary period during the Pliocene when the major uplift of the Andes mountains occurred, 2–8 million years ago. The Guiana Shield and the Brazilian Shield, to the north and south of the Amazon valley respectively, also were uplifted during the Tertiary when a strong erosion surface developed in these areas and an existing sedimentary cover over the crystalline basement rocks was largely removed. Details of the Tertiary history in the Amazon basin are not known because of the lack of marker horizons and index fossils in the deposits of that period (Alter do Chāo or Barreiras Formation). Probably the sediments were laid down in large swamps and marshes, and on vast flood plains. Fluvial-continental clastic sediments were transported westward into the sub-Andean basin of the upper Amazon region as well as eastward into the Atlantic Ocean. The formation of a large Amazon delta during the late Miocene and rapid delta progradation across a pre-existing carbonate shelf during the Pliocene reflect the period of major uplift and erosion of the Andes when the Amazon River began to carry a huge load of clastic sediments into the Atlantic Ocean. Vertical movements in the Andes continued during the Pleistocene into Recent times, as indicated by uplifted river terraces, tilted or faulted gravel and fan deposits, and strong earthquakes.

Three regionally important tectonic features (structural 'highs') cross the Amazon depression and at depth connect the Guiana Shield with the Brazilian Shield. These subsurface highs extend in a NNW–SSE direction and have been designated (from west to east) Iquitos Arch, Purús Arch, and Gurupá Arch (see, e.g. Bigarella 1973). They originated during Palaeozoic and Mesozoic times probably through an accelerated subsidence of the basins between these arches (which themselves lagged behind the regional subsidence) rather than as narrow zones of orogenic uplift. Because of our ignorance as to the details of the stratigraphy of the Tertiary strata in Amazonia, a possible influence of these old tectonic arches on the palaeogeography and sedimentation in Amazonia during the last 60 million years remains unknown. During the early Tertiary, the western arches possibly caused large portions of western Amazonia (the Solimões region) to form a low lying erosion area or gentle watershed which was 'drowned' and covered by the Iquitos-Pebas formation during the middle and late Tertiary (review by Grabert 1983; Beurlen 1970). The present position of the Tertiary Alter do Chão Formation in Amazonia varies between 100 and 350 m indicating large scale gentle (epeirogenic) movements after the deposition of these strata. The strata form a slightly southward-dipping structural high crossing the Amazon between the mouth of the Rio Tapajós and that of the Rio Xingu (Beurlen 1970, p. 331).

Data for a reconstruction of the history of tropical America during the Quaternary period, i.e. during the last two million years, come from several different sources: from geological investigations of Pleistocene and post-Pleistocene deposits, from geomorphological interpretations of extant landforms, from analyses of the pollen content of Quaternary strata and from palaeoclimatological interpretations. Field data are available from only few study areas in the vast tropical lowlands of South America. The results indicate that Pleistocene climatic–vegetational fluctuations severely affected also the neotropical lowlands and not only the higher levels of tropical mountains. Recent summaries regarding the vegetational history of equatorial America, Africa and Indo-Malesia during the Quaternary are those of Livingstone and van der Hammen (1978); Flenley (1979); Peterson, Webb, Kutzbach, van der Hammen, Wijmstra, and Street

(1979); Street (1981); and Whitmore (1981); see also Graham (1973a, b, 1979), van der Hammen (1979, 1983) and several contributions in Prance (1982b) and Smith (1982).

Geoscience data are insufficient so far to map the changing distribution of forest and non-forest vegetation during the various climatic periods of the Pleistocene. Many more intensive regional studies are needed to establish a comprehensive interpretation. However, the evidence which is available does permit the formulation of a hypothesis or a model to be tested by future investigations and to be compared with the results of analyses in other fields of inquiry such as the biogeographical studies of vertebrates, insects, and plants described later in this book.

Geology

Sediments deposited off northeastern South America between latitudes 20°N and 10°S during the last ice age (Wisconsin–Würm) include arkosic sands which contain 25–60 per cent feldspar (Damuth and Fairbridge 1970). This is much higher than the 17–20 per cent found in Recent sands of this region. Feldspars occur commonly in igneous and metamorphic rocks of the South American shield areas and of the Andes. Under conditions of climatic aridity, these minerals are fairly stable during the weathering process but they decompose rapidly when the climate is humid. Because of the high feldspar content of these glacial sands, Damuth and Fairbridge concluded ' . . . that an arid to semi-arid climate dominated large portions of equatorial South America during the Pleistocene glacial phases in complete contrast to the present-day and Pleistocene inter-glacial humid tropical climate'. They suggested ' . . . that the arkosic sands were derived from the Brazil and Guiana Shields near the ocean at a time when the climate was much drier than today'.

This interpretation was challenged by Milliman, Summerhayes, and Barretto (1975) and Irion (1976) who discussed the possibility of an Andean or central Amazonian origin of the arkosic sands. In the case of an Andean origin, it was felt that rapid erosion in the mountains left insufficient time for decomposition of the feldspars and the transport in suspension by Amazon waters permitted the feldspars to reach the Atlantic Ocean, to be deposited in the sands off northeastern South America. Irion (1976) speculated that during the Wisconsin glacial phase of lowered sea-level unconsolidated Tertiary deposits of the Alter do Chão (Barreira) Formation in central Amazonia may have been eroded in this way furnishing the arkosic sands. Either of the above alternative interpretations would diminish their palaeoclimatic significance. However, studies of seabottom cores taken in the Caribbean Sea and their content of foraminiferal assemblages as well as of aeolian biogenic detritus also indicate climatic aridity during the glacial phases in South America (Bonatti and Gartner 1973; Prell 1973; Parmenter and Folger 1974). This additional evidence supports the interpretation of Damuth and Fairbridge.

Irion and Absy (1978) and Absy (1979) studied lake sediments from the central Amazon valley near Manaus that were deposited during the latest part of the last glacial stage and mainly in post-Pleistocene (i.e. Holocene) time. Mineralogical–chemical, sedimentological, and pollen-analytical methods did not uncover indications of regional changes in the distribution of forest and savanna vegetation during the time of deposition. Probably fairly humid conditions prevailed continuously in the broad valley of the Amazon River during the Holocene. Strata of Pleistocene (rather than Holocene) age should now be sampled in terra firme regions between the large Amazonian rivers in order to contribute relevant data for the discussion of palaeoenvironments in Amazonia during glacial and interglacial phases. Sediments exposed along the Rios Juruá and Acre in W Brazil and along the Rio Beni in NE Bolivia are also Holocene in age and indicate large-scale flooding and deposition in SW Amazonia probably resulting from the sudden draining of glacial lakes in the high Andes of Peru and Bolivia toward the end of the Pleistocene (Campbell *et al.* 1984, 1985). This interpretation implies the destruction and flooding of forests over vast lowland regions in this portion of Amazonia during a certain interval of the late Quaternary lasting several thousand years.

World sea-level was about 80–100 m lower than today during the last glacial period (Wisconsin-Würm, 13 000 to 18 000 years BP). During Quaternary periods of raised sea-level (interglacials) many coastal lowlands of South America were flooded and large portions of the Amazon valley were converted into huge inland brackish and freshwater lakes extending west into eastern Peru. Terraced Pleistocene deposits are widely present in the lower Amazon valley (Klammer 1971, 1984) and consist of several sequences of fluvial sands partly cemented by descendent iron solutions. Five different terrace levels have

been found between 10 and 100 m above local water level and represent remnants of fill-in terraces resulting from cyclic erosion and aggradation. High stands of interglacial world sea-level compared to present sea-level decreased from about +180 and +100 m during two interglacials of the latest Pliocene, to +60, +30, and +17 m during successive Pleistocene interglacials, and to +3 m (4000 a BP) and finally +1 m (2300 a BP) during the Holocene (see Fittkau 1974, Table 2; note that this table erroneously lists the dátes as BC rather than BP).

Geomorphology

Geomorphical processes such as soil formation, stream-bed cutting, and erosion patterns occurring in modern environments yield characteristic surface features distinguishing the landscape of areas developed under different climatic conditions. In humid climates rock weathering is predominantly chemical whereas in arid climates it is predominantly mechanical. Mechanical weathering and erosion of largely unforested land surfaces under arid climates lead to the formation of usually steep mountain slopes surrounded by gently dipping erosion surfaces (pediplanes) and/or gravel beds (pediments). Rivers become choked with debris and show braided patterns. A pavement of coarse alluvium develops on the surface under arid conditions due to aeolian removal of the fine fractions of the soil and the absence of perennial runoff. Palaeopavements are seen in outcrops of Quaternary strata as 'stone lines' which indicate the previous action of an arid climatic cycle. In humid regions, by contrast, streams are broad and meandering and hill slopes are less steep and lack extensive gravel beds around them. Because of the thick cover of vegetation, runoff is reduced and weathering of the rocks is due to chemical action of organic acids rather than mechanical forces (Fig. 1.5; Bigarella, Mousinho, and da Silva 1969; Bigarella and Becker 1975).

If the landforms at present found in a given region are consistent with its climate and vegetation then it may be assumed that climate was stable over a long period of time. On the other hand, the modern vegetation sometimes covers a landscape or soils obviously produced under a different climatic regime from that currently found. In these cases analyses of the landforms and of soil profiles permit inferences to be made about past climates and the changes which have since taken place.

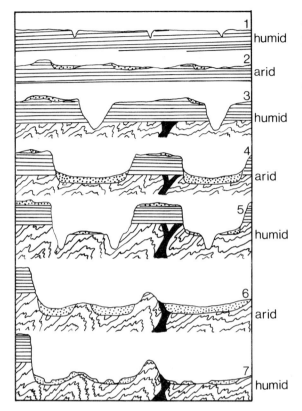

Fig 1.5. Schematic cross-section profiles illustrating the geomorphic development of the Rio Caroní area, southern Venezuela (C on Fig. 1.6). (After Garner (1974).) Humid climatic periods characterized by fluvial incision alternated with arid periods when extensive aggradation occurred.

In several areas of South America tropical rain forest grows on surface sediments composed of coarse, mechanically fragmented detritus (alluvium) containing rather unstable rock and mineral fragments, e.g. feldspars. These sediments resemble alluvium formed in deserts or generated by glaciers (Garner 1975). Forests had certainly been displaced far away from these areas when the surface debris was accumulated under arid climatic conditions. These situations have been described from southeastern Brazil (Bigarella and de Andrade 1965), southern Venezuela (Fig. 1.5; Garner 1966, 1967, 1974), and lower Amazonia (Tricart 1974; Journaux 1975). The reverse situation is also known. Well-preserved fossil forest soil occurs under dry caatinga vegetation of northeastern Brazil indicating the former more widespread occurrence of forests in this presently arid region (K. S. Brown, personal communication).

In southeastern Brazil, several pediments along the base of the forested Serra do Mar and on the coastal lowlands indicate the repeated change from humid to arid conditions. The landscape development in the Rio Caroni drainage system in southern Venezuela (Fig. 1.5) also points to planation-aggradation under one or several severely arid climatic phases. The desert erosion mechanisms that emplaced the gravels in this area, now occupied by rain forest and savanna, led to the deposition of gravel beds in stream channels and valleys sometimes to the point of actual burial of these depressions by alluvium. Resumption of humid conditions in more recent times has led to the chaotic channel networks due to alluvially clogged depressions (Garner 1974, 1975).

Interpretations of surface landforms based on radar imagery have been published for selected areas in the Amazon valley by Tricart (1974) and Journaux (1975). These authors observed soil formations, dissected, and gullied land surfaces and crusts (stone lines) characteristic of areas with sparse vegetation thus pointing to a period of drier climate than today and displacement of rain forest from portions of Amazonia during the late Quaternary. The slope development indicated continued dissection of the land surface during a transition from semiarid to humid conditions.

In the llanos plains of the Orinoco River (Venezuela and eastern Colombia), Tricart (1974) discovered, through interpretation of LANDSAT-1 satellite photography, an extensive dune system. The dunes are composed of alluvial sand which spread from the northern Andes during the last cold-arid climatic phase (Wisconsin-Würm). They are partly fossilized and were probably formed by strong action of the trade winds. As to northern Venezuela see data summarized by Ochsenius (1983).

Klammer (1981, 1982) sees evidence in southeastern Brazil for only one strong period of aridity with desert conditions inland and more moderate dryness on the coast during the early Quaternary prior to the onset of cyclic glacial–interglacial fluctuations when, in his opinion, the climate turned increasingly humid.

Ab'Sáber (1967, 1982) gathered geomorphological data on pediments, fluvial terraces, stone lines, palaeosols, and iron crusts in Brazil for many years. An interpretive map (Fig. 1.6) which he published recently (Ab'Sáber 1977, 1982; Brown and Ab'Sáber 1979) depicts, on the basis of geomorphological data, the distribution of major vegetation types in South

America during the last Wisconsin–Würm glacial phase (13 000–18 000 years BP). The lowland tropical rain forest is shown restricted to gallery forests along the main river courses and to a number of isolated areas ('refugia') in Amazonia and along the wind-exposed outer slopes of mountains in central and eastern Brazil. Non-forest vegetation, i.e. caatinga, cerrado, savanna, is shown to have replaced the forest in most areas of central South America. This map thus illustrates the statement by Garner (1975) ' . . . that vast portions of the [South American] lowlands . . . had been arid during higher-latitude glacial episodes'. Ab'Sáber (1982) presented a more detailed account of the criteria he used in the delineation of each of the forest enclaves mapped. He confirmed that the placement of most of the Brazilian forest refuges as shown on the map was based on abundant geomorphological data which he had collected personally during his fieldwork. For the refugia in southeastern Colombia and along the Rio Juruá in western Brazil Ab'Sáber used radar images and other indirect data. The refugia shown in the Guianas, Venezuela, and western Colombia are extrapolations based more on topography, exposure to prevailing winds at that time, and published geomorphological information. No biogeographical data on faunal or floral endemism entered this geomorphological reconstruction of environmental conditions in South America during the last glacial period. As Prance (1981) has pointed out, it is unlikely that large areas of Amazonia developed a uniform vegetation type as dry as caatinga 15 000 years ago as is suggested by the Ab'Sáber model. Other vegetation types notably transition forest (especially liana forest and bamboo forest), besides more open vegetation, probably separated the rain-forest refugia and fragmented the distribution of many deep-forest species. Bibus (1983) confirmed Ab'Sáber's (l.c.) observations through pedological studies in the lower Rio Tapajós region, also concluding that in this area extensive soil erosion and redisposition occurred in the recent geological past under an open vegetation and in the absence of dense rain forest.

Garner (1974) and Douglas (1978) have reviewed the recent advances in the field of tropical geomorphology emphasizing the importance of Quaternary climatic changes for interpretations of landscape development.

In summary the studies reviewed above indicate that, throughout Amazonia, the present forest is in a dynamic state, and its modern distribution reflects neither an old nor a stable pattern (Bigarella and Ferreira 1985).

Palaeoclimatology

Regional variation of annual rainfall in tropical South America is mainly determined by the surface relief. This originated during the late Tertiary although some changes through continued uplift of the Andes and of the shield areas, as well as through erosion of high relief areas, probably occurred during the Quaternary also. Precipitation is highest near the Atlantic coast, on the exposed northeastern slopes of the Guianan mountains and near the eastern base and along the slope of the Andes. By contrast, total annual rainfall decreases on the leeward (western) side of mountains and in a zone across lower Amazonia (Fig. 1.3). Even though climatic patterns during Pleistocene glacial and interglacial periods probably differed from the present, in general, areas that now receive high rainfall were probably still wet relative to other areas. Tropical rain forest is likely to have persisted in these areas but disappeared from those intervening areas where rainfall dropped below 1500 mm per year during arid phases.

Damuth and Fairbridge (1970) suggested that a strong westerly air-flow against the southern Andes produced heavy snow in the high areas, lowering the snow line by about 1000 m, and, warmed by a sudden descent (Föhn), must have led to a rain shadow and desiccation east of the Andes. Toward the equator, this same wind system would have tended to swing round to become southeasterly but would be very dry, even in Amazonia. The winds from the Atlantic Ocean probably moderated the effect of these southeasterly air currents in central South America where certain areas, especially near mountains or in hilly regions, remained humid even during arid periods. In reconstructing isolated pockets of humid climate and hence of rain forest (Fig. 1.6), Brown and Ab'Sáber (1979) postulated that, during the last glacial phase, the predominant July wind direction was more from the north than today.

Recently the July sea-surface temperature has been mapped for 18 000 BP (i.e. during the Wisconsin–Würm Glacial) on the basis of counts of plankton (Foraminifera, Radiolaria, Coccolithophoridae) preserved in deep-sea sediments (Climap project members 1976; Gates 1976). The varying composition of the biota was used to estimate the positions of currents and surface water-masses based on standards developed from Recent planktonic assemblages and their correlation with the surface water-masses of the present oceans. The underlying assumption, that the

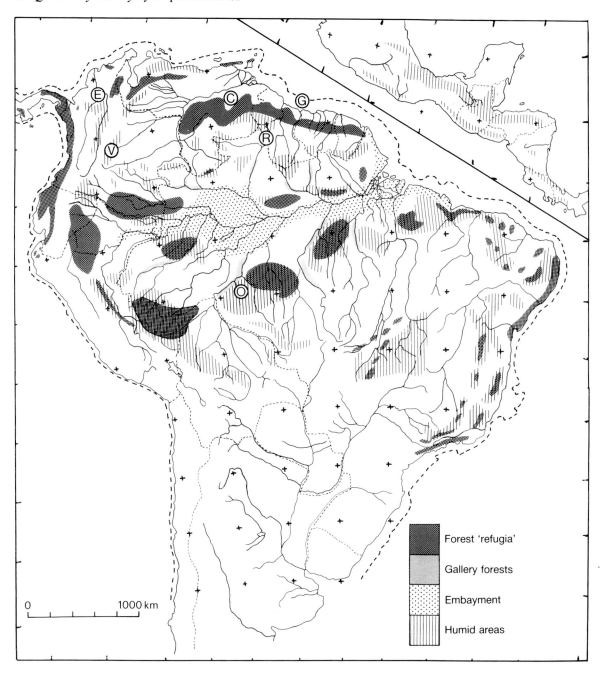

Fig. 1.6. Tropical forests ('refugia', solid) and generally humid areas (hatched) in tropical South America during the last cold dry Pleistocene climatic phase (18 000–13 000 years ago; Wisconsin–Würm) as reconstructed on the basis of geomorphological data (Ab'Sáber 1977) and palaeoclimatic-pedological considerations (Brown 1979; Brown and Ab'Sáber 1979). Solid – tropical forest, in favourable lowland areas and along escarpments where winds were forced to rise giving off part of their moisture. Light grey–gallery forests. Broken line follows the probable glacial coastline of lowered sea-level. Stippled area outlines the estimated Günz/Mindel interglacial Amazonian embayment (sea-level raised by about 60 m above present situation; areas of impeded drainage flooded during most of the year in those times were very extensive in central Amazonia). Letters designate locations mentioned in the text.

planktonic assemblages of 18 000 BP reacted in a similar manner to different water-masses and currents as the equivalent assemblages of today, seems valid. Marked cooling of surface waters in equatorial regions of up to 6 °C is found at 18 000 BP and this is in part due to increased upwelling of cool waters to the surface. However, the temperature reduction of surface waters was less than 2 °C in the Gulf of Mexico, the Caribbean Sea and in the western equatorial Atlantic and the average for equatorial surface ocean waters was 2.3 °C.

Palynology

Analyses of the pollen content of Quaternary lake and bog deposits from several areas of South America have revealed patterns of floristic changes which reflect climatic–vegetational fluctuations during the Pleistocene and post-Pleistocene. Detailed studies have been made in the highlands of the Eastern Andes of Colombia, documenting the immigration of an upland flora during the final uplift of this mountain range toward the end of the Tertiary and repeated vegetational changes during the following Ice Ages (van der Hammen, Werner, and van Dommelen 1973; van der Hammen 1974; Flenley 1979). By contrast, pollen profiles from the tropical lowlands permit interpretations mainly for the post-Pleistocene period. We urgently need palynological studies of dated Pleistocene strata, especially from Amazonia and other forested areas of tropical America during the last glacial phase or during an earlier cold phase. The results of such studies would permit mapping the changing vegetation of the tropical lowlands and thereby lead to the confirmation, modification, or disproval of the suggestions of vast Pleistocene environmental changes in these areas which climatological and geomorphological data suggest. Palynological studies may enable palaeobotanists to chart the location of areas of persistent forest or savanna during arid and humid climatic periods of the Pleistocene, respectively.

The results of one study of the pollen content of Pleistocene to Holocene sediments in southern Amazonia (Fig. 1.7(a)) tend to corroborate the notion of extensive vegetational changes in the tropical lowlands during the Quaternary. This composite pollen profile of bore hole samples from Capoeira and Katira, central Rondônia, western Brazil (O of Fig. 1.6) indicate that, during a prolonged period of the Quaternary, grass savanna had replaced

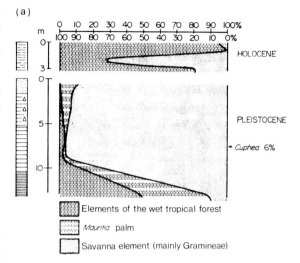

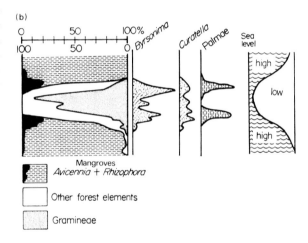

Fig. 1.7. Pollen diagrams from central and northern South America. (a) From Rondônia, western Brazil; O on Fig. 1.6. Vertical scale (m) indicates length of sediment cores; lithology shown on left. (b) Scheme of the succession of vegetation types (as reflected in pollen diagrams) during an interglacial-glacial–integlacial cycle in the present coastal plain of Guyana and Surinam, northeastern South America (G on Fig. 1.6). Vertical axis represents time/depth (no scale). (After van der Hammen (1974) and Absy and van der Hammen (1976).)

the rain forest which today occupies the sample region. The studied section could not be dated exactly but most probably is late Pleistocene to Holocene in age. The diagram shows the widespread occurrence of wet tropical forest and swamp forest when the yellow-brown clay of the basal section was deposited. The younger clays contain predominantly pollen of savanna elements (mainly grasses, but also herbs like *Cuphea*) indicating the disappearance of the forest from this region. Pollen of forest elements dominate most of the younger (Holocene) section, taken at Capoeira, but the abundance of savanna elements in the samples from c.2 m depth indicate the temporary reduction of forests.

Wijmstra (1971) and van der Hammen (1974) studied several core sections from the coastal lowlands of Guyana and Surinam in northeastern South America (G of Fig. 1.6). Some of the profiles are over 100 m in length and known to be Pleistocene and Holocene in age on the basis of ^{14}C-dated samples. The lithology of the sediments and their pollen content reflect the low and high sea-level stands of glacial and interglacial periods, respectively, as well as the associated changes of vegetation at the study sites. A schematic summary diagram (Fig. 1.7(b)) combines the results of several pollen diagrams from the coastal Guianan lowlands and illustrates the succession of vegetation types during one interglacial–glacial–interglacial cycle. Several such cycles occur in some cored sections thus reflecting the repeated glacial–interglacial cycles of the Pleistocene. The sites were covered with the shallow waters of the shelf sea and the coastline lay southward during the high sea-level stands of the interglacial periods, i.e. near the bottom and top of the summary diagram. The pollen spectrum for the periods of high sea level is dominated by mangrove pollen, especially of *Rhizophora* species, which produce abundant pollen that is easily transported by coastal currents. Mangrove trees of the genus *Avicennia* are poor pollen producers and the pollen is mainly found inside the mangrove belt itself. Therefore, the two intervals in the diagram characterized by up to 20 per cent *Avicennia* pollen mark the periods when the mangrove belt passed through the study sites. The coastline was located seaward during the intervening period of low sea-level (glacial phases) when savanna elements (grasses) dominated the vegetation. Scattered shrubs and low trees of the genera *Byrsonima* and *Curatella* grew on the savannas, their pollen forming a conspicuous percentage of the total spectrum. The occurrence of savannas in the large study area during glacial periods, or at least during a part of each glacial period, cannot be explained by the influence of edaphic factors alone. It is probable that a comparatively dry or strongly seasonal climate prevailed in these regions during the glacial phases (van der Hammen 1974).

Pollen profiles from lake deposits in the grasslands of eastern Colombia and interior Guyana (respectively V and R on Fig. 1.6) demonstrate that a dry forest, or closed savanna woodland, preceded the open savanna found in these regions today (Wijmstra and van der Hammen 1966; van der Hammen 1974). The pollen diagrams from these study areas (Fig. 1.8) document repeated changes through time of the vegetation cover. In the region of Laguna de Agua Sucia, eastern Colombia it can be seen that a grass savanna existed prior to 4000 BP (Fig. 1.8(a)). The peat layer which permitted the dating of this part of the section as 4000 to 3800 a BP was formed during a period of low lake level, i.e. dry climate. The climate became more humid and the lake level rose again after 3800 BP when *Byrsonima* and various forest elements invaded the region. Later on, savanna

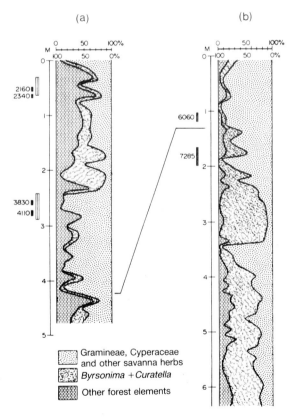

Fig. 1.8. Pollen diagrams of the Late Pleistocene and Holocene from savanna plains in (a) eastern Colombia and (b) Guyana. (After van der Hammen (1974).) Four digit numbers are absolute dates of samples (years BP).

elements again gradually increased and another peat layer formed around 2200 BP when the lake level was very low (dry climate). The final sharp increase of savanna elements during the last 2000 years may reflect the influence of burning by man.

The pollen diagram from Lake Moreiru in the Rupununi savanna of Guyana (Fig. 1.8(b)) is partially dated by two peat samples from layers deposited at 6000 and 7300 BP respectively. Here, an extensive *Byrsonima* woodland existed prior to 7300 years ago but was replaced by a grass savanna during a dry climatic period at a depth of 340 cm which corresponds to a dry phase probably of the last glacial period.

In the Lake Valencia region, northern Venezuela, xeric herbaceous vegetation existed under dry climate during the late Pleistocene and was replaced by arboreal plant communities under a more humid climate at about 10 000 years BP (Bradbury, Leyden, Salgado-Labouriau, Lewis, Schubert, Binford, Frey, Whitehead, and Weibezahn 1981, Binford 1982).

Holocene pollen profiles from central Amazonia (Absy 1979, 1985) indicate the occurrence of periods of reduced effective precipitation around and shortly after 4000 BP, around 2100 BP, and around 700 BP. These dates correspond to somewhat drier periods indicated in the profiles discussed above and so probably indicate regional phenomena. The studies in central Amazonia did not yield palynological evidence for a major extension of savanna vegetation during the portion of the Holocene represented by the studied profiles. The analysed changes of vegetation were mainly due to local processes of sedimentation and changes of water level. One important succession reflected in the diagrams is from open water via floating meadows (mainly of grasses) and *Cecropia* stands to swamp forest. Wijmstra (1967) observed similar Holocene fluctuations of local vegetation in the wet lowlands of the lower Magdalena valley of northern Colombia (E on Fig. 1.6). The climatic fluctuations which can be inferred from these pollen profiles are shown on Fig. 6.10 (p. 173) and discussed further in Chapter 7. As to Middle America, Graham (1982) and Toledo (1982) discussed palaeobotanical evidence documenting climatic-vegetational fluctuations in Mexico and adjacent areas during the Cainozoic (see also Lewin 1984).

In conclusion these palynological studies give direct evidence of vegetational changes in tropical America during the Quaternary associated with changes in climate. The vegetational fluctuations during the post-Pleistocene were comparatively restricted. Much larger changes in distribution of lowland forests and savannas probably occurred earlier, during the climatic reversals of the Pleistocene, for which time interval, however, only very limited palynological evidence is as yet available.

A model of fluctuating climate and vegetation

The geoscientific field evidence, indicating an unstable climate and corresponding vegetational changes in lowland tropical America during the late Pleistocene and Holocene, has been adduced from only a limited number of widely separated study sites. However, because of the regional nature of the climate over the tropics, vegetational changes at these scattered locations indicate that the climatic pattern throughout the tropical lowlands has changed repeatedly. The estimate for the July climate of 18 000 years BP, i.e. during the Wisconsin–Würm glacial (Gates 1976; Climap project members 1976) indicated that over tropical land areas the simulated ice-age cooling averaged about 5 °C compared with an average cooling of the equatorial surface ocean waters of only 2.3 °C. As a consequence grassland, steppes, and deserts spread at the expense of forests which regained lost ground and coalesced during humid interglacial periods. Nevertheless, tropical forest and non-forest vegetation certainly are old, having originated during the Cretaceous and Tertiary. In the areas of their uninterrupted occurrence forests and savannas continuously offered ecologically stable and equable conditions to tropical animals and plants, thus explaining the long history of co-evolution and co-adaptation. It is the overall geographical distribution of these vegetation types that has changed dramatically in response to the changing world climates of the Pleistocene and in this way setting the stage for widespread organic differentiation at the subspecies and species level during the recent geological past. No similarly effective climatic–vegetational events are known in South America during the preceding Tertiary period when world climates were fairly stable or changing more slowly and tropical conditions extended far beyond their current northern and southern limits. However, major tectonic events probably had climatic impacts especially in the region of the Andes.

Geoscientific data gathered over the past twenty years demonstrate rather convincingly that the earlier notion of the 'stable tropics' which offered constant environmental conditions over large continental areas throughout the Quaternary period needs revision. The tropical lowlands experienced alternating cool-dry and warm-humid climatic phases which correlate with the glacial and interglacial periods respectively of high latitudes. Although temperature fluctuations were not strong enough to eliminate the tropical biotas in equatorial regions variation in the total annual precipitation and/or its annual distribution caused vast changes in the distribution of tropical forest and non-forest floras and faunas. During arid climatic periods, rain forests were restricted to 'refuge' areas with a favourable orographic and hydrographic setting, such as along major rivers (broad gallery forests) and near the base or along the slopes of mountains which intercepted moisture-laden winds. The surface relief was a major factor in determining the location of remnant forests during dry climatic periods, through local rainfall maxima and favourable soil conditions. Since the present relief of tropical America originated mainly during the late Tertiary and earlier, and therefore is older than the Quaternary climatic fluctuations, several favourable areas probably served repeatedly as forest refugia during successive Pleistocene arid periods. Other refugia, unrelated to relief factors, probably changed

their location due to differences in climatic conditions in the successive dry periods. Details inevitably remain speculative. We may consider the scenario reconstructed by Ab'Sáber (1977, 1982; see also Brown and Ab'Sáber 1979; Brown 1979, Bigarella and Ferreira 1985) and here summarized in Fig. 1.6 as a geoscientific model with which biogeogrpahical data may be compared.

Climatic-vegetational cycles during the Quaternary have affected tropical Africa to a similar degree as the lowlands of tropical South America (Moreau 1963; Livingstone 1975; Livingstone and van der Hammen 1978). '. . . The redistribution of Kalahari sand over most of the present Congo forest area, towards the end of the mid-Pleistocene indicates that the forest must have been vastly reduced and pushed into edges and corners, presumably with concurrent extinctions' (Moreau 1966, p. 161). Lowland forest refugia during arid periods existed along the coast of the Gulf of Guinea (West Africa), and in the areas of Gabon and eastern Zaire as well as along the coast of East Africa (review by Hamilton 1976). There is also evidence for formerly more arid climates in the Malay archipelago and tropical Australia (Walker 1972, 1982; Webster and Streten 1978; Verstappen 1975), though as yet no evidence that the present day rain forest areas were substantially reduced (Whitmore 1981).

THE HIGHLANDS

The Pleistocene climate in the highlands of tropical America alternated between cooler conditions than at present, with or without an effective increase in precipitation during glacial periods, and warmer, humid or dry conditions during interglacial periods. In general, highland and subtropical areas of South America may have had very different climatic variations during the Quaternary from lowland tropical regions. A correlation of climatic periods in these various areas is still problematic. The snow line in the mountains was lowered considerably and the glaciers extended far beyond their present limits. Numerous lakes formed in the altiplano region of the Peruvian and Bolivian Andes and in many other intermontane basins of the northern Andes. Lacustrine sediments of one of these glacial lakes in the Colombian Andes ('Lake Humboldt', Sabana de Bogotá) have been studied by van der Hammen and Gonzalez (1960, 1964), van der Hammen *et al.* (1973), and van Geel and van der Hammen (1973) and Hooghiemstra (1984).

The extensive pollen analytical work of these authors permitted analysis of the altitudinal movements of montane vegetation belts in response to the Pleistocene climatic fluctuations in northern South America. An interesting discovery is the fact that the climate in the Colombian Andes was very dry during the peak of the last glaciation (20 000–13 000 BP). This agrees with findings from the neotropical lowlands reviewed above. Other dry phases are marked on the diagram from Laguna Fúquene in the Eastern Andes of Colombia by fluctuations of the lake level (Fig. 1.9). These dry phases occurred earlier than the long arid peak and also later (9500–11 000 BP, 'El Abra' interval). The shift of vegetation belts through the area of Laguna Fúquene as documented by pollen preserved in the sediments is illustrated in the diagram (Fig. 1.9) by a heavy black line. Páramo and subpáramo vegetation occurred during the last glacial period and forest invaded the area later during the Holocene.

Two or three glacial periods are recognized in

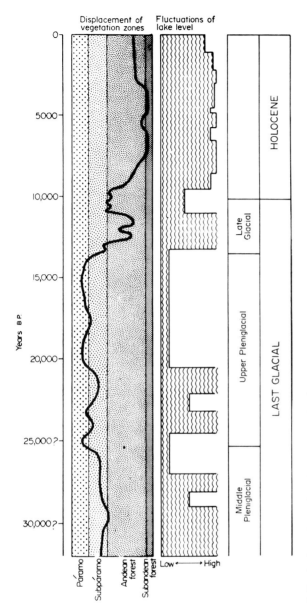

Fig. 1.9. Displacement of vegetation zones and fluctuations of lake level, Laguna de Fúquene, Eastern Andes, Colombia during the Late Pleistocene and Holocene. (After van der Hammen 1974.)

reasonable in most cases. Remains of the early Pleistocene glaciations are missing in some areas of the northern Andes (Sierra Nevada de Santa Marta, Sierra de Perijá) and may indicate that these mountains were still too low to be glaciated at that time.

The advancing glaciers compressed the altitudinal climatic and vegetation zones along the mountain slopes, as the glacial temperature depression probably decreased toward the lowlands which remained tropical. The lowland temperature reduction probably was only 4-5 °C (Emiliani 1972; Gates 1976) but reached 7-8 °C, or even 11 °C, at elevations of 2000 to 3000 m. Similar compression occurred throughout the tropics (Flenley 1979). The lower limit of the Páramo Zone, Temperate Zone, and Subtropical Zone in the Colombian Andes during cold phases may have been located at approximately 2000, 1400, and 500 m, respectively. Accurate determinations are still lacking and regional variation is to be expected, within broad limits, depending upon slope exposure, precipitation, and other local conditions. No pollen data are available so far to trace the changing distribution of the Andean forests in Ecuador during humid and dry climatic phases of the Pleistocene. A *Podocarpus* forest grew along the eastern slope of the Ecuadorian Andes at 1100 m elevation during the Wisconsin-Würm glaciation, i.e. at least 700 m lower than it does today (Liu and Colinvaux 1985).

It follows that the montane forests of the Andes were probably more continuous than today when depressed during cool, humid glacial periods, when the Temperate and Subtropical Zones were located along the less dissected lower mountain slopes. Conversely, the upward shift of the climatic zones during interglacial periods must have led to an increasing dissection of montane vegetation.

Pleistocene climatic conditions similar to those in the Andes probably prevailed in other mountains of the Neotropical Region. For example, in the Chirripó massif of Costa Rica, Weyl (1956) reported landforms which indicate glaciation of the high Middle American mountains at least during the last glacial period. The evidence for glacial activity in the Serra do Mar and Itatiaia mountains of southeastern Brazil is controversial. Possibly the cooling was insufficient to have caused the formation of glaciers on these mountains (only 1000-2000 m and 2800 m elevation respectively).

No geological data are available for the Pleistocene period in the Guiana Highlands (1000-3000 m elevation). The higher elevations of the table moun-

most mountain ranges on the basis of successive levels of moraines left by the advancing glaciers. A correlation with the Kansan (Mindel), Illinoian (Riss) and Wisconsin-Würm glacial periods appears to be

tains in this region probably were cooler during the glacial periods resulting in a lowering of the montane vegetation zones similar to the situation in the Andes. The increased area of montane zones would have allowed the development of patches of montane tropical forest on lower peaks now covered by lowland forest. Similarly, the small mountains 500–900 m in elevation located in southeastern Colombia may have been capped by montane forest thus forming a highly discontinuous climatic–vegetational 'bridge' between the Andes and the Guiana Highlands (see Fig. 4.18).

Simpson-Vuilleumier (1971), Simpson (1975, 1979, 1983). Salgado-Labouriau (1979), Schubert (1979), Flenley (1979), and Clapperton (1983) have summarized the present knowledge of Pleistocene environmental changes in the highlands of South America and Livingstone and van der Hammen (1978) have done so for Middle America and Mexico.

2

SOILS AND VEGETATION

SUMMARY

The soils which underlie tropical rain forest at the present day are briefly characterized (Table 2.1) and mapped. The distribution of the major soil units in the Brazilian Amazon as revealed by recent aerial and ground surveys is shown (Fig. 2.1). Consideration is given to the kinds of soil and topography most likely to have supported humid forest continuously through past drier climatic periods (Table 2.2, Fig. 2.2). Some of these soils are clothed today in mosaics of dense and open forests. It is believed that these mosaics

TABLE 2.1. Nomenclature of soil units of the neotropical forests (Brown 1979)

No.	Brazilian (translated)	FAO/UNESCO	Soil Taxonomy
1	Eutrophic Structured 'Terra Roxa'	Eutric Nitosol	Rhodic Paleudalf
2	Eutrophic Red-Yellow Podzolic Soil, Reddish Brunizem	Ferric Luvisol, Chromic Luvisols, Luvic Phaeozem	Oxic Tropudalf, Oxic Tropustalf, Paleustoll, Argiudoll
3	Dystrophic Structured 'Terra Roxa'	Dystric Nitosol	Rhodic Paleudult
4	Dystrophic or Alic Red-Yellow Podzolic Soil	Orthic Luvisol, Orthic Acrisol	Oxic Tropudult, (Alic Tropudult)
5	Eutrophic Dusky Red Latosol, Eutrophic or Mesotrophic Red-Yellow Latosol (not sandy)	Rhodic Ferralsol, Orthic Ferralsol	Eutrorthox, some Haplorthox
6	Eutrophic Cambisol	Eutric Cambisol	Eutropept
7	Eutrophic Lithosol (in mosaics)	Lithosol (eutric mosaics)	Lithic (Eutric) Troporthent
8	Eutrophic Alluvial Soil	Eutric Fluvisol	Eutric Tropofluvent
9	Eutrophic Humic Gley	Eutric Gleysol	Eutric Tropaquept, Eutric Tropaquent
10	(Andosol)	Andosol	Andept
11	Dystrophic Cambisol	Dystric Cambisol	Dystropept
12	Dystrophic Lithosol	Lithosol (dystric mosaics)	Lithic (Dystric) Troporthent
13	Plinthic Podzolic Soil	Plinthic & Gleyic Acrisol, Plinthic Luvisol	Plinthic Paleudult, Plinthaquic Paleudult
14	Humic or Plinthic Latosol	Humic Ferralsol, Plinthic Ferralsol	Plinthumox, Plinthic Haplorthox
15	Alic Red-Yellow Latosol, Yellow Latosol, Dark Red Latosol	Acric Ferralsol, Xanthic Ferralsol	Haplorthox, Acrorthox
16	Sandy Dystrophic Podzolic Soil	Acrisol	Psammentic Paleudult
17	Grumosol	Vertisol	Tropudert
18	Dystrophic Alluvial Soil, Dystrophic Humic Gley	Dystric Fluvisol, Dystric Gleysol	Dystric Tropofluvent, Dystric Tropaquept, Dystric Tropaquent
19	Hydromorphic Podzol	Podzol	Tropaquod, Tropohumod
20	Red-Yellow Quartz Sand	Ferralic Arenosol, Acric Ferralsol	Oxic Quartzipsamment
21	Quartz Sand	Arenosol, Regosol	Quartzipsamment
22	Lateritic or Concretionary Soil, Planosol	(Petric Phase), Planosol	(Petroferric Phase), Plinthaquult, Albaquult

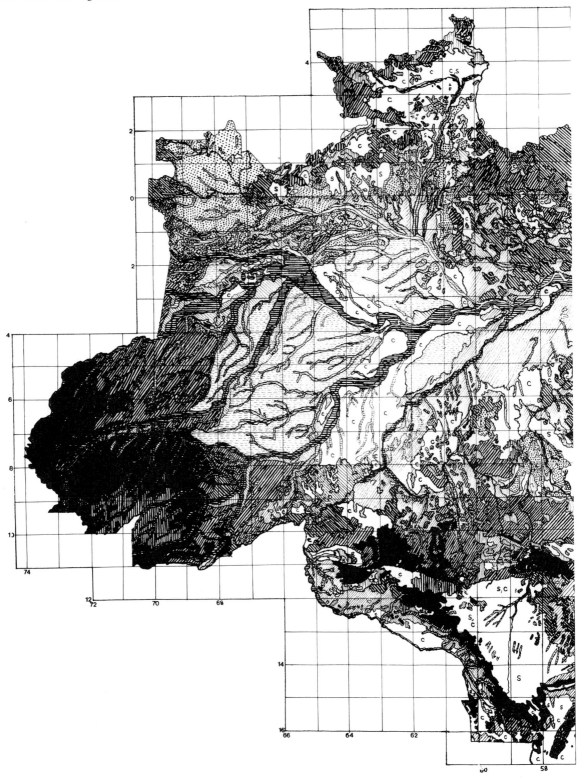

Fig. 2.1. The major soil units in the Brazilian Amazon, based on recent survey. See text and Table 2.2.

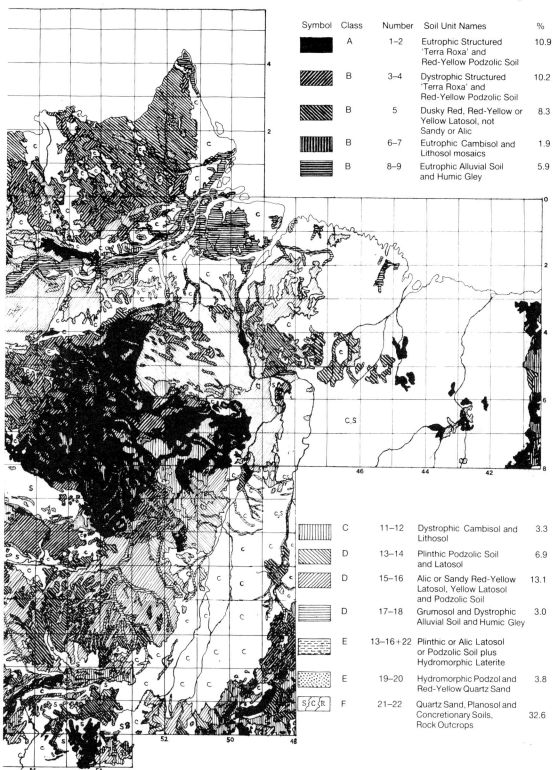

Symbol	Class	Number	Soil Unit Names	%
	A	1–2	Eutrophic Structured 'Terra Roxa' and Red-Yellow Podzolic Soil	10.9
	B	3–4	Dystrophic Structured 'Terra Roxa' and Red-Yellow Podzolic Soil	10.2
	B	5	Dusky Red, Red-Yellow or Yellow Latosol, not Sandy or Alic	8.3
	B	6–7	Eutrophic Cambisol and Lithosol mosaics	1.9
	B	8–9	Eutrophic Alluvial Soil and Humic Gley	5.9
	C	11–12	Dystrophic Cambisol and Lithosol	3.3
	D	13–14	Plinthic Podzolic Soil and Latosol	6.9
	D	15–16	Alic or Sandy Red-Yellow Latosol, Yellow Latosol and Podzolic Soil	13.1
	D	17–18	Grumosol and Dystrophic Alluvial Soil and Humic Gley	3.0
	E	13–16+22	Plinthic or Alic Latosol or Podzolic Soil plus Hydromorphic Laterite	
	E	19–20	Hydromorphic Podzol and Red-Yellow Quartz Sand	3.8
	F	21–22	Quartz Sand, Planosol and Concretionary Soils, Rock Outcrops	32.6

TABLE 2.2. Characteristics, typical vegetation, and classification of soil units of the neotropical forests.

No.	Soil unit name (Brazilian nomenclature)	Fertility	Drainage	Weathering Stage	Typical Relief	Micro-climate diversity	Typical vegetation	Root develop-ment	Humid micro-climates*	Vegetation in drier climates	Class
1	Eutrophic structured 'Terra Roxa'	4	G	A	Und	3	hill rain forest (dense and open)	++ H/V	3	hill rain forest (open), liana and semi-deciduous forest	A
2	Eutrophic Red-Yellow Podzolic Soil, Brunizem	4	G	A	Und	3	hill rain forest (dense and open)	++ H/V	3	semi-deciduous forest	A
3	Dystrophic structured 'Terra Roxa'	3	G	A	Und	3	hill rain forest (liana transition forest) (open)	++ H/V	3	hill rain forest (open), semi-deciduous forest, (open) transition forest	B
4	Dystrophic Red-Yellow Podzolic Soil	3	G	A	Und	2	hill rain forest (transition forest) (open)	++ H/V	3	hill rain forest	B
5	Dusky Red or Red-Yellow Latosol not sandy or alic	2–3	VG	AA	GnUnd	2	lowland and hill rain forest (dense)	++ H/V	2	lowland and hill rain forest (dense and open) transition forest	B
6	Eutrophic Cambisol	3	VG	FE	Und	2	hill rain forest, bamboo transition forest	+, H	3	montane forest, semideciduous transition forest	B
7	Eutrophic Lithosol (in mosaics)	3	EX	E	Mount	2	hill rain forest (dense), montane forest	+, H	3	montane forest, semideciduous transition forest	B
8	Eutrophic Alluvial Soil	3	M	FE–A	Level	1	floodplain forest (open and dense)	+, H/V	3	gallery forest (open and dense)	B
9	Eutrophic Humic Gley	3	P	FE	Level	1	inundated forest to savanna	+, H	3	inundated forest to savanna	B
10	(Andosol)	3	VG	FE	Level-Mount	2	transition rain forest	++ H/V	2	semideciduous transition forest	C
11	Dystrophic Cambisol	2	VG	FE	Level-Und	2	lowland and hill rain forest transition forest	+, H	2	transition forest, cerrado, savanna	C
12	Dystrophic Lithosol	1	EX	E	Mount	1	montane forest cerradão	+, H	3	semideciduous transition forest, cerrado	C
13	Plinthic Podzolic Soil	1	IMP	A	GnUnd	2	lowland rain forest, Amazonian savanna	+, H	3	transition forest, Amazonian savanna	D

No.	Soil	Fertility	Drainage	Weathering	Relief	Microclimate diversity	Vegetation	Root development	Probability of conservation*	Vegetation	Class
14	Plinthic or Humic Latosol	1	M	A	Level	1	lowland rain forest, Amazonian savanna	+, H	2	transition forest, savanna. cerrado	D
15	Alic Red-Yellow Latosol, Yellow Latosol, Dark Red Latosol	1	VG	AA	GnUnd	2	lowland rain forest, transition forest, Amazonian savanna	++, H	1–2	semideciduous transition forest cerrado, caatinga	D
16	Sandy Dystrophic Podzolic Soil	2	VG	A	GnUnd	1–2	lowland rain forest, transition forest	++, H/V	2	transition forest cerrado, savanna	D
17	Grumosol	3	M	A	Level	1	lowland rain forest, cerrado, savanna	+, H	2	cerrado, caatinga, savanna, desert	D
18	Dystrophic Alluvial Soil, Dystrophic Humic Gley	1–2	P–VP	E	Level	1	transition forest, inundated forest, inundated savanna	+, H	3	gallery forest in savanna	D
19	Hydromorphic Podzol	1–2	P	A	GnUnd-Level	1	campina forest, campina	+, V	1	open campina white sandforest	E
20	Red-Yellow Quartz Sand	1	EX	A	Level	1	white sand campina forest, lowland rain forest	++, H/V	1	open campina white sandforest, savanna, caatinga	E
21	Quartz Sand	1	EX	A	Level	1	lowland rain forest, cerradão, campina, Amazonian savanna	++, H/V	1	cerradão, savanna	F
22	Lateritic or Concretionary Soil, Planosol	1–2	P	A	Level	1–2	lowland rain forest, Amazonian savanna	+, H	1	Amazonian savanna	F

*Probability of conservation of a humid microclimate within the vegetation when macroclimate becomes dryer (3 = highest, 1 = lowest).

Abbreviations: Fertility: 4 = highest, 1 = lowest. Drainage: EX = excessive, VG = very good, G = good, M = moderate, IMP = imperfect, P = poor, VP = very poor. Weathering stage: AA = very advanced, A = advanced, FE = fairly early, E = early. Typical relief: Und = undulating, GnUnd = gently undulating, Mount = mountainous. Microclimate diversity: 3 = highest, 1 = lowest. Vegetation classes as those in Table 2.3; for more detail see Brown (1982a). Root development: H = horizontal, V = vertical.

may prove to have the greatest biotic endemism in addition to appreciable species diversity.

The principal vegetation types of lowland tropical South America (Table 2.3) are briefly described. Recent surveys make it possible to provide a much more accurate map than heretofore of the vegetation types of the Brazilian Amazon (Fig. 2.4). Terra firme *rain forest* occupies only about half of the area in perhumid or variably seasonal climates without serious water deficiency in the soil. Its most striking feature is an amazing species diversity. Most of these species have a poor capacity for dispersal. Within this same climate there are also extensive areas of edaphically controlled closed or open *forest on white-sand* soils, as well as some *Amazonian savannas*. In the regions where the climate is humid but distinctly seasonal or where there are soils inimical to rain forest development, there is a tendency towards *transition forests* which occur in part in a belt from a few to over 100 km wide between the rain forest and the various kinds of *savanna and savanna forest*. The driest climates have caatinga, a kind of *tropical thorn scrub*.

Several major forest formations occur. About 13 per cent of the Brazilian Amazon is clothed by periodically or permanently inundated vegetation types and one per cent by montane rain forests. The now extensive savanna and savanna forest including the cerrado, mainly occupies the planalto region of central Brazil. It has much local endemism. In various climates throughout the region there occurs Amazonian savanna, which is open grassland with some resemblances to (but important differences from) open parts (campo limpo) of the cerrado. The largest (Roraima-Rupununi) lies along the Brazil–Guyana frontier; the Venezuelan Gran Sabana and Colombian llanos are also extensive. Many species of Amazonian savanna are adapted to long-distance dispersal and the endemism varies greatly from little to much depending upon edaphic factors. Transition forests, of which various quite distinct formations have recently been discovered, are penetrated by gallery forest and interspersed with islands of cerrado and Amazonian savanna. There is no strong correlation between the various sorts of transition forest and soil type.

It is likely that in past times of drier climate transition forest was much more extensive than today and that this, rather than cerrado or Amazonian savanna, replaced much of the terra firme rain forest. Within the white sand forests the more open kinds were probably commoner but this distinctive forest did not extend beyond its specialized substrate. Transition forest and gallery forest are the kinds being felled most rapidly for agriculture at the present day. In order to interpret past oscillations in Amazonian lowland vegetation it is crucial to have detailed studies of them. They are not yet even accurately mapped.

Finally, a map is presented, Fig. 2.8, based on a synthesis of the vegetational, palaeoclimatic, pedological, and geomorphological evidence, which shows areas where humid forest probably persisted during the last, Wisconsin–Würm, ice age.

SOILS

Soils represent a fundamental factor in the development and persistence of characteristic biological domains. Under a given climate regime and without human interference, each major soil class tends to support a distinct and recognizable type of vegetation.

A number of different soil types occur under tropical forests in the neotropics today. These are often mapped as certain types of latosols and podzolic soils, or sometimes as mosaics of these with cambisols, lithosols, or alluvial soils (Brazilian nomenclature). In this section the characteristic soil units often found under various kinds of forests in a humid tropical climate are identified as far as possible. Each soil unit is then classified with relation to the vegetation which usually occurs on it in more arid or seasonal climates.

The principal soil units of the neotropics have been mapped in recent publications by FAO/UNESCO (1971, 1976). The soils of a large part of the Brazilian Amazon have been much more thoroughly mapped by the Projeto RADAMBRASIL (Projeto RADAM 1973–75, Projeto RADAMBRASIL 1975–83). The 22 soil units commonly encountered under neotropical forests are listed in Table 2.1 and these are grouped into six classes A–F and characterized in Table 2.2. The distribution of the soil units in the Brazilian Amazon is shown in detail in Fig. 2.1 and of the six classes in Fig. 2.2. A less detailed map for the entire

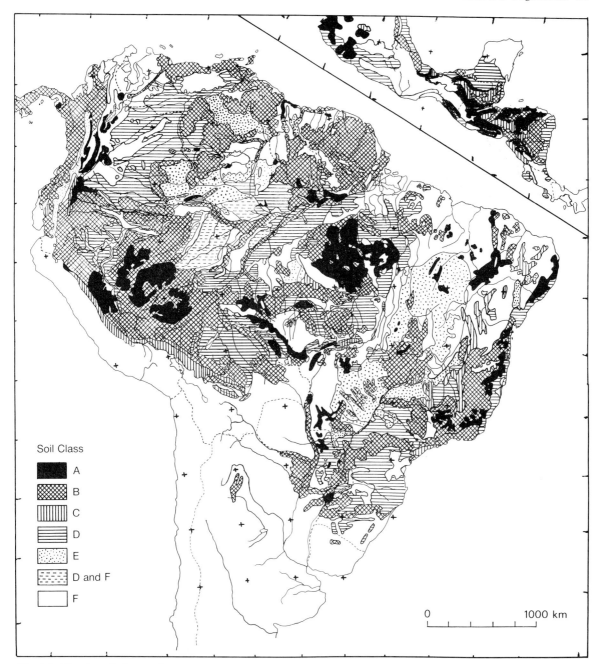

Soil Class

■	A
▨	B
▥	C
▤	D
▦	E
▦	D and F
□	F

0   1000 km

Fig. 2.2. The six soil classes of tropical America. These are progressively less likely to have remained forested during past dry epochs. See text and Table 2.2.

neotropics was published in Brown (1979, 1982*a*).

It should be noted that the soil units are in fact very fine mosaics of different soil types, in which the dominant soils may determine the principal veg-etation above them, but are not uniformly distributed throughout the mapping units. This is shown in Fig. 2.3 (taken from Brown 1979) for the region of Jaru, Rondônia (Brazil) which is possibly the richest area

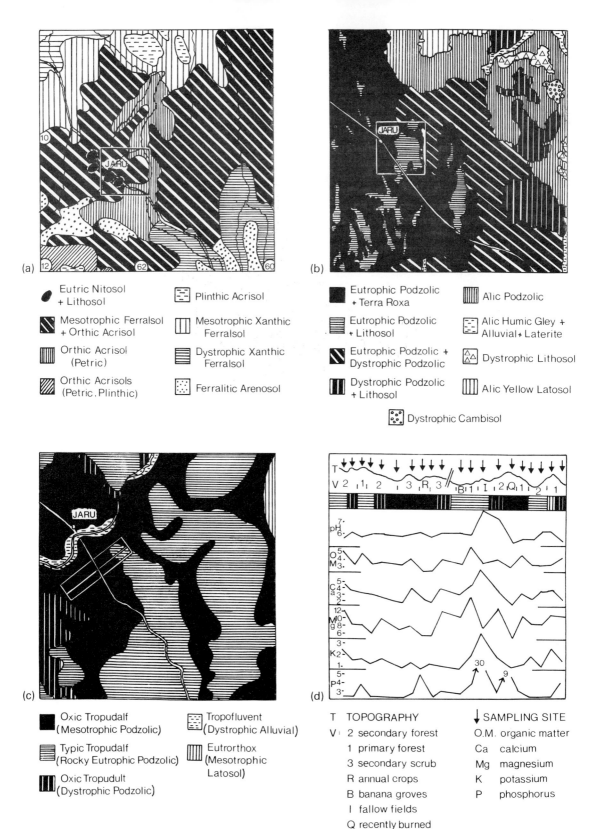

Fig. 2.3. Soil units in the region of Jaru, Rondônia, Brazil (10° 27′S, 62° 27′W), in four different classifications and scales; the small box in each of maps a, b, and c is enlarged in the following map.

(a) FAO/UNESCO, 1:5 000 000. (b) RADAMBRASIL, 1:1 000 000. (c) CEPLAC (Carvalho Filho and Leão 1976), 1: 200 000. (d) Sampling transect by Silvia Helena S. Arruda, and K. S. Brown Jr, every 200 m, 1:50 000 (soil units on the band coded map c). Note the point which had obviously been fertilized shortly before the sampling, and the drop of all values except P in the burned area (it had not rained since).

in the world for insects and plants (Brown 1976*c*; Prance as quoted in Absy and van der Hammen 1976). Four successively more detailed scales down to a transect of sampling points every 200 m show the progressively finer subdivision of homogeneous units into mosaics of distinct soil types. The occurrence of such mosaics, which is an important facet of tropical forest ecology (see van Wambeke 1978), is taken into account in the present analysis, which is, however, focused on broader biogeographical regions where single soil units may be dominant.

The nomenclature of tropical soils is variable and confusing, and exact equivalents in different classifications are impossible, since they are based on different criteria (Beinroth 1975; Sanchez 1976). The units are thus first named here in a generalized, very near Brazilian, nomenclature and the equivalents in the FAO/UNESCO and Soil Taxonomy classifications are given in Table 2.1. Of the three systems the Brazilian classification corresponds best with ecological reality in the neotropics, but the other two systems are also very useful and widely employed for comparisons.

The question of age and stability of soils under tropical forest is still debated by pedologists; suggestions for the rate of endogenous formation or exogenous accumulation of a metre of latosol in the Amazon basin vary over four orders of magnitude. For present purposes, a simple division can be made into soils according to the stages of weathering, which may or may not correspond to age (probably of the order of hundreds versus tens of thousands of years).

In Table 2.2 five measures are given of the capacity of each soil unit to make water and nutrients available to the rhizosphere of a forest. These measures are typical natural fertility, drainage, weathering stage, relief, and microclimatic diversity associated with topography. The next column indicates the vegetation most regularly associated with the soil unit on the RADAMBRASIL maps, in areas of less than three months' dry season in the Amazon Basin. The following two columns give additional criteria considered important for the preservation of this vegetation under a more rigorous climate in which water deficiency is more marked, due to higher temperatures, lower rainfall, or a longer dry season. These criteria are root penetration and conservation of humid microclimates within the vegetation during periods of drier macroclimate. There follows a column giving the vegetation which is usually found on the soil unit today where it occurs under less humid or more seasonal climates,

at the borders of the Amazon Basin.

In the final column an estimation is made of the efficiency of the soil unit in conserving tropical forest through a long dry period, such as that which terminated the Wisconsin-Würm glaciation (Chapter 1, Fig. 1.6). Six classes are recognized. Class A (soil units 1 and 2) is considered highly likely to have preserved tropical rain forest under such a climate, since it bears at least seasonal forests (but evergreen in ravines) under highly unfavourable climates today in the southeastern Amazon. Class B is usually favourable; Class C is sometimes favourable; Class D would only preserve large areas of forest under regimes at least as humid as today. Class E is considered mainly unfavourable, and Class F always unfavourable as to its efficiency in maintaining tropical forest. Both Classes E and F often bear savanna in the present.

Classes A–F are distributed in such a way that some regions are more likely to have had a continuous rain forest cover during climatic vicissitudes than are others. There are considerable areas where it is likely that a slight decrease in water availability would have led to the degradation of closed rain forest to transition forest or more open vegetation, becoming semi-deciduous in the more seasonal climatic regimes.

Much of the tall, heterogeneous, dense forest of the Amazon Basin subsists on a sandy and infertile soil (Sombroek 1966; Stark 1970, 1971*a, b*). It may only persist under climates at least as favourable as those of the present. In some cases, this forest may even represent a fairly recent recolonization from various different forest types of surrounding, more folded land – which may help to account for its apparent high species diversity.

It is also noteworthy that many of the more base-rich and fertile soils, which it is postulated should have supported the most drought-resistant tropical forests over long periods of climatic fluctuation (Table 2.2), are covered by mosaics of dense forest in ravines and around water-springs and open forest on higher rolling land. Such mosaics may prove, with further studies, to be the most stable of all neotropical forest systems, with the largest values for endemism in the overall biota and fairly high species diversity (if perhaps not necessarily in canopy trees), showing characteristics of selection for permanence in carefully organized and 'saturated' communities. The geomorphological evidence suggests a long history for this kind of ecological system as does its persistence in areas of seasonal climate in the southern Amazon today.

A number of subsoil characteristics have also been taken into account in drawing up Table 2.2. Any soils with reworked stone-lines in upper horizons have been regarded as unfavourable to long-term forest stability (class F) since these stones seem to have an exogenous origin in nearby rock or laterite formations, spread out on the surface under a sparse vegetation in the late Pleistocene (Chapter 1; Ab'Sáber 1977, 1982). Other unfavourable characteristics such as salinity, high content of toxic elements such as aluminium, hardpans or concretions, regional plinthite, peat, coarse sandy texture with excessive drainage, very low fertility or impeded drainage are all regarded as suggestive of the elimination of soils from consideration as supportive of long-term species-rich tropical rain forests. Soils at an early weathering stage or on unstable surfaces (alluvial soils, humic gleys, cambisols, and lithosols) have been considered favourable to the persistence of forest only when highly fertile or in mosaics with more stable and favourable soils, to which they could be contributing nutrients.

In summary, the soils which may be presumed to be good indicators of long-term stability of tropical rain forests through periods of climatic fluctuations, are the more fertile and fine-grained 'tropical forest soils'. These soils are usually formed slowly under dense vegetation in a warm, humid climate and are continually replenished with nutrients. They are often rapidly degraded when exposed to direct rainfall or strong wind action under sparse vegetation (Brown and Ab'Sáber 1979; Brown 1982a).

Details of the interpretations of the FAO/UNESCO and RADAMBRASIL soil maps, including the classification of all the soil units according to the criteria in Table 2.2, have been presented elsewhere (Brown 1979). Further comments may be found there and also in Brown and Ab'Sáber (1979) and Brown (1982a).

VEGETATION

Present-day vegetation

To understand the changes in vegetation cover in the past it is necessary to be familiar with at least the major, present-day vegetation types of the lowland neotropics and to have an explanation of terms such as savanna, cerrado, caatinga, and rain forest which have already been used in Chapter 1 and in this chapter in connection with soils. The contemporary vegetation of most of the Brazilian Amazon is shown on a map (Fig. 2.4) which is based on the recent survey by side-looking airborne radar (SLAR) plus air photographs in true and false colour and extensive flights and ground checks (Projeto RADAM 1973-5, Projeto RADAMBRASIL 1975-1983). Life zone maps have been published for Colombia (Espinal and Montenegro 1963), Peru (Tosi 1960), and Venezuela (Ewel, Madriz, and Tosi 1976) and a vegetation map for the whole of South America (Hueck and Seibert 1972). The most important vegetation types are summarized below but there are many subdivisions and transitional types which cannot be considered in detail here. The most commonly used summary of Amazonian vegetation is the classic work of Ducke and Black (1953, 1954). More up-to-date summaries are those of Pires (1973) and Prance (1975, 1978), and especially the reports of the RADAMBRASIL radar survey of Amazonia, of which 32 volumes covering all of Brazilian Amazonia are now available (Projeto RADAM 1973-5, Projeto RADAMBRASIL 1975-83), and the recent radar survey of Colombia (Proyecto Radargrametrico del Amazonas 1979).

The principal vegetation types are summarized in Table 2.3. As everywhere in the tropics vegetation is especially influenced by rainfall and its seasonality. However, in lowland Amazonia other factors such as soil, altitude, and topographic variation play an important role. It is only possible to arrange the vegetation types approximately according to the degree of water stress experienced, and a rigid scheme such as the Holdridge life-zones does not work well over a lowland area as large and varied as Amazonia.

Many classifications of tropical vegetation types especially those of the Old World tropics have been based on the classification of Schimper 1898, 1903) which recognized four main types, see also Champion (1936), Champion and Seth (1968), Burtt Davy (1938), Keay (1959), Richards (1952), Baur (1968), and Whitmore (1975). The groups 'Tropical thorn forest', 'Savanna and savanna forest', 'Tropical seasonal forest', and 'Tropical rain forest' do not apply well to the neotropics where there is a much greater diversity of vegetation and a much less distinct correlation between vegetation type and climate.

For example, major savanna areas within high

TABLE 2.3. Principal vegetation types of lowland tropical South America (see Fig. 2.4)

1. Caatinga (tropical thorn scrub)

   a. Arboreal
   b. Shrubby
   c. Parkland

2. Cerrado (central Brazilian savanna and savanna forest)

   a. Cerradão = savanna forest
   b. Campo cerrado = isolated tree savanna
   c. Campo limpo = parkland savanna

3. Amazonian savanna

   a. Savanna on terra firme
   b. Inundated savanna (campo de várzea)

4. Transition forests

   a. Semideciduous forest
   b. Mixed forest with palms
   c. Babassu forest (pure *Orbigyna*)
   d. Liana forest
   e. Bamboo forest
   f. Floodplain open mixed forest

5. Amazonian rain forests on terra firme

   a. Lowlands, relatively flat relief, dense forest
   b. Undulating and submontane relief (<100 m) hill forest

6. White sand formations

   a. Campina
   b. Campina forest (campinarana or caatinga Amazônica), and Wallaba forest
   c. Campina rupestre (sandstone rock outcrops)
   d. Coastal restinga (beach forest)

7. Inundated forest types

   (i) Swamp forest

   a. Permanent swamp forest (whitewater)
   b. Permanent igapó (black or clear water)

   (ii) Periodically flooded forest

   c. Mangrove forest (seawater inundation)
   d. Tidal várzea (tidal inundation, fresh water)
   e. Seasonal várzea (white water)
   f. Seasonal igapó (black or clear water)
   g. Flood plain forest (flash flooding)

8. Gallery forests of savanna areas

9. Montane forest types (over 700 m altitude)

   a. low montane (to c. 1000 m)
   b. cloud forest (upper montane forest)
   c. open montane habitats – subalpine vegetation
   d. rocky outcrops (Inselbergs)

10. Secondary forest and agricultural areas

rainfall and aseasonal areas include the edaphic savannas of southern Venezuela, the coastal savannas of Amapá, Brazil, the Humaitá and Puciari savannas between the Rio Madeira and Purus, the Ilha do Marajó in Pará, and the savannas of the upper Rio Negro region. In contrast the large lower rainfall, seasonal region, termed the Aw bridge in Chapter 1, includes a large block of dense terra firme forest between the Guianas and the Amazon River in the boundary region between the states of Pará and Amazonas.

Reason for this disparity must include the fact that tropical Asia is much more broken up into land-water strips which together with monsoon and typhoon effects could easily cause a more generalized climate/vegetation situation. South America is a very broad continent in which only the rivers break up the land and fewer mountain ranges occur east of the Andes. The different history of the two regions where the cooler periods of the Pleistocene had a more marked effect on the vegetation have also probably contributed to the disparity between vegetation type and climate patterns.

The vegetation types of tropical South America vary from the arid types where there is a very strong dry season (such as caatinga of northeastern Brazil and the thorn scrub of northern Venezuela and Colombia) to rain forest, occurring where there is no or virtually no dry season and very little water stress. It must be emphasized, however, that although it is convenient for discussion and comparison between different parts of the tropics to divide the vegetation types into major groups the exact boundaries are somewhat arbitrary. In reality vegetation is not always so sharply bounded and mixed forest is frequent in Amazonia. Thus, in Table 2.3 transition

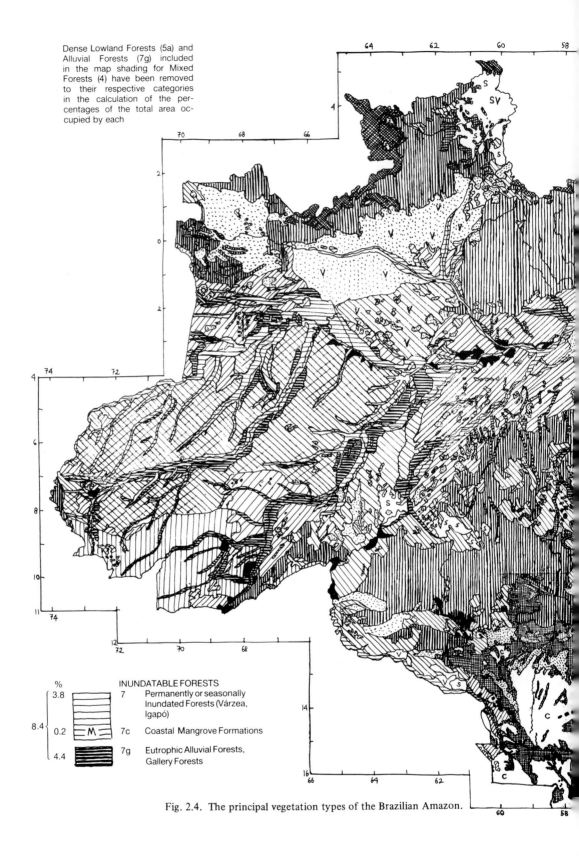

Dense Lowland Forests (5a) and
Alluvial Forests (7g) included
in the map shading for Mixed
Forests (4) have been removed
to their respective categories
in the calculation of the per-
centages of the total area occupied by each

INUNDATABLE FORESTS

%			
3.8		7	Permanently or seasonally Inundated Forests (Várzea, Igapó)
8.4 { 0.2	M	7c	Coastal Mangrove Formations
4.4		7g	Eutrophic Alluvial Forests, Gallery Forests

Fig. 2.4. The principal vegetation types of the Brazilian Amazon.

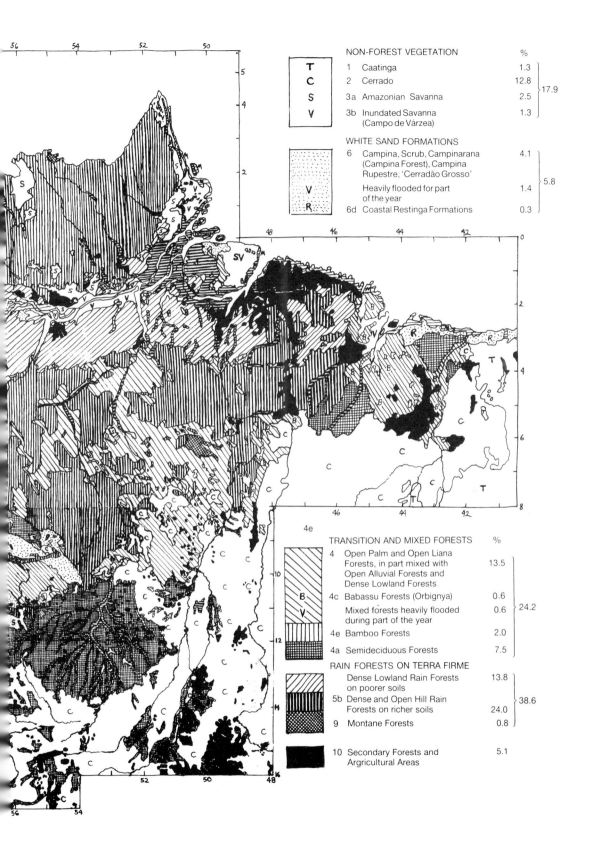

NON-FOREST VEGETATION %

1	Caatinga	1.3
2	Cerrado	12.8
3a	Amazonian Savanna	2.5
3b	Inundated Savanna (Campo de Várzea)	1.3

17.9

WHITE SAND FORMATIONS

6	Campina, Scrub, Campinarana (Campina Forest), Campina Rupestre, 'Cerradão Grosso'	4.1
	Heavily flooded for part of the year	1.4
6d	Coastal Restinga Formations	0.3

5.8

TRANSITION AND MIXED FORESTS %

4	Open Palm and Open Liana Forests, in part mixed with Open Alluvial Forests and Dense Lowland Forests	13.5
4c	Babassu Forests (Orbignya)	0.6
	Mixed forests heavily flooded during part of the year	0.6
4e	Bamboo Forests	2.0
4a	Semideciduous Forests	7.5

24.2

RAIN FORESTS ON TERRA FIRME

	Dense Lowland Rain Forests on poorer soils	13.8
5b	Dense and Open Hill Rain Forests on richer soils	24.0
9	Montane Forests	0.8

38.6

10	Secondary Forests and Argricultural Areas	5.1

forest is mainly found in seasonal climates and is to be considered mainly as the neotropical representative of tropical seasonal forest of the other parts of the tropics (see Schimper 1898, 1903) but has one type, bamboo forest much of which occurs where the climate is continuously wet.

Since climate, and especially seasonality, have a definite influence on the distribution of vegetation today, climatic changes, such as those which occurred in the Pleistocene (Chapter 1), will have caused changes in the vegetation. There will also have been an interaction with soil, as has already been discussed earlier in this chapter and is shown on Table 2.2. We shall return to the subject in the concluding section of the chapter. First, the principal vegetation types of Amazonia will be briefly discussed, following the arrangement of Table 2.3. It must be borne in mind that there is still a lack of accurate and detailed phytosociological data because much of the survey has produced forest inventories based largely on local names of the trees and often lacking in botanical precision.

1. **Caatinga**. This tropical thorn scrub occurs today mainly east of Amazonia in northeast Brazil in the region subject to the arid type B climate of the Köppen classification. Rainfall is low, often below 200 mm and rarely above 1000 mm per annum (Fig. 1.2, p. 4). The caatinga is a low, arboreal, deciduous scrubland which produces leaves and flowers during the short rainy season and is leafless and dormant for much of the year. Caatinga has been divided into three types by RADAMBRASIL. **Arboreal caatinga** (*Caatinga arborea*) is dense, tall scrub forest. **Shrubby caatinga** (*Caatinga arbustiva*) is lower and dominated by shrubs of a uniform height. **Parkland caatinga** (*Caatinga parque*) is an open savanna-like vegetation with scattered trees and many Cactaceae (e.g. *Cereus*, *Leocereus*, *Pilocereus*) and other plants, such as spiny legumes, adapted to xeric habitats.

Typical genera of the Brazilian caatinga include *Bromelia*, *Caesalpinia*, *Capparis*, *Euphorbia*, *Jatropha*, *Manihot*, *Mimosa*, and *Opuntia* and several endemic genera of Cactaceae such as *Leocereus*, *Tacinga*, and *Zehntnerella*.

The Brazilian caatinga is interesting because of its physiognomic similarity with the vegetation of the arid regions of northern Venezuela/Colombia and Central America. The distribution of many genera and species in these three separate areas of arid vegetation is suggestive of a more continuous distribution in the past. For example, the genus *Melocactus*

(Cactaceae) is common in both Venezuela and north-eastern Brazil, and *Licania rigida* of Brazil is most closely related to *L. arborea* of northern Colombia and Central America. These last two species form a vicariant pair. Long-distance dispersal of their drupaceous fruit seems unlikely and this also indicates an historical connection.

Typical genera of the Venezuela–Colombia arid region include *Acacia*, *Bulnesia*, *Bursera*, *Caesalpinia*, *Capparis*, *Cercidium*, *Cereus*, *Croton*, *Jacquinia*, *Lemaireocereus*, *Opuntia*, and *Prosopis*.

Sarmiento (1975) studied the arid vegetation types of South America and, using numerical methods, showed a close generic relationship between the Brazilian caatinga and the dry vegetation of the Caribbean area. There are 41 genera in common, a large number considering the depauperate flora.

Even within the extremely arid caatinga region, there are a few patches of forest on small hills of over 500 m altitude. These forest covered hills are called *brejos* locally. They have been described in some detail by Andrade-Lima (1982). The forest exists because of clouds attracted by the elevation. These small islands of forest contain many Amazonian forest species such as *Apeiba tibourbou*, *Virola surinamensis*, and *Norantea guianensis*. The brejos are present day refugia for Amazonian species within the caatinga biome. A more detailed study of these contemporary refugia would be helpful to reveal the characteristics and dynamics of small forest refugia.

2. **Cerrado** is savanna and savanna forest. It covers an area of about 1 500 000 km^2, dominates the planalto region of central Brazil, and borders the southern limit of present-day Amazonia. Cerrado is characteristic of a region with a markedly seasonal rainfall of 1500–2000 mm (Fig. 1.2). It mainly occurs at altitudes above 500 m but also at lower altitudes in some areas, especially in Mato Grosso.

Cerrado has been subdivided into many different types of which three are important. **Cerradão** is a low, dense, evergreen savanna forest (in the sense of Schimper 1903) in which the trees are 5–15 m tall and are close but do not form a continuous canopy, and where there are no open spaces. There is a distinct shrub layer, the ground has tufts of grass and terrestrial Bromeliaceae are frequent. **Campo cerrado** is a more open savanna with frequent, but isolated, trees up to 2 to 5 m in height, which are scattered more or less evenly throughout. **Campo limpo** is open grassland with very few trees and is commonest in areas which are permanently swampy, and on high hill slopes.

The trees of the cerrado are tortuous and often have a thick, fire-resistant bark. Suffrutices with underground trunks are common. These plants produce shoots which are burnt off in fires but sprout again quickly from the underground trunk. Such adaptations indicate that the cerrado is adapted to sporadic natural fires. The best summary of work on the cerrado region is that of Eiten (1972).

The cerrado has some characteristic tree species which are widespread throughout, such as *Byrsonima verbascifolia*, *Bowdichia virgilioides*, *Caryocar brasiliensis*, *Curatella americana*, *Dimorphandra mollis*, *Hancornia speciosa*, *Qualea grandiflora*, and *Stryphnodendron barbatima*. There is also much local endemism. Many species are restricted to small areas and each range of hills tends to have a distinct flora. This suggests that the cerrado is an old formation. Most important to a discussion of palaeovegetation are the numerous small 'islands' of forest which occur scattered throughout the cerrado region. Most of the mesophytic forest in the cerrado region occurs along the water courses as gallery forest, but in some areas of richer soils and more constant water availability, forest extends up slopes and into more open areas. The species of the gallery forest and forest islands are much more closely related to those of Amazonia than to those of the adjacent cerrado and these mesophytic forests also form fragmented present-day refugia in the planalto region for various forest species.

3a. **Amazonian savannas** are the open grasslands which occur scattered throughout the rain forest region. Some of the larger savannas occur within the drier and more seasonal parts of Amazonia, in the regions with under 2000 m of rainfall (Fig. 1.2). Amazonian savannas have a general physiognomic and floristic similarity with cerrado, especially campo cerrado, but are distinguished in several ways (Eiten 1978). They occur predominantly in lowland areas and have less local endemism and species diversity, few suffrutices, and less tortuous trees. They have several species in common with cerrado, for example *Byrsonima verbascifolia*, *Curatella americana*, *Hancornia speciosa*, *Palicourea rigida*, *Qualea grandiflora*, and *Salvertia convalariaeodora*. The distribution of such species has been used as evidence of previous more continuous distribution of savannas, but caution must be used here, for, in marked contrast to those of rain forest, many savanna species are adapted to long-distance dispersal (see Kubitzki, 1983). Bird- and wind-dispersed diaspores are much more common among savanna species than among rain forest species and this facilitates distribution between isolated patches of savanna.

The largest Amazonian savanna is that of Roraima-Rupununi which covers 54 000 km^2 at the Brazil-Guyana frontier. Others occur in subcoastal Amapá on the Suriname-Brazil frontier in the Tumucumaque region (the Sipaliwini savanna); the Rio Paru region of Pará and the Rio Trombetas-Cumina region of Pará; south of the Amazon around Santarém; near Humaitá in the Madeira river basin (3416 km^2, Braun and Ramos 1959); near the curve of the Rio Machado at the Mato Grosso/Amazonas/Rondônia intersection; the Rio Cururu region south of Serra de Cachimbo; and the Tarapoto region and Pampas del Heath in Peru. There are also many savannas in Amazonian Venezuela. These have been well defined by Huber (1982) who recognized three types: (i) the grassy llanos and llanos-type savannas confined to the north of the region; (ii) grassy inundated savannas of the Manapiare-Parucito basin; and (iii) the true Amazonian savannas of the central and western part of Amazonas Territory. The Amazonian savannas of Venezuela are characterized by a high endemism and the presence of strictly Amazonian floristic elements. Huber proposed that the Amazonian savannas, mainly of edaphic origin are centres of pre-Quaternary origin, whilst the llanos and the inundated savannas are relict areas resulting from the Pleistocene arid periods.

The savannas vary greatly in their botanical diversity and degree of endemism. The Roraima, Sipaliwini, and Humaitá savannas have rather few endemic species and Descamps, Gasc, Lescure, and Sastre (1978) stated that there is no endemism in the savannas of French Guiana. By contrast, a few savannas have an extremely rich and diverse flora, for example those of the Rio Cururu region in southern Pará which occur on sandstone. The sandstone savannas of the Guayana region of Venezuela are also richer in endemism than those of central Amazonia. In general, edaphic savannas which occur on special sites are richer than those which occur on terrain with conditions similar to those of the surrounding forest.

On the northern fringes of Amazonia other extensive savannas are the llanos in Colombia and the contiguous Gran Sabana in Venezuela. To the southwest of Amazonia there is an extensive upland savanna in the Departments of San Martín and Madre de Dios in Peru. Savannas also occur in southern Amazonian Brazil along the Rio Guaporé and there are extensive savannas in lowland Bolivia. These savannas occur on clay soils or yellowish soils in contrast to other open vegetation types such as campina (discussed below) which occur on white sand soils.

3b. **Inundated savanna.** Closely related to the upland savannas discussed under (2) and (3a) are the periodically inundated savannas, or *campos de várzea* as they are called in Brazil. The largest area of flooded savanna is that which occurs on Marajó island in the Amazon delta, but other flooded savannas also occur inland, for example along the Amazon and between the Rios Xingu and Tapajós, especially in areas of recent sediments and on eutrophic humic gley soils. One of the main differences from upland savanna is the greater abundance of Cyperaceae, e.g. *Cyperus giganteus*, which tend to replace the grasses, although there are still many species of grasses present e.g. *Panicum* spp and *Paratheria prostata*.

4. **Transition forests**, which in part form a belt between the high rain forest typical of Amazonia and the areas of cerrado and savanna, have several important types. They occur more frequently in areas with medium but markedly seasonal rainfall. Transition forests run along much of the boundaries of Amazonia and the Brazilian and Guiana shields and vary greatly in width from a few kilometres to over 100 km (see Fig. 2.4). Similar transition forest types occur to a lesser extent around the larger savannas especially those south of the Amazon river. The transition forests are taller and/or with a more closed canopy than cerradão and lower and more open than typical Amazonian terra firme rain forest. There are areas of open forest.

There is much **semideciduous forest** in the transition region to the south of Amazonia as well as large patches of other transition forest types of which the two most widespread are **Babassu palm forest** and **liana forest**.

Babassu palm forest is an open forest which is abundant in the southeastern border of Amazonia in Maranhão, Goiás, and Pará. It is a mixture of tall palms, mainly *Orbignya phalerata* and evergreen tree species which are widely spaced and do not form a closed canopy. Tree height varies from 10–25 m, and there are many clusters of *Orbignya* in pure stands. This type of forest has low species diversity and few endemics. It is probable that it covered wider areas during periods of drier and more seasonal climate.

Liana forest (called *cipoal* or *mata de cipó* in Brazil) is, as the name implies, rich in woody climbers. It is open forest with the trees well-spaced and often

completely entwined by the lianas. The tree species are some of those commonly found in typical Amazonian rain forest, but there is much less diversity. Lianas of the families Bignoniaceae, Malpighiaceae, and Menispermaceae are so abundant that it is impossible to walk through the forest without cutting through the tangled stems. Liana forest was little-known until the work of RADAMBRASIL drew attention to its extent in southern Amazonia, especially in the area between the Rios Tapajós, Xingu, and Tocantins. However, liana forest also occurs in other regions, for example in small patches in Roraima (Brazil) near to the savanna area and also in the southern parts of Amazonian Peru. There are no detailed studies of this vegetation type and no adequate explanation of the factors which cause it to occur. It is described briefly in Velloso, Japiassu, Goes Filho, and Leite (1974). Among the large, characteristic trees of the liana forest of the Xingu region are: *Astronium gracile, A. lecointei, Bertholletia excelsa* and *Elizabetha paraensis*. The small trees include: *Acacia polyphylla, Bauhinia* sp, *Cenostigma, Orbignya* sp, *Poecilante effusa*, and *Sapium marmieri*.

Falesi (1972) made a study of the soils of the liana forests along the Transamazon Highway from Marabá to Altamira. He observed that liana forest occurs on all types of soil found in the region, both on the poor soils and the more fertile 'terra roxa'. No correlation was found between soil and the occurrence of liana forest.

One of the most interesting features of the transition forest belt is the contact between the tropical rain forest floras of Amazonia and of the Atlantic coastal region of southern Brazil. Pires (personal communication) studied an area near Sarare (59° 30'W, 15°S in Mato Grosso) and found such typical Amazonian elements as forest species of *Enterolobium schomburgkii, Hernandia guianensis, Parkia, Schizolobium amazonicum, Simaruba amara,* and *Spondias lutea* mixed with others obviously from the São Paulo forest region as *Aspidosperma, Myroxylon,* and *Poeppigia procera*. He also found *Amburana* a genus characteristic of northeastern Brazil.

There is a group of species known from the belt of transition forest but which are absent from the forest of central Amazonia. Many of these species occur in an area from Peru around the south of the region, for example the mahogany *Swietenia macrophylla* and the genus *Myroxylon*.

There are also a number of species which occur both in the cerrado and in the open transition forest types; for example, *Apuleia leiocarpa*, *A. molaris*, *Erythrina ulei*, *Plathymenia foliosa*, *P. reticulata*, and *Physocalymma scaberrimum*.

The transition forest has been termed a 'zone of ecological tension' by the RADAMBRASIL vegetational survey. In places transition forest is interspersed with small islands of cerrado and elsewhere itself occurs as small patches in the cerrado. Within Amazonia it occurs mainly in the drier, more seasonal parts, especially in the region between the Rios Tapajós and Xingu where there is just such a cerrado-transition forest mosaic. Here active change is taking place with forest slowly replacing cerrado. Certain aspects of transition forests of this region are discussed in the phytosociological researches of Eiten (1975) and of Ratter, Richards, Argent, and Gifford (1973) for northern Mato Grosso.

A much greater extent of this mixture of transition forest and cerrado is likely to have occurred during the Pleistocene dry periods, and transition forest itself was probably more extensive.

Transition forests are, in broad terms, found in regions with a seasonal climate but, as elsewhere in the tropics, there are complex interactions between periodic drought and other facets of the environment. Rain forests extend further into seasonal climates on soils where periodic water stress is compensated by good water retention from either suitable soil physical properties or from topographical position, for example along rivers as gallery forests which penetrate a long way into the cerrado region. It is likely that soil nutrient status interacts such that rain forest occurs on the most fertile soils. The interdigitation probably works in the other direction too and it is likely that seasonal forests extend into areas with no or slight dry season on the most easily droughted sites, for example ridge crests and coarsely textured soils, and also on oligotrophic soils. Thus it is not to be expected that there will be an exact match between areas with a dry season and those carrying transition forest. In fact, the rain forest-seasonal forest ecotone remains very little studied in the neotropics. Indeed, the extent and diversity of transition forests has only recently been discovered. The transition area is an extremely complex mixture of open forest with or without dominance by palms, liana forest, patches of cerrado and gallery forest. It is still not mapped exactly because of its reticulate nature. Ecological investigation of this ecotone will be extremely rewarding and of great practical importance because of the present extensive disturbance and

destruction of many economically important transition forests (especially those including mahogany, Brazil nut, and babassu palm as dominant species).

Bamboo forest is an open transition forest type of limited distribution in a large area from Amazonian Peru across Acre and into Bolivia. It is particularly abundant in the State of Acre where it dominates large areas. Species of bamboo are abundant in the understory and even reach the canopy of 30 m, in places spreading to a crown of 10 m diameter. Bamboos are particularly abundant in openings along stream margins and in clearings. The bamboo forest was studied by Soderstrom and Calderón (personal communication). It contains few tree species, but those which occur are the typical species of terra firme rain forest. The bamboos never form pure stands, but are interspersed with trees which lend support to them. Some are thorny which make this forest extremely hard to penetrate. Soderstrom and Calderón collected three species of bamboo, two of *Bambusa* subgenus *Guadua* and one of *Merostachys*. One *Bambusa* (*B. superba*) forms dense clumps along the edges of rivers and streams and the other (an undescribed species) produces more open clumps and has culms that reach high into the trees for support. Soderstrom measured a culm which he cut down as 29.77 m in length. From the upper nodes the branches reach out into the trees for support. A side-branch measuring 9.69 m was recorded which in turn bore a branch of 4.96 m. The result is that forest bamboos spread far out into the canopy. Apart from the recent work of Soderstrom and Calderón and their brief report for Projeto RADAMBRASIL (1976) the only other publication which refers to any details of the bamboo forest is that of Huber (1906) where he has a section entitled 'quelques observations sur les Bambusees du Rio Purus'.

5. **Rain forest on terra firme.** This is the single most extensive forest type of present-day Amazonia (see Fig. 2.4). It occurs in areas where the climate is without a marked dry season on plateaux above the flood-level of the rivers, and occupies about 51 per cent of the region (over 90 per cent according to Pires (1973) but now realized to be an overestimate following the RADAMBRASIL survey).

The rain forest on terra firme is characterized by its closed canopy with some emergent species and its height, usually about 25–35 m but attaining 50 m in some areas. The biomass is much greater than in any other type of forest. The most striking feature of this rain forest is its amazing species diversity. All detailed studies of the forest have shown this. For example, Prance, Rodrigues, and da Silva (1976) found 179 tree species of over 15 cm diameter on a hectare of forest near Manaus and 236 of 5 cm diameter and over. Klinge (1973) found a total of 502 woody species per hectare in his studies around Manaus. However, in spite of its species richness the forest is remarkably uniform in its physiognomy throughout the region. Some variation with topography was detected by Takeuchi (1960). The floristic composition is by no means uniform and the Amazon has been divided into various phytogeographical regions. For example Rizzini (1963) recognized eight regions, and Ducke and Black (1953, 1954) also divided it into eight, rather different, regions. Prance (1977) (see Fig. 2.5) made minor modifications to the eight regions of Ducke and Black. These major divisions have their explanations in the history of the region and in the present-day climate. For example, the wetter Atlantic coastal region, the part of eastern Amazonia with a more seasonal climate with lower rainfall, and the much wetter, less seasonal, western region near to the Andes all fall in different phytogeographical regions. Amazonian forest has a mixture of widespread and almost ubiquitous species e.g. *Caryocar glabrum* (Fig. 2.6) and of extremely local endemics which are specially abundant in some areas, as will be elaborated upon in Chapter 3. The distribution of the numerous species within the rain forest is also dependent upon altitude, topography (Takeuchi 1960), soil, drainage, and climatic seasonality. Some species show very distinct habitat preferences and occur only in small areas.

Most trees of the terra firme rain forest have seeds which are initially dispersed by gravity and only move a short distance. However, in many cases secondary dispersal agents on the forest floor, such as rodents and other vertebrates, move them further. In marked contrast to those of cerrado and campina vegetation, terra firme forest species are mostly not adapted to long-distance dispersal. This must have an effect on the rate of expansion of the forest into potentially colonizable habitat although accidental and secondary dispersal give many gravity-dispersed species a faster rate of dispersal.

There is as yet very little evidence for structural and physiognomic differences in terra firme forest between the slightly different climates which exist in different parts of its range (compare Figs. 1.2, rainfall; 1.3, seasonality with Fig. 2.4), but the matter remains largely unexplored. Baur (1968), as a result of observations in tropical rain forest throughout the world, pinpointed slight differences clearly and for the first time. He described two different dry land lowland tropical rain forest formations and considered that both formations occur in tropical South America. These are subdivisions of the terra firme rain forest. Tropical lowland evergreen rain forest occurs in the least seasonal parts of the tropics and is replaced by tropical semievergreen rain forest in parts where the climate is slightly more seasonal. The former is the main rain forest formation of southeast Asia, the latter of Africa.

The recent publications of RADAMBRASIL have distinguished two main types of forest on terra firme associated with the topography, firstly forest on lowlands with relatively flat relief, and secondly hill forest on undulating land between 250 and 700 m elevation.

5a. **Lowland forest** (or dense lowland forest) occurs mostly on the area of Tertiary sediments between the major rivers and mainly below 250 m. It is characterized by the presence of emergent species up to 50 m tall and a canopy height usually of about 30 m. The understorey has few shrubs and consists mainly of young individuals of the canopy trees.

Lowland forest is generally on poorer soils than **hill forest** (5b) which replaces it at over c. 250 m altitude. Hill forest has a lower canopy and emergents only reach c. 35 m tall. The canopy is more open and thus the understorey is often much thicker. The physiognomy of hill forest is much more varied than that of lowland forest with marked differences between the slopes and valleys. The species content is also rather different. Hills between 250 and 500 m often bear denser forest which varies from 10-15 m in height on the spurs and up to 25 m in the valleys.

There is a large area of terra firme forest in western Brazilian Amazonia that is much more interspersed with open forest (Fig. 2.4) than that of central and eastern Amazonia. This area, mainly south of the Rio Solimões and bordered to the south by the bamboo forest of Acre, is a complex mosaic of tall closed rain forest on terra firme, and more open forest and forest on alluvial areas. It has been separated on Fig. 2.4 from the rest of the terra firme forest because of its more open nature and its physiognomic similarity to some of the transition forests.

The Amazon–Orinoco hylaean forest is by far the largest part of the neotropical rain forest. There are also two other much smaller, detached blocks. These

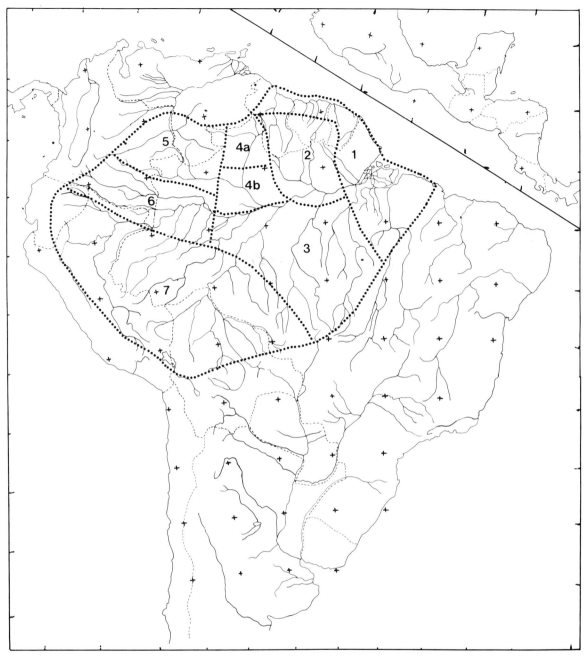

Fig. 2.5. The seven major phytogeographic regions of Amazonia. 1. Atlantic Coast; 2. Jari–Trombetas; 3. Xingu–Madeira; 4. Roraima–Manaus; 5. Northwest–Upper Rio Negro; 6. Solimões–Amazonas west; 7. Southwest (Prance 1977.)

are mainly terra firme rain forest though both have small areas of other types especially of inundated forest. West of the Andes rain forest occupies the Pacific coast of Colombia and extends south to the extreme north of Ecuador and north into parts of Central America, namely most of Panama, the Osa Peninsula of Costa Rica, and extending in the Caribbean drainage as far as southeastern Mexico. This

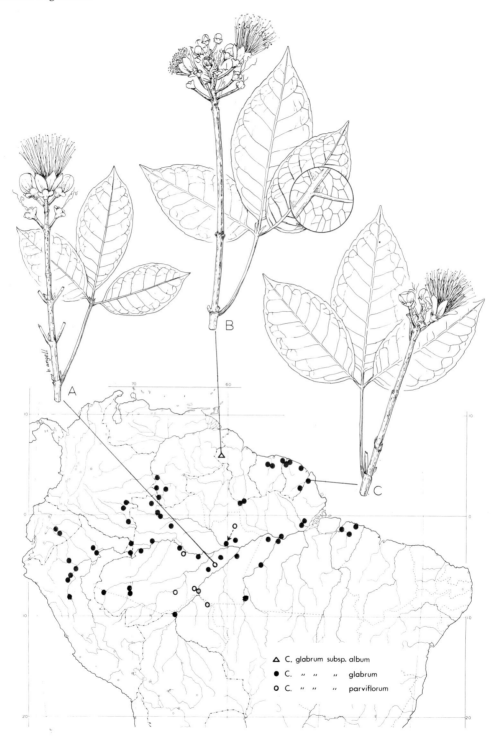

Fig. 2.6. The distribution of *Caryocar glabrum*, showing the morphological variation of the three varieties.

second block now persists in Middle America only as relict fragments of which the three biggest are Darién (Panama), Mesquitia (Honduras/Nicaragua), and Chiapas (Mexico). On the Atlantic coast of Brazil from Pernambuco south to Santa Catarina the flanks of the Serra do Mar and the coastal plain have the third rain forest block. This too has become greatly reduced and fragmented.

6. **Vegetation on white sand soils.** Within Amazonia there are various areas where podzols of leached white quartz sand soils or regosols occur. This nutrient-deficient habitat supports distinctive vegetation which varies from areas of open savanna to closed low forest. This is the formation called heath forest in the Eastern tropics and which is virtually absent from Africa. A reduced number of species with a tendency towards dominance is characteristic of all white sand areas. In some parts there is local endemism. All vegetation types on white sand have a tendency towards small and scleromorphic leaves and gnarled trunks, but there is otherwise great diversity in forest structure and physiognomy between types. An excellent summary of the white sand vegetation is given in Anderson (1981).

The white sands have a variety of origins. The belts of sand in the Guianas are former sea beaches, those of central Amazonia are former river beaches or beaches created by the inland sea at the time of maximum embayment of Amazonia. Other sandy areas are derived from sandstone rocks.

Wallaba forest. In the Guianas there are both forests and open savannas on sand. The densest forest is dominated by *Eperua falcata* and two other species of *Eperua* locally called Wallaba.

Campina (forest) (campinarana of RADAMBRASIL; sometimes called Amazonian caatinga; see Lisboa 1975; Klinge and Medina 1979; Anderson 1981). In Brazil there is much forest on white sand in the upper Rio Negro region. This forest is over 18 m tall and the trees tend to be of small diameter, rather tortuous and not forming a closed canopy. The understorey and ground vegetation is extremely rich and contains many endemics in such families as Bromeliaceae, Marantaceae, and Rapateaceae. This forest type was described and analysed in some detail by Rodrigues (1961*b*). The campina forest of the lower Rio Negro around Manaus is dominated by such species as *Aldina heterophylla*; *Humiria balsamifera* (which is also common on the Guianas and the coastal restinga) and *Glycoxylon inophyllum*. The trees are twisted and are loaded with large numbers of epiphytic orchids, Araceae, Gesneriaceae such as *Codonanthe* (always growing in ant gardens), and pteridophytes. In the lower Rio Negro the campina forest is lower and poorer in species, and many open areas occur which are locally called campinas. The open areas of sand in campinas are covered by the blue-green alga *Stigonema tomentosum* and clumps or islands of shrubs and low trees occur with abundant lichens (*Cladonia* sp) around their perimeters. It is possible that some of this area was opened up from campina forest by Indian activity about 800 years ago (Prance and Schubart 1978). The vegetation of open campinas is described in some detail in Anderson, Prance, and de Albuquerque (1975).

The campina forest and campinas were also described and discussed by Klinge and Medina (1979) who related them to the heath forests of Asia and listed some of the characteristic white-sand species based mainly on the incomplete lists of Takeuchi (1960). They draw attention to the extensive area of white sand vegetation in the Rio Negro region of Brazil and Venezuela. A survey of the species which occupy some open campinas (Macedo and Prance 1978) showed that three-quarters of the species are adapted to long-range dispersal by birds, bats, or wind. Thus, as in the case of savannas, caution must be used in drawing conclusions from the distribution of species of the white sand areas, as they too have a much higher capacity for long-range dispersal than the species of the rain forest.

One of the most interesting areas of open vegetation on sandy soil is that of Serra do Cachimbo on the Pará-Mato Grosso border in Brazil. This range of hills, 400–500 m in altitude, is predominantly sandstone. The dominant vegetation is extremely similar in its physiognomy to that of the campinas of the lower Rio Negro region (Lleras and Kirkbride 1978). Previously, this vegetation was classified as cerrado (see for example Projeto RADAM 1974) because the area around the airport was the focus of study but in fact that small area is a savanna on sandstone, whose vegetation is physiognomically similar to cerrado. Actually most of Cachimbo is covered by campina which has an interesting mixture of species characteristic of Amazonian campinas and those of the cerrado. For example, such typical cerrado species as *Caryocar brasiliensis* and *Parinari obtusifolia* occur. To-day, the populations of cerrado species are isolated from the cerrado zone to the south by a belt of Amazonian rain forest. Both the species mentioned have large and heavy fruits which are unlikely to have been dispersed across an area of continuous rain forest and this isolated occurrence is indicative of a continuous connection at some stage. Campina forest also occurs in other isolated areas of sandy soil, with a similar physiognomy and appearance to that of the Rio Negro, but with a rather different species content. For example, there are quite large patches in the Chapada dos Parecis in Rondônia where many of the same genera occur as in the Rio Negro, such as *Abolboda, Clusia, Humiria, Paepalanthus, Retyniphyllum, Syngonanthus, Ternstroemia, Tovomita,* and *Xyris*. Another large area occurs northwest of Cruzeiro do Sul in Acre, between the Rio Moa and Ipixuna and was described in Projeto RADAM-BRASIL (1977).

Campina rupestre is the term given to the campina-like vegetation which occurs on open rocky or degraded sandstone rocky areas. Physiognomically and in species composition this is similar to campina on sandy soil. Particularly common in campina rupestre in northern Amazonia are the genera *Clusia* and *Retyniphyllum* and various Theaceae. Cactaceae and xerophytically adapted orchids such as *Cyrtopodium* can also occur in these areas. Campina rupestre is abundant on the Serra do Cachimbo in southern Pará.

Restinga is a low scrub which occurs along the coast of Brazil. It is found on the sandy soil of former sand dunes beyond the influence of salt water, inland from sandy beaches and sometimes from mangrove forest too. It is not extensive in Amazonia where it occupies a narrow, interrupted belt running eastwards from near the Amazon delta in northern Pará and Maranhão. Towards the east restinga sometimes resembles caatinga or cerrado in physiognomy. In the Amazon it is more open than in eastern coastal Brazil where it ranges as far south as Santa Catarina. Restinga is rather uniform and is best characterized by the presence of the shrub *Chrysobalanus icaco*. Amongst the herbs *Bulbostylis capillaris* and *Ipomoea pes-caprae* are abundant.

It is hard to imagine that the various kinds of vegetation on sandy soil ever had a much wider distribution in past drier periods, since they are tied to the specialized habitat of white sand. The disjunct ranges of the campina species can be understood in light of the predominance of long-distance dispersal (Macedo and Prance 1978) and does not require the explanation of a previous, more continuous distribution.

7. **Inundated forest types**. The periodic heavy rainfall and consequent rise and fall of river levels creates substantial areas of both periodically and permanently flooded forest. Inundated forest has less species diversity than forest on terra firme, and also has less regional endemism, although some striking endemism occurs such as *Polygonanthus amazonicus*, known only from flooded beaches around Maués in central Amazonia.

Several kinds of inundated forest can be recognized, divided between permanently and seasonally flooded types. They have been described in detail by Prance (1979).

(i) Swamp forest
a. **Permanent swamp forest (whitewater)**
b. **Permanent igapó (black or clear water)**
Both types of permanent swamp forest have a low species diversity and cover only limited areas of Amazonia. The soil under the forest is a eutrophic humic gley. In some areas of dystrophic humic gley palm swamp occurs. In Amazonia this is usually dominated by species of *Mauritia*, but *Euterpe* is also common in palm swamp along with several other genera of palms.
(ii) Periodically flooded forest
c. **Mangrove forests** develop along the coasts flooded twice daily by silt-laden seawater. In the Amazon estuary the dominants are species of *Avicennia, Laguncularia,* and *Rhizophora.*
d. **Tidal várzea**. This forest occurs in the delta region up to 100 km from the sea. It is flooded twice daily by fresh water as the result of back-up caused by tide. The forest is similar in species composition to seasonal várzea.
e. **Seasonal várzea**. This is forest periodically flooded by white-water rivers. It is tall and physiognomically similar to upland forest but is less diverse in species and has more lianas.
f. **Seasonal igapó** is forest periodically flooded by black- or clear-water rivers. It is lower with fewer species than that of white-water rivers and many of the trees show scleromorphic adaptations. This type of forest is described in Rodrigues (1961*a*). Many of the species are different from those of seasonal várzea.
g. **Flood plain forest** develops on areas which are flooded at irregular intervals by quickly draining flash floods. This type of forest occurs mainly in the upper reaches of the rivers and has greater diversity of species than other inundated forest types since many of the species of terra firme can survive the short periods of inundation.

The inundated forests are important because they can persist, although with decreased area, under drier climate regimes. Although many tree species from the terra firme forest do not survive flooding, they can exist on the fringe of inundated forests in the same way that forest species survive today in the outer edge of the gallery forest of the cerrado region of central Brazil.

8. **Gallery forest**. Today all major savanna areas such as the Colombian llanos, the Roraima-Rupununi savanna, and the cerrado of central Brazil have abundant gallery forests along the watercourses. Gallery forests exist within a drier vegetation type because of the availability of water. They are not usually subject to long periods of flooding but only to flash flooding of the forest nearest to the watercourses. The forest is present because near to the streams the watertable is nearer to the surface, and consequently water is available to the trees which are often rather deeply rooted. With the greater availability of water, plants can produce a greater biomass than in the nearby savannas, and as a result forest grows rather than savanna. The gallery forest of the cerrado contains some species which are characteristic of Amazonian terra firme rain forest. Figure 2.7 shows the distribution of *Cariniana estrellensis* a common forest species of the coastal forest around Rio de Janeiro and São Paulo which extends its range to Amazonia through the gallery forests of the cerrado region. Similar continuous distribution of forest species must have occurred in Amazonia in the times of reduced forest cover. Thus, while many species were isolated into several populations in refugia, some others continued to have a more continuous distribution through the gallery forest. The gallery forests of the cerrado region contain a mixture of forest species characteristic of Amazonia and of the Atlantic coastal forests. Gallery forest also have another important function, as a refuge for animals during the burning of the cerrado. The role of gallery forest in the past drier periods has been underestimated by workers on refugia.

9. **Montane forests**. Within Amazonia these occur at altitudes of over 700 m, and mainly around the northern and southern fringes. Depending on the height of the mountains the vegetation consists of **lower montane rain forest** on the slopes (characteristically between 700 and 1000 m) with **upper montane rain forest**, often called cloud forest, above. This is a very humid elfin forest which is full of epiphytic bryophytes and lichens. Lowland species do not generally reach as high as the cloud forest. Lower shrubby or savanna **subalpine vegetation** occur at the higher altitudes, as in the Guayana highlands, Pico da Neblina, and on the eastern slopes of the Andes. The distribution of species between the different areas of upper montane rain forest is of interest and is profoundly influenced by the type of rock of the mountain. There is much more endemism on the sandstone tops of the Guayana highlands than granite mountains of the Guianas.

Rocky outcrops have a specialized unique vegetation which, in many areas, includes plants adapted to withstand seasonal drought. There are rock outcrops (inselbergs) within Amazonia. Cactaceae and orchids with large pseudobulbs such as *Cyrtopodium andersonii* are quite common. Many of the species characteristic of rock outcrops are widespread and occur on many different outcrops widely separated geographically.

Montane forest types are common outside Amazonia on both sides of the Andes, the coastal cordillera in Venezuela, Sierra Nevada de Santa Marta, Panama and Central America, and in the highlands of Atlantic coastal Brazil. The considerable variation of these different montane forests is not discussed here.

10. **Secondary forest** is of increasing importance in Amazonia owing to the amount of felling of primary forest which is taking place. Obviously, the species which now dominate secondary areas also had a natural distribution prior to the occurrence of man-made secondary forest. A few secondary forest species are found throughout the tropics, others grow in open spaces beside rivers, on old landslides, or where forest has been felled by severe storms. Because the seeds and stumps have not been destroyed by fire, regenerated vegetation of naturally cleared areas is different from that on burned areas. Secondary forest species do not generally occur in the transition regions between forest and savanna and were probably of no importance in the natural changes of vegetation which took place during Pleistocene climatic variations. Modern vegetation maps show an increasingly large amount of a new man-made vegetation and also of agricultural areas which it is not necessary to discuss here.

Vegetation history

This brief summary has shown that the most extensive vegetation in Amazonia is lowland terra firme rain forest which occupies about half of the total area. In

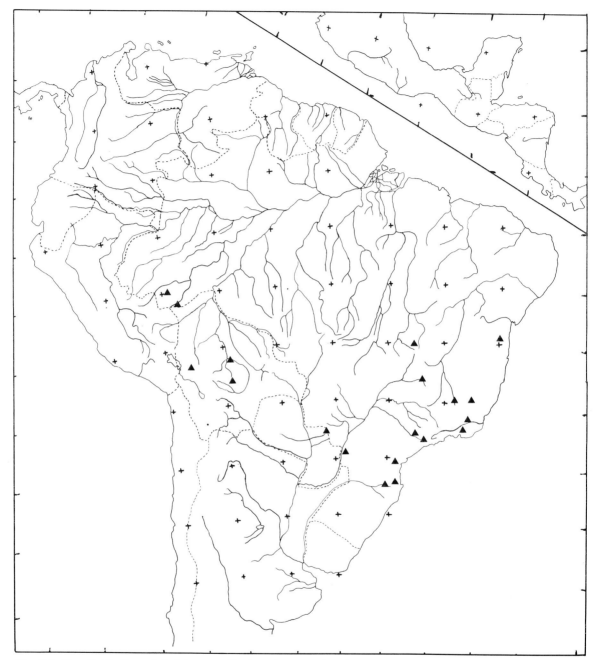

Fig. 2.7. The distribution of *Cariniana estrellensis*, a forest species of southern Brazil and Amazonia connected through gallery forest.

addition many other kinds of vegetation occur interspersed with the rain forest because of local variations in climate, soil, topography, and water conditions. We now turn to consider the historical changes in balance between these different vegetation types.

The climatic changes of the past undoubtedly did cause variations in the area covered by the different vegetation types. Today the wetter, less seasonal, parts of the region have a more continuous covering of rain forest. Cerrado, Amazonian savanna, and the various sorts of transition forest (especially Babassu palm and liana forest) are all much more extensive in the parts of Amazonia with a lower and more seasonal rainfall, in a belt which extends from the Gran Sabana region of Venezuela, through the Roraima-Rupununi savanna region, the Rio Trombetas–Rio Paru region, Santarém and the area between the Tapajós, Xingu, and Tocantins rivers and then to the cerrado in northern Goiás (see the vegetation map Fig. 2.4). This area is referred to as the 'Aw climatic bridge' in Chapter 1 (Fig. 1.3) and Fig. 1.2 shows the 2000 mm isohyet. It can be seen that there is some correlation between seasonality, annual rainfall and the distribution of vegetation types.

On the other hand, it is significant that certain contemporary savannas occur in regions of much higher present-day rainfall. Eden (1974) studied two such savannas in Amazonian Venezuela, La Esmeralda with an annual rainfall of 2900 mm and Santa Barbara with 2700 mm. He concluded that these isolated tracts of savanna within the wetter climatic regions are relics of the more widespread savanna which existed in drier periods. He found that the distribution of savanna is not necessarily correlated with adverse soil conditions, or with the distribution of soil types. He concluded that burning is one of the reasons for the persistence of these savanna areas in a wetter period, but that they were not created by fire. The presence of these savannas is good evidence for the former more widespread distribution of this vegetation type. The range of certain of the terrestrial savanna animals, e.g. the South American rattlesnake *Crotalus durissus* (see Müller 1972, 1973) provides corroboration. It is improbable that this and other non-forest animals dispersed through rain forest to reach their presently isolated non-forest stations.

It is significant that the region which falls in the belt of Aw climate is not covered predominantly by savanna, but by a mixture of various types of transition forest such as liana forest and Babassu palm forest and even rain forest north of the Amazon River (see Figs. 1.3 and 2.4). In most papers discussing past vegetation fluctuations emphasis has been placed on the dramatic change from tall species-rich rain forest to savanna. However, it is probable that the drier climates of the Pleistocene did not cause nearly the whole region to be turned into savanna, but instead much of it probably developed more open or lower types of forest. In some cases

these could have served as a barrier between refugial areas of denser forest often preserved by favourable soils. Obviously though, the area of savanna, and possibly the northeastern area of caatinga, did increase enormously, especially adjacent to already existing tracts where the change could take place with facility.

The distribution of vegetation types at the last cold dry phase of the Pleistocene, 18 000–13 000 BP (the Wisconsin–Würm glaciation) as deduced entirely from geoscientific data by Ab'Sáber (1977) was shown in Fig. 1.6. Between the isolated humid areas where he deduced that rain forest persisted would have lain a complex mosaic of the different vegetation types of seasonally dry climates which would have been an effective barrier to the dispersal of most rain forest plants and animals, though some would probably have kept a more continuous distribution by persisting in gallery forest, some of the types of transition forest, and small 'islands' of rain forest within cerrado, just as seen at the present day.

The only direct biological evidence for the past distribution of the different lowland vegetation types comes from the few, widely scattered pollen diagrams. This palynological evidence was reviewed in Chapter 1. It does indeed show that the vegetation of seasonal, relatively dry climates has been more extensive in the past, but the data are far too scanty to be useful for mapping past vegetation.

Even in the absence of direct evidence it is instructive to consider likely details in the past changes in vegetation. The rain forest at the present day is interspersed with blocks of varying size of the different kinds of more xeric vegetation (Fig. 2.4 and discussion above). Xeromorphic forests on white sand are now known, from the recent RADAM survey, to be much more extensive than was formerly believed. They include closed forests (campina forest or Amazonian caatinga) and more open forests (campina and campina rupestre). Observations on white-sand areas near to Manaus have revealed that succession is taking place from open to closed forest with a concomitant increase in biomass and species richness (Anderson *et al.* 1975). It is thought that this change is an adjustment to a more humid climate and that, on white sand, past climatic fluctuations were accompanied by oscillations between open and closed forest. It is unlikely that this group of forests ever extended beyond the white sand because they are edaphically determined and their flora is restricted to this type of site. Ab'Sáber (1977) expressed the opposite view but did not present any evidence.

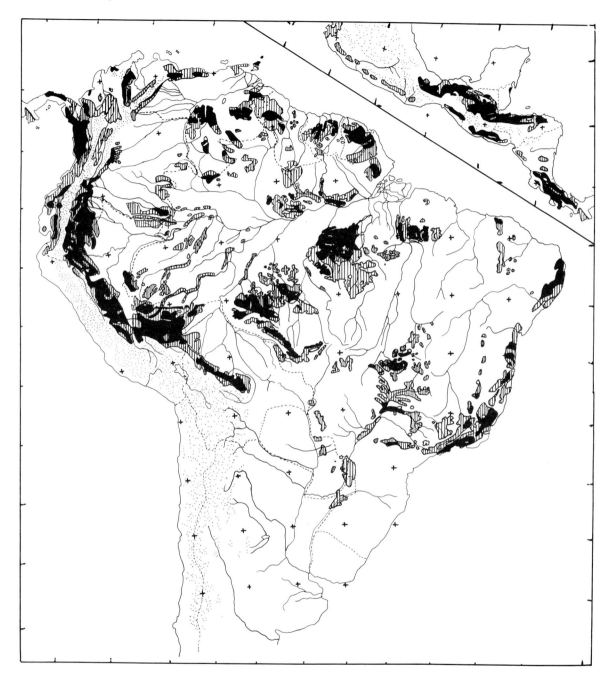

Fig. 2.8. Areas where tropical humid forest probably persisted in the last (Wisconsin–Würm) ice age deduced from all the different lines of geoscientific evidence, see text. Sixty per cent (hatched) and 80–100 per cent (solid black) likelihood of persistence are shown based on overlap of positive evidence. (After Brown 1979, Fig. 132).

Natural reafforestation at the present day has also been reported from the Guianas by Descamps, Gasc, Lescure, and Sastre (1978), in this case on rocky outcrops from which certain characteristic species of open dry areas are disappearing, e.g. *Clusia* spp, *Pitcairnia geyskesii*, and *Sipanea pratensis* var. *dichotoma*. Descamps and his colleagues have also suggested that rocky outcrops and certain savannas with xeromorphic species are good evidence for drier palaeoclimate. These areas occur today as isolated refuges of xeric flora and fauna surrounded by rain forest. They believe that the present-day landscapes of the llanos savannas of Colombia resemble the landscape of the Guiana plateau during the drier periods.

The relationships between Amazonian savanna and cerrado also throw light on past vegetation. These are two distinct vegetation types despite similarities in physiognomy and some species in common (e.g. the ubiquitous *Curatella americana*). Most of the characteristic species of cerrado are absent from the Amazonian savannas, e.g. *Caryocar brasiliensis* and *Diospyros hispida*, whose flora is much poorer. These disjunct savannas within the Amazonian rain forest have progressively fewer typical cerrado species the farther north they are away from the cerrado region (Eiten 1972). One would expect a much stronger cerrado floristic element in these savannas if the cerrado had encroached into Amazonia during the Pleistocene. Its absence suggests that instead the rain forest was replaced by transition forest, or perhaps in places by caatinga as Ab'Sáber (1977) has suggested.

If transition forest was indeed at times more extensive, then species peculiar to it would have had wider ranges and be today relatively restricted and refugial. Such patterns have been found to occur in climbing Bignoniaceae which are restricted to the liana type of transition forest (Gentry 1979).

It can be seen that in order to interpret past oscillations in Amazonian lowland vegetation it is crucial to make detailed studies of transition forest and also of gallery forest. At present not even their distribution is accurately mapped. Yet these are the forests being felled most rapidly for agriculture so the need is urgent. Once species composition has been ascertained it may be possible to interpret pollen profiles from earlier times to indicate these vegetation types. At present the limited pollen diagrams available from the lowlands show the presence or absence of grass pollen which indicates savanna or cerrado, but not the transition forest types. Much of the shrubby caatinga also lacks grass and future pollen studies could indicate the extent of the expansion of the caatinga.

The ranges of individual plant species give much more substantial evidence for changes in vegetation and hence refugia and to this we shall turn in Chapter 3.

AREAS WHERE HUMID TROPICAL FOREST PROBABLY PERSISTED

A direct summation of palaeoecological evidence for and against continuity of integrated humid tropical forest systems during the late (Wisconsin–Würm) ice age has been undertaken by Brown (1979). Humid forest is used here as a general inclusive term for all those forests adequately humid to include rain forest organisms, that is rain forests in the limited sense of the previous pages plus parts of the other broad forest formation groups. In the summation one point each was assigned to the tropical forest refuges based on palaeoclimatic, pedological and geomorphological considerations (Ab'Sáber 1977; Brown and Ab'Sáber 1979; Fig. 1.6), favourable areas for continued high rainfall under the modified palaeoclimatic regime (Brown 1979; Fig. 1.6), favourable vegetation formations (Fig. 2.4), and usually favourable soils (class B, Table 2.2). Two points were given to always favourable soils (class A, Table 2.2) and a point was subtracted for unfavourable soils (classes E and F, Table 2.2) and those areas of Amazonian savannas with high endemism (Fig. 2.4). The total possible was thus five points; the summation produced a map for probability of humid forest stability according to superposition of the four different lines of palaeoecological evidence. The regions of higher probability are shown at two levels in Fig. 2.8. With the inclusion of negative elements in the summation this model represents a strong averaging in which concordant data are reinforced and discordant data eliminated. This model was used by Brown (1979) for independent correlation with the biogeographical patterns detected in butterfly subspecies. It resembles all of its contributing data-sets somewhat, but none closely.

3
BIOGEOGRAPHY OF NEOTROPICAL PLANTS

SUMMARY

The neotropics, especially Amazonia, remain one of the most undercollected parts of the world. If allowance is made for the artefacts due to unequal sampling (Fig. 3.1), range-maps show some areas have far more species than others. The botanical data support the existence of forest areas with high endemism. These are interpreted to have remained stable through drier climatic periods. In Mexico and Central America eight refugia have been recognized (Fig. 3.2). Present-day species ranges in the Andes show clear evidence of the influence of past vertical movements of the vegetation belts as the climate has fluctuated (Fig. 3.3). In the mountains east of the Andes localized endemism is strongly influenced by the great variety of soils and habitat types (Fig. 3.4). In the terra firme rain forests of lowland South America twenty-six centres of endemism can be recognized (Fig. 3.6). Eight of these lie in the Pacific coast and Caribbean forest (numbers 1–8). Three (24–26) lie in the Atlantic coast rain forest and the remaining fifteen (9–23) lie in the great block of the Amazonian–Orinoco–Guayanan rain forest. These centres occur within the continuous mantle of the rain forest. It is also possible to detect centres of endemism in dry forest patches, for example in Bignoniaceae, and in the isolated humid forest patches known as brejos associated with small mountains in the dry caatinga of northeast Brazil.

It has been suggested by Morley that *Mouriri*, a genus of rain forest trees, evolved in response to habitat specialization not to refugia, but critical re-examination shows that the species of *Mouriri* which inhabit terra firme rain forest have ranges which fit the recognized refugia.

Little is known about speciation in refugia but it is believed to have been an important cause of the species-richness of the rain forest, in addition to other previously recognized modes of species evolution. Speciation must now be viewed against continual environmental instability not stability as was formerly believed. For example, the occurrence of twelve *Eschweilera* species in a single hectare of rain forest near Manaus is believed to result from allopatric speciation in refugia when the climate was drier, then migration as the forest recoalesced following the return to more humid climatic conditions. Likewise, the complex, taxonomically intractable, polymorphic variation of so-called ochlospecies could have arisen in refugia. The type of differentiation which actually has occurred in woody plants in refugia will, however, only be known after a lot more experimental work has been done.

INTRODUCTION

The refuge theory postulates large changes in vegetation cover and distributions of plant species during the Pleistocene. The theory for Amazonia was developed by the zoologist Jürgen Haffer (1969) and since then many zoologists have commented on it and have added their ideas (see Chapter 7). Although the implications of the refuge theory for botany are important, botanists have produced very few papers about refugia based on plant distributions and vegetation types. This is an unfortunate lack because knowledge of the changes in plant distributions and vegetation types is basic to the theory, and information pertaining to as large a range of plant species as possible is needed. The paucity of botanical comments is probably largely due to the inadequate specimen collections from the region. Only two botanists have discussed Pleistocene vegetation in any detail. The highland areas of South America have been studied by B. B. Simpson (1975, 1979) and the lowland refugia of Haffer (1969) were discussed and modified slightly in the light of plant distributions by Prance (1973). There are, however, several discussions of refugia for smaller areas, e.g. that for Mexico and adjacent Central America by Toledo (1976) and for Venezuela by Steyermark (1979).

Until recently botanists were explaining plant speciation in the lowland tropical rain forests on the assumption that they had remained stable over a long period of geological time (e.g. Fedorov 1966; Ashton 1969; Richards 1969). Botanists have come to recognize recently that this stability of the forest was only presumed (Flenley 1979) and that changes in climate during the Quaternary and Holocene must be considered as a factor and their role in speciation must be worked out. Although changes in the forests have been known to have occurred for a decade now for South America, and longer still for Africa (Moreau 1963, 1966), there has been relatively little work of a reinterpretative nature by botanists.

This new emphasis on the instability of rain forest in the Pleistocene and Holocene does not necessarily eliminate the importance of other modes of plant speciation. For example, in the Amazonian rain forest habitat there is an enormous number of niches available and the niche specialization emphasized for example by Ashton (1969) and Richards (1969) is likely to be important for the speciation of forest trees and climbers. Competition and interaction with pollinators has led to phenological separation. Examples are the separation of closely related species into different strata of the forest or into different forest types, for example inundated forest and forest on terra firme. The danger of discussing any one theory is that other putative engines for speciation may be ignored. It now appears that Pleistocene vegetation changes have been one of the important factors influencing plant speciation and therefore species diversity of Amazonia, but the discussion below assumes that other methods are also important.

Simpson and Haffer (1978) reviewed the various methods used by biologists to determine the speciation patterns and location of palaeoecological refugia in Amazonia. They point out five modes of analysis which have been employed to define refugia:

1. The mapping of distributional ranges of taxa as circumscribed by the taxonomist to pinpoint centres of endemism.

2. The extrapolation of the centres of dispersal from the mapping of numerous organisms. These are regions where many groups have overlapping ranges.

3. Mapping combined with studies of areas in which closely related forms come into contact, as employed by Haffer and other ornithologists (see Chapter 5).

4. The statistical analysis of geographical variation as carried out by Vanzolini (1970) for a group of lizards. Some areas showed a larger than normal amount of variation, and others a lack of variation. The latter are interpreted as areas which have been stable over a long period of time.

5. The genetic analysis of hybrid zones which establishes the hybrid nature of intermediate forms found in areas of secondary contact. This has been used for insects by Brown (1977a, 1979, Chapter 4), Turner (1976), and others.

Botanists have mainly used the first method because the lowland neotropics remain one of the most undercollected regions of the world. It is not possible to distinguish hybrid zones for the majority of plants which have a long generation cycle, and the sampling is too sporadic to work out geographical variation in a way similar to that produced for the lizards by Vanzolini. The botanical papers discussed here are all concerned with present-day centres of endemism and with disjunct distribution patterns of forest species as circumscribed by the various taxonomists concerned.

In addition to these five lines of evidence for refugia, botanists have used a few other facts as indicators for climate changes. For example, D. R. Simpson (1972), in a paper about Peruvian refugia based on evidence from the Rubiaceae, pointed out some of the xeromorphic features of trees of the rain forest of Peru. He suggested that xeromorphic traits must have evolved in a dry climate and proposed that species with these adaptations are relicts from gallery forests and forest islands which formerly existed in the midst of savannas. This is an interesting idea but there are many present-day habitats in Amazonia where xeromorphic adaptations are an advantage such as campina, seasonal black-water igapó, and rocky outcrops which still exist as permanent or periodically dry habitats in lowland Amazonia under present-day climates (see Chapter 2). These could also produce xeromorphic adaptations which have later been retained after migration of the species into the rain forest. It is considered, however, that such migration is not as likely as that caused by the greater disturbance of climate changes.

To give an idea of the sample which botanists are using Fig. 3.1 illustrates the collecting density of species by degree square of the tree genus *Hirtella*

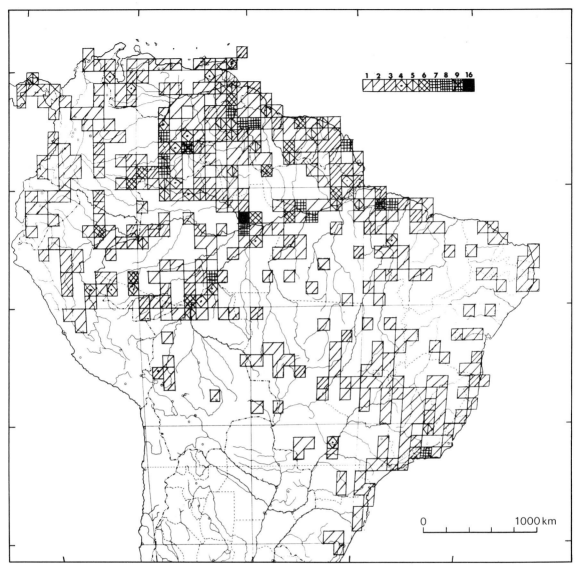

Fig. 3.1. The collecting sample of *Hirtella* (Chrysobalanaceae), showing number of species collected per degree square of latitude and longitude. This indicates well-collected areas such as around Manaus and Belém, and is useful to separate artefacts of collecting from the true presence of endemism.

(Chrysobalanaceae). It is immediately apparent that clusters of species appear around the regions which have been more intensively collected, such as Manaus with 18 species in its degree square and Belém with eight species. It is probable that an equal number of species of *Hirtella* occur in many other degree squares but have not yet been collected there. This means that it is not possible to pinpoint centres of endemism and to know which disjunct distributions are likely to be real and which are probable artefacts. This is one reason why botanists have lagged behind zoologists in their methodology for the definition of refugia. It does not detract from the importance of plants which should be the primary indicators of refugia because refugia are the direct result of changes in the vegetation cover.

This review of the botanical evidence is arranged geographically treating briefly Mexico and Central America first, followed by the Andean Highlands. The lowland tropical rain forest area is discussed in more detail since it is the region of focus of the refuge theory. Palynology has already been reviewed in Chapter 1.

BOTANICAL CENTRES OF ENDEMISM AND REFUGIA

Mexico and Central America

One of the most detailed botanical studies is that of Toledo (1976, 1982) for Mexico and adjacent northern Central America. Toledo pinpointed five Pleistocene refugia mainly from evidence of centres of endemism and the distribution of endemic species in a similar way to methods used in Amazonia. He also used other lines of evidence:

1. The abnormal distribution of temperate elements as relicts within areas of tropical rain forests, for example *Quercus**, *Pinus*, *Podocarpus**, *Myrcia*, *Cerifera* (see also Gómez-Pompa 1973).

2. The distribution of xeromorphic species in tropical rain forest, for example *Erythroxylum bequaertii*, *Guettarda gaumeri*, *Lysiloma bahamensis*, and *Neomillspaughia emarginata* (cf. Simpson 1972 on Peruvian Rubiaceae discussed above).

3. The large tolerance to drought of certain species of the tropical rain forest, as is evident from their geographical distributions.

4. Distribution of tropical rain forest species into the cooler climate zones of today, e.g. *Guarea grandifolia* (Meliaceae, cited as *G. chichon*), a species common in the Guianas, western Amazonia, Chocó, and Panama.

5. The differing distribution patterns of the dominant species of the tropical rain forest showing the recolonization capacities of different trees.

6. The latitudinal distribution of tree species.

7. The study of leaf shape and morphology (e.g. the ratio of different types of leaf margin) as an indicator of climate type.

On the basis of these data Toledo proposed five primary refugia (Fig. 3.2) where rain-forest species were protected from simultaneous general lowering of temperature and rainfall during the Pleistocene cool-dry climatic periods.

In addition Toledo (1981) recognized northeast of

*It should be noted that although Toledo considers *Quercus* and *Podocarpus* temperate they range widely in the tropics in the Old World (editors).

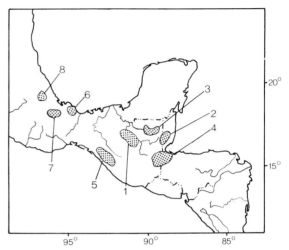

Fig. 3.2. The Mexican and Central American refugia of Toledo (1976, 1982). 1–5, Primary refugia: 1, Lacandona, Chiapas, Mexico; 2, Maya mountains, Toledo district southeast Belize; 3, Petén, Guatemala, near Tikal and Flores plus part of Cayo district Belize; 4, Lake Izabal and environs, Guatemala; 5, Soconusco, Chiapas, Mexico; 6–8, Secondary refugia: 6, Los Tuxtlas; 7, Sierra de Juarez; 8, Córdoba.

these primary refugia three of what he termed secondary refugia in which the rain forest species were protected from either the lowering of temperature or of precipitation, but not both (Fig. 3.2).

The methods of designation of refugia by Toledo show a thorough study with aspects which could be applied to Amazonia but since they are outside Amazonia they are not discussed in further detail here. The refugia of Central America could have been important in providing a route for northern species into the northern part of South America when the climate became warmer and wetter and recoalescence of the forest occurred. There are some species of Central America with disjunctions well into South America, e.g. *Licania arborea*, common in Central America and the extreme north of Colombia is also in Amazonian Peru. This type of distribution is probably the result of the Pleistocene disruption of the continuous rain-forest cover.

Graham (1982) has also discussed refugia in Mexico. His pollen studies in Veracruz State indicate that during the late Miocene the area was covered by oak–*Liquidambar* forest rather than rain forest. The change in the vegetation to rain forest at the present day is consistent with data from the Amazon and also with Toledo's view that forest refugia occurred in Mexico and Central America.

The forests of Central America have been so disrupted by the effect of early, Pre-Colombian civilizations that the history is obscure when the main criterion for refugia is based on endemism.

The Andes

In addition to the palynological evidence from the Andean region reviewed in Chapter 1 some of the earliest botanical evidence for Pleistocene climate changes came from the study of the highland páramo vegetation. B. Simpson was the first botanist to comment on the refuge theory in South America (Simpson-Vuilleumier 1971). This paper was a general review of geological and palynological evidence and it also presented considerable details about glaciation in the Andean highlands and the southern tip of South America. The review was based mainly on the speciation patterns of high Andean plants and birds but it presented no really new botanical evidence. Attention was drawn to the relationship of the flora of the Venezuelan highlands with that of the plateau of central Brazil, and to the distribution pattern of the genus *Polylepis* (Rosaceae) in the Andes. The distribution of this genus, which forms distinct isolated patches of woodland at higher elevations in the Andes, is consistent with a sequence of humid–arid changes in climate along the central Andean slopes where the tree line fluctuated. Later the same author presented her botanical evidence in detail (Simpson 1975). This paper also dealt exclusively with the high tropical Andes and described the floristic changes to the northern Andean páramo flora over 3000 m elevation, during the Pleistocene, the puna of the altiplano in Peru, the upper Andean forests, and the dry desert scrub of the high intermontane valleys. Simpson also included a good review of the history of the uplift of the Andes and the gradual availability of the various Andean habitats for plant colonization. Since most of these habitats have become available only in the Quaternary or late Tertiary, their flora is intimately connected with that of the lowlands. Simpson found an altitudinal and

latitudinal variation in the way plant species moved into the Andean habitats, the manner of differentiation during the Pleistocene, and the time of immigration. Speciation appears to have taken place mainly by divergence following geographical isolation, especially during the Pleistocene. With the exception of the altiplano most species expanded their ranges when the lowering of the high-altitude habitats occurred during the Pleistocene. At lower elevation in the northern Andes of the eastern cordillera direct migration was possible during glacial periods. The several interglacial periods when vegetation belts were raised were times of isolation and differentiation. In contrast, in the altiplano the glacial periods were times of population fragmentation (because the snow line was lowered) accompanied by differentiation and/or speciation.

Simpson distinguished two elements in the páramo

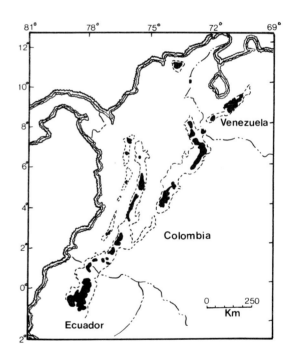

Fig. 3.3. The distribution of páramo 'islands' in the northern Andes (dark areas), determined by tracing the 3500 m contour lines, except for Tama, Perija, and Faralla where the 3000 m contour was used. These dark areas include both páramo and the mountain top glacier region above the vegetation zone. The dotted line is the 2500 m contour to show the effect of a lowering of the vegetation zones during glacial times. Many páramos then become contiguous instead of isolated islands. (After van der Hammen (1974).)

flora, species groups which are not closely related to lowland groups, and others which are. It is the latter that interest us here. An analysis of the flora of the northern páramos of the high Andes shows that the area was colonized during glacial periods in an analogous way to oceanic islands. There are significant correlations between páramo size, their distances from source areas, and the number of plant taxa which now inhabit them. There is an even stronger correlation between richness and páramo size in the glacial period which suggests that most colonization occurred then when plant propagules were able to disperse more easily because of increased size of the páramos.

These observations together with much palynological work have proved indisputably that there were considerable changes in the South American flora during the glacial periods. The changes in páramo in the extreme highlands (Fig. 3.3) meant changes in cloud forest and mountain slope forest at lower altitudes. The importance of the slope forest as a possible migration route for plant species must be considered in discussion of the lowland forest. The details of the lowland flora have not been worked out in such detail as those of the highlands. An interesting part of the highland work is the comparison with, and use of some of the concepts of, island biogeography which has been little discussed by botanists.

Steyermark (1979) indicated fifteen different montane centres of endemism in the western Andean part of Venezuela based on the distribution of endemic species of plants.

Highlands outside the Andes

A few other highland areas have been discussed in terms of refuge theory and the effect of the Pleistocene climate changes on the vegetation. Steyermark (1979) treated the Guayana Highland area under the name Pantepui in which he included both the sandstone mountain tops (tepuis) and the nearby lowland savanna area, the Gran Sabana. Pantepui includes several places which had been designated as forest refugia by previous workers, for example, the Roraima, Ventuari, and Imerí refugia of Brown (1976a) and also various lowland edaphic savannas and igneous lava formations in southern Venezuela. Steyermark stressed the edaphic variability and history of the region as a cause of much of the endemism that he used to define 'refugia' and dispersal centres. It is interesting that the lowland

edaphic savannas of this region have rather a high endemism in contrast to the savannas scattered throughout much of lowland Amazonia. Steyermark pointed out that only 39 genera, or 8.5 per cent, of the 459 genera, are endemic to the summits of the Pantepui, and emphasized the number of summit species which also occur on the slopes. He considered, contrary to previous proposals, that migration from the lowlands to the summits has been much more important than the reverse. Within or adjacent to the Pantepui region Steyermark found six centres of endemism (see Fig. 3.4).

In summary, the amount of endemism in the Pantepui region is very high but localized and is correlated with the great variety of soils and habitat types. The presence of many high peaks likely to attract cloud cover suggests some present-day habitats may have remained humid during past dry periods. However, only a few of the centres of plant endemism are coincident with likely lowland forest refugia, namely Imataca and Imerí. These are discussed further below.

Further to the east Sastre (1976) made a study of

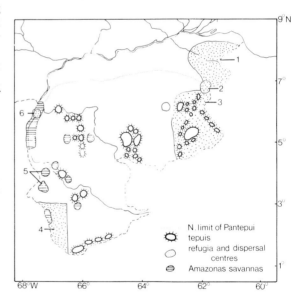

Fig. 3.4. Southern Venezuela to show the Pantepuí region as defined by Steyermark (1981) and its endemic centres: 1, Imataca centre (which lies just north of Pantepuí; 2–6, the Pantepuí centre; 2, Venamo dispersal centre; 3, The Gran Sabana dispersal centre; 4, The Rio Negro lowland forest centre; 5, the Amazonas savannas near the Rio Guainía (sandy soils); 6, the Atures dispersal centre (igneous rock).

the open vegetation areas of the Guianas with particular attention to the species of savannas and the open areas of mountain tops. He found that, whereas the Guiana savannas show no endemism, the sandstone mountains over 1000 m altitude do show considerable endemism. He estimated that 40 per cent of the plant species of granite outcrops also occur in the lowland savannas, and that 55 per cent are confined to mountain tops. Of these 55 per cent, 40 per cent are specific to granite. He observed that upland mountain savanna species are usually divided into populations separated by 300 km or more of forest. Long-distance dispersal by birds can explain some of these distributions, others can be explained by hypothesizing the spread of savanna in dry periods. He also recognized the Guiana mountains as a centre of differentiation for species of open habitats.

The lowland forests of South America

Various botanists have commented on individual centres of endemism of the lowland tropical rain forest region of South America. The only treatments of the entire region are those of Prance (1973, 1982a). These will be used here as a basis for discussion of all the botanical evidence. Figure 3.5 is an early attempt to locate lowland centres of endemism (Prance 1973) and Fig. 3.6 a refinement based on much more data (Prance 1982a). The changes between the two maps are small and are mainly in areas which remain poorly collected botanically. The botanical evidence is primarily the concentrations of endemic lowland forest species. Other evidence for past drier climates, such as disjunct distribution of dry forest species and the occurrence of xeromorphic adaptations in rain forest species, helps to highlight the effects of drier climate, but is of little use for the pinpointing of centres of endemism.

At present it is possible to delimit at least 26 centres of endemism in the lowlands (Fig. 3.6). When more collections become available some of these could possibly be split up further. For example, Steyermark (1979) has suggested five separate centres within the two listed here for coastal Venezuela.

The correspondence between the centres of endemism discussed below and at least a few species of Chrysobalanaceae is shown in Fig. 3.7 which also indicates the main edaphic adaptations of various species.

The twenty-six areas which stand out as centres of forest endemism are discussed below.

1. *Darién*. The tropical rain forests of Panama are by far the richest in Central America (Gentry 1978) and recent collections have shown that the forests of Darién, for example Cerro Jefe, are rich in endemic species. The centre of endemism extends upwards into the cloud forest. Some species are also found in the Chocó region to the south, discussed below, and Gentry considers that it was a source for some species found in Darién.

There are also species in common with the Guianas to the east, but disjunct, for example *Licania affinis* and *Caryocar nuciferum*. The latter species has been known from the Guianas since the eighteenth century but was only found in Darién and Chocó in the later 1970s. The palm *Phytelephas* occurs in an area which spans the Darién and Río Magdalena centres of endemism and is disjunct between there and East Peru (Moore 1973). Further study is likely to reveal additional species common to Darién and other South American centres of endemism but it is unlikely that the Mexican and Central American centres to its north have much in common with it.

2. *Chocó*. One of the richest and most distinctive centres of plant endemism, and one recognized by all workers, is the lowland Pacific coastal forest of Colombia including Chocó province and the extreme northwest of Ecuador. Figure 3.7 shows, for example, that there are several species of Chrysobalanaceae endemic to the region.

Three groups of palms occur in Chocó and are disjunct in other areas; *Wettinia* in Chocó and East Peru; *Chelyocarpus* in Chocó and both South Peru and Rondônia; and *Orbignya* section *Spirostachys* in Chocó and around Leticia in Amazonian Colombia (Moore 1973). These strong affinities, with the same or related species found both in Chocó and across the Andes in the western Amazon region, are also found in many other groups. The two regions undoubtedly had stable forest refugia in which such groups persisted during the Pleistocene. The species were probably isolated by the Tertiary uplift of the Andes rather than by Pleistocene climatic changes and have not since been able to coalesce. Chocó is important as a Pleistocene refuge, but has remained much more isolated than other centres of endemism east of the Andes and is rather unimportant as a place from which plant species have spread out to other lowlands, with the exception of Central America. The presence of great species diversity in Chocó indicates an area which has been stable for a long time and certainly remained an area of rain forest during the

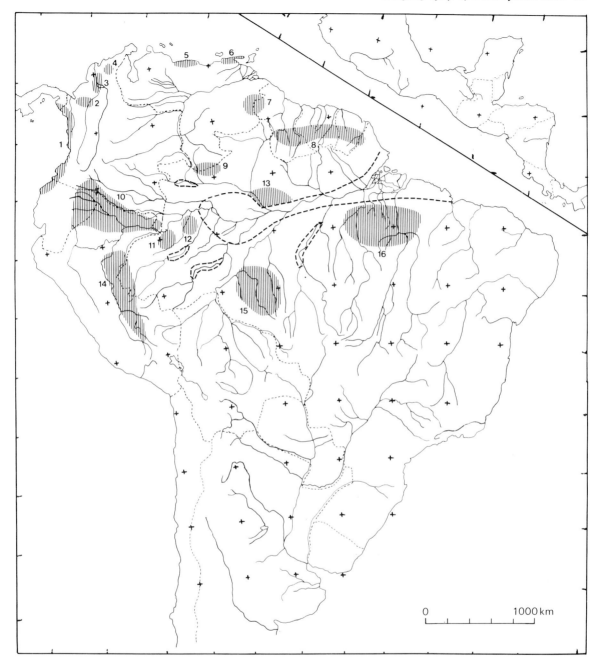

Fig. 3.5. Forest refugia proposed by Prance (1973), based on studies of four families of flowering plants, Caryocaraceae, Chrysobalanaceae, Dichapetalaceae, and Lecythidaceae: 1, Chocó. 2, Nechi; 3, Santa Marta; 4, Catatumbo; 5, Rancho Grande; 6, Paria; 7, Imataca; 8, Guiana; 9, Imerí; 10, Napo; 11, Olivença; 12, Tefé; 13, Manaus; 14, East Peru; 15, Rondônia–Aripuanã; 16, Belém–Xingu.

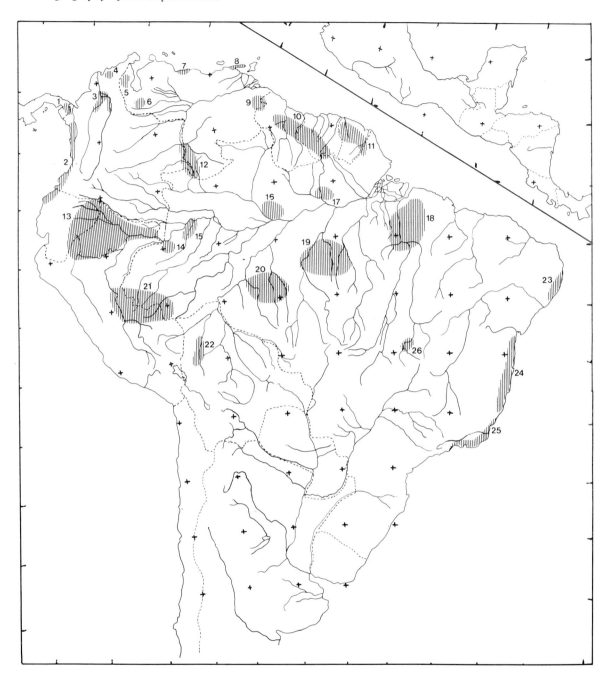

Fig. 3.6. Forest endemism centres proposed by Prance (1979), based mainly on distribution of five woody Angiosperm families. This is a refinement of Fig. 3.5. 1, Daríen; 2, Chocó; 3, Río Magdalena; 4, Santa Marta; 5, Catatumbo; 6, Apure; 7, Rancho Grande; 8, Paria; 9, Imataca; 10, W. Guiana; 11, E. Guiana; 12, Imerí; 13, Napo; 14, São Paulo de Olivença; 15, Tefé; 16, Manaus; 17, Trombetas; 18, Belém; 19, Tapajós; 20, Aripuanã; 21, E. Peru–Acre; 22, Beni; 23, Pernambuco; 24, Bahia; 25, Rio-Espírito-Santo; 26, Araguaia.

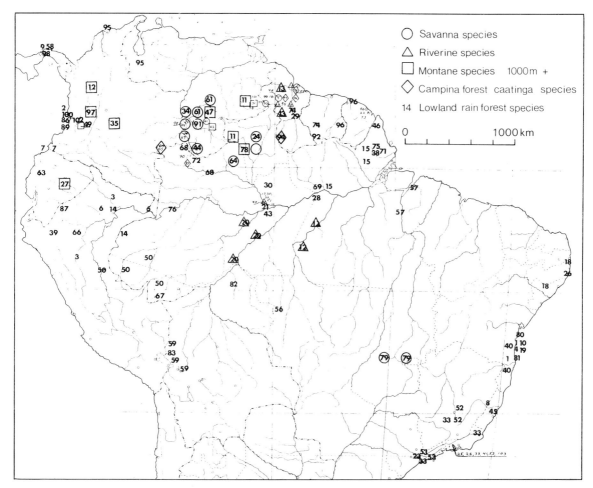

Fig. 3.7. The distribution of 103 locally endemic species of Chrysobalanaceae: Open number = lowland rain forest species. 1, *Parinari alvimii*; 2, *P. chocoensis*; 3, *P. klugii*; 4, *P littoralis*; 5, *P. maguirei*; 6, *P. parilis*; 7, *P romeroi*; 8, *Exellodendron gracile*; 9, *Maranthes corymobsa*; 10, *Couepia belemii*; 11, *C. canescens*; 12, *C. contracta*; 13, *C. comosa*; 14, *C. dolichopoda*; 15, *C. excelsa*; 16, *C. exflexa*; 17, *C. glabra*; 18, *C. impressa*; 19, *C. insignis*; 20, *C. krukovii*; 21, *C. marlenei*; 22, *C. martinii*; 23, *C. meridionalis*; 24, *C. multiflora*; 25, *C.·parvifolia*; 26, *C. pernambucensis*; 27, *C. recurva*; 28; *C. reflexa*; 29, *C. sandwithii*; 30, *C. spicata*; 31, *C. steyermarkii*; 32, *C. stipularis*; 33, *C. venosa*; 34, *Hirtella adderleyi*; 35, *H. adenophora*; 36, *H. angustifolia*, 37, *H. angustissima*; 38, *H. araguariensis*; 39, *H. aramangensis*; 40, *H. bahiensis*; 41, *H. barrosoi*; 42, *H. caduca;* 43, *H. conduplicata*; 44, *H. cordifolia*; 45, *H. corymbosa*; 46, *H. couepifolia*; 47, *H. cowanii*; 48, *H. dorvalii*; 49, *H. enneandra*; 50, *H. excelsa*; 51, *H. fasciculata*; 52, *H. floribunda*; 53, *H. glaziovii*; 54, *H. guyanensis*; 55, *H. insignis*; 56, *H. juruensis*; 57, *H. lancifolia*; 58, *H. latifolia*; 59, *H. lightioides*; 60, *H. longifolia*; 61, *H. longipedicellata*; 62, *H. orbicularis*; 63, *H. pauciflora*; 64, *H. pimichina*; 65, *H. scaberula*; 66, *H. standleyi*; 67, *H. subglanduligera*; 68, *H. subscandens*; 69, *H. tenuifolia*; 70, *H. vesiculata*; 71, *Acioa guianensis*; 72, *A. schultesii*; 73, *A. somnolens*; 74, *Licania albiflora*; 75, *L. amapaensis*; 76, *L. angustata*; 77, *L. apiculata*; 78, *L. aracaensis*; 79, *L. araneosa*; 80, *L. bahiensis*; 81, *L. belemii*; 82, *L. bellingtonii*; 83, *L. boliviensis*; 84, *L. buxifolia*; 85, *L. cabrerae*; 86, *L. calvescens*; 87, *L. cecidiophora*; 88, *L. chiriquiensis*; 89, *L. chocoensis*; 90, *L. compacta*; 91, *L. cordata*; 92, *L. couepiifolia*; 93; *L. crassivenia*; 94, *L. cuprea*; 95, *L. cuspidata*; 96, *L. cyathodes*; 97, *L. durifolia*; 98, *L. fasciculata*; 99, *L. foldatsii*; 100, *L. fuchsii*; 101, *L. furfuracea*; 102, *L. glauca*; 103, *L. glazioviana*.

Pleistocene. In terms of plant species the isolation or differentiation is far greater than that between the centres of endemism east of the Andes. The Andes mountains have formed an effective barrier to plant dispersal so that the contribution of the plants of the Chocó centre to lowland Amazonia has been far less than for animals which have greater mobility.

3. *Río Magdalena-Cauca*. There are a large number

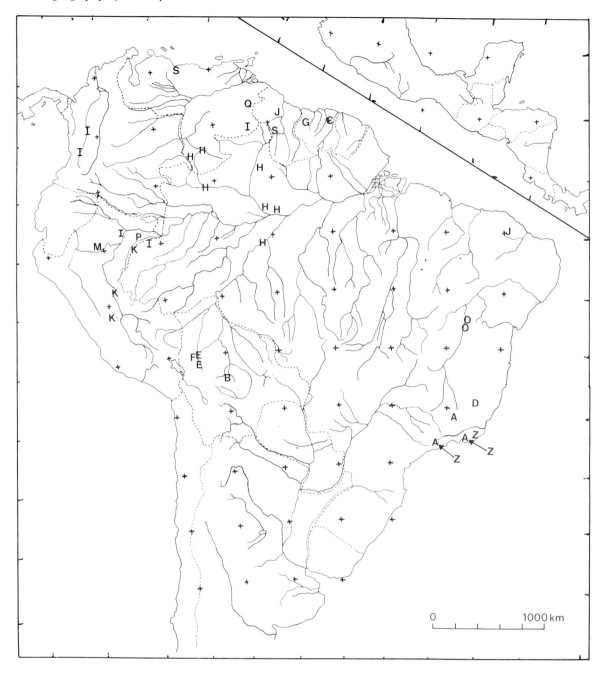

Fig. 3.8. Distribution of locally endemic taxa of *Trigonia* (Trigoniaceae):

A, *T. paniculata*; B, *T. boliviana*; C, *T. candelabra*; D, *T. eriosperma* subsp. *simplex*; E, *T. echitifolia*; F, *T. floccosa*; G, *T. coppenamensis*; H, *T. sprucei*; I, *T. sericea*; J, *T. villosa* var *macrocarpa*; K, *T. killipii*; M, *T. macrantha*; O, *T. nivea* var *fasciculata*; P, *T. prancei*; Q, *T. bracteata*; S, *T. costanensis*; Z, *T. rytidocarpa*.

of plants endemic to the valleys of the Magdalena and Cauca rivers in Colombia, especially the latter. Botanically, it is hard to distinguish separate areas within this region but the area termed the Nechí refuge by Haffer (1969) and recognized by other later authors appears to be the centre of endemism, although many plant endemics extend much further south in the Cauca valley, such as *Licania glauca* (Fig. 3.7) and *Trigonia sericea* (Fig. 3.8). The region of the Nechí centre of endemism still remains very poorly collected botanically and the forests have been largely disturbed or destroyed. Subdivisions probably exist within the area marked on Fig. 3.6, but it is impossible to delimit them accurately from present botanical evidence. Cauca was designated as a separate centre from Nechí by Brown (1977*b*).

4. *Santa Marta*. This area on the Caribbean coast near the Colombia/Venezuela border contains the Sierra Nevada de Santa Marta (5775 m). In addition to the distinct and endemic flora of the high mountain itself the slope forest must be regarded as an endemic centre in terms of the species of plants found there. This area was probably of lesser importance as a refuge because species diversity of the slope forest, despite high endemism (e.g. *Stephanopodium aptotum*, Fig. 3.9) is rather low.

5. *Catatumbo*. This area on the Colombia/Venezuela border southwest of Lake Maracaibo has been proposed as a centre of endemism by most authors beginning with Haffer (1969). Steyermark (1979) has shown that Catatumbo is an important centre of plant endemism, and the study of many plant families indicates that this is so. The rain forest here has recently been largely destroyed.

6. *Apure*. The area designated as the Apure centre although close to Catatumbo, is separated by the easternmost cordillera of the Andes, a high, dry barrier. Apure was not mentioned in some of the earlier publications such as Haffer (1969) and Prance (1973) though Brown (1977*b*), Steyermark (1979), and others recognize it. This area is still poorly known botanically, but the increasing collection sample from Apure indicates strongly that there is a centre of forest endemism in the region marked in Fig. 3.6. *Mouriri barinensis* is an example of an Apure endemic. Steyermark (1979, 1982) has subdivided Apure into several centres of endemism in the Andes (Ayari, Barinitas) and San Camilo. Of these the San Camilo area (located in Apure between the Río Uribante and the Río Arauca and extending eastward to the State of Barinas between the Ríos Caparo and

Apure) has besides endemics some species with relatives in other endemism centres and in the western Amazon, for example *Licania latifolia* and *Dichapetalum latifolium*.

7, 8. *Rancho Grande and Paria*. Centres of endemism on the coast of Venezuela have been proposed by various workers beginning with Haffer (1969) who indicated a single northern Venezuelan centre in the western part of the coastal cordillera. Steyermark (1979) proposed five centres of diversity, based on his thorough examination of the plant species. He pointed out the rather close relationship between the plants of the coastal cordillera and the forest of the south of Venezuela, and hypothesized that these two areas must have had contact, perhaps at the height of the Pleistocene humid periods or through gallery forest. He cited *Froesia venezuelensis* and *Stephanopodium venezuelanum* (Fig. 3.9) as good examples of northern refugial species of predominantly southern genera. The area of Steyermark's five coastal centres includes the Rancho Grande and Paria centres of Prance (1973), Brown (1977*b*), and others. The two most important areas within northern Venezuela in terms of the plant families studied are the area of Aragua around Rancho Grande, and the Paria peninsula. There are also many species common to both areas such as *Mouriri rhizophorifolia* which also extends into the Imataca centre. Based on the present state of knowledge it seems preferable to recognize only the Rancho Grande and Paria centres of endemism rather than the five suggested by Steyermark.

9. *Imataca*. This area on the Venezuela/Guyana frontier which also includes the Altiplanicie de Nuria is a centre of much botanical endemism in terra firme lowland rain forest and is also one of the insect endemism centres of Brown (1979) but was not recognized by Haffer (1969). Figure 3.7 shows the cluster of eudemic Chrysobalanaceae in the region, and Fig. 3.8 *Trigonia bracteata*, an Imataca endemic. Figure 3.7 shows that endemism extends into Guyana where it is also strong. However, the two areas must be separated because there are many species confined to one only. The amount of endemism in this region is further increased by the number of edaphic endemics. For example, Fig. 3.7 shows that on the Guyana side there are many savanna and riverine endemics, although on the Venezuelan side most of the endemics are forest species, although there are certainly also many savanna endemics too. Steyermark (1979) centres this area just north of Venamo and included the Altiplanicie de Nuria.

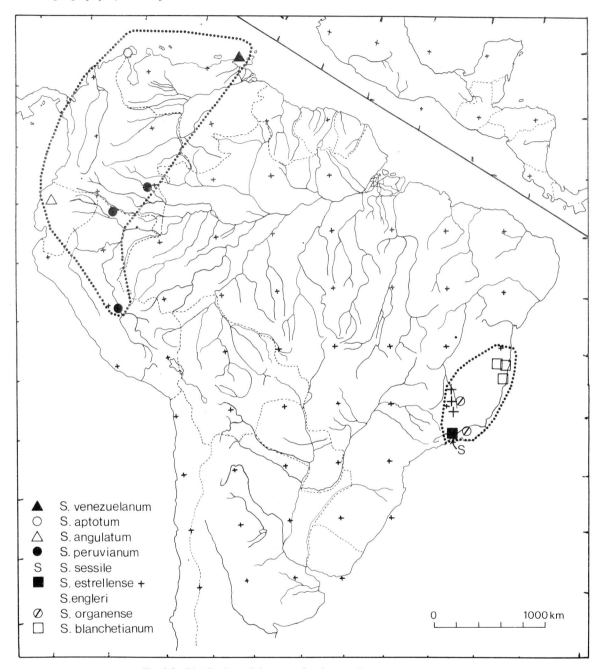

Fig. 3.9. Distribution of the genus *Stephanopodium* (Dichapetalaceae):

Steyermark (1979, 1982) has stressed the striking difference between the area designated as the Imataca centre and the mountainous Pantepui area discussed above which is almost contiguous to the south. Steyermark (1982) provided a long list of species which in Venezuela are restricted to Imataca. This list is slightly misleading since many of these species have extensive ranges outside Venezuela.

However, there is no doubt that Imataca does have an extremely high concentration of lowland rain forest endemics, and Steyermark's list includes, for example, *Licania latistipula*, *Piranhea longepedunculata*, *Ocotea subalveolata*, and *Dilkea magnifica*.

10, 11. *West and East Guiana*. Most workers have indicated one large centre of endemism in the Guianas (Haffer 1969; Prance 1973, 1979). Brown (1979) distinguished both a Manaus/Guiana centre extending into southern Guyana and an Oyapock centre in French Guiana and Amapá and analysis of the numerous plants now recorded also shows that there are at least two distinct patterns of rain-forest endemism, one centred in Guyana and the other in French Guiana, as well as a large number of edaphic endemics, confined to such habitats as the white-sand Wallaba forest and the various sandstone and granite mountains and savannas discussed by Sastre (1976). The West Guiana centre of endemism extends diagonally across Guyana into southwest Suriname and the eastern one from the east of Suriname through French Guiana well into Amapá, Brazil. The refuge was probably located well inland from the coast because of former higher sea levels. There is strong endemism near the coast, shown for example by Chrysobalanaceae in Fig. 3.7, but this might be partly an artefact resulting from the concentration of botanical collecting in the vicinity of Cayenne.

Descamps *et al.* (1978), who worked on the plants and animals of the savannas and rock outcrops of French Guiana, divided the Guianas into three biogeographical subregions based on the distribution of various forest species. They concluded that speciation took place in more than one centre and that during the times of dry climate the Guianas were broken up into at least three refugia. They suggested that the easternmost refuge was located north of the Tumucamaque mountains around Saül between the Tampoc and Camopi rivers, and between the Comté and the Approuague. This is a little farther northwest than the Oyapock centre of Brown (1977*b*).

Granville (1978, 1982) has made the most detailed study of the part of the East Guiana centre in French Guiana. He agreed with Descamps *et al.* (1978) that this centre of endemism is located well north of the Tumucumaque mountain area where it is very dry even today and proposed that the Pleistocene refugium bordered to the south by the Inini-Camopi range and to the north by Kaw mountain and that it extended past the Oyapock river into Amapá, Brazil to the east and into Suriname to the west. Granville's

deductions were based on geomorphology, soil, and vegetation. He cited a number of endemic species which have been collected around Saül such as the new genus *Elephantomene eburnea*. He also cited a number of species which occur in the Guiana centre of endemism and disjunct in other centres such as *Adiantum cordatum* from Saül and the Panama-Darién refuge, *Heliconia lourteigii* from Saül and Camopi and also the Napo region. The data of Descamps *et al.* (1978) and Granville (1978) as well as all the other botanical evidence indicate that at least the eastern part of the Guiana centre is well north of the area where it was placed by Haffer (1969) and Vanzolini (1970). Granville listed the various areas in French Guiana which were unlikely to have been refugia, such as the recent coastal formations and white-sand areas. He also commented that the coastal savannas and the inland rock outcrops function as present day refugia for plants of more xeric habitats.

12. *Imeri*. The Imerí centre of endemism is centred around Serra Imerí and southeast to Cerro Neblina. Small lowland forest patches have probably persisted through this area owing to the presence of the high mountains and their associated rainfall. Much plant endemism is evident (see, for example, Fig. 3.7). As in the Guiana area there is much edaphic adaptation to the igneous outcrops, the savannas, the mountains, and to campina forest on white sand, but there are also a large number of lowland terra firme rain forest endemics. The endemism is extremely complex and even Steyermark's (1979) account does not give the full details for the Venezuelan part of the region. The very high level of endemism suggests that this region has been stable for a long time. Botanical collections in 1978-9 by Nilo T. Silva and Umbelino Brazão from the foothills of Cerro Neblina and within the Imerí area contained an exceptionally large number of lowland endemics. The Imerí refuge is likely to have comprised several different patches of forest, but present knowledge is insufficient to locate them exactly.

13. *Napo*. This large area, extending from the Andean foothills eastwards to the Trapecio Amazonico of Colombia, is widely recognized as a centre of endemism. The whole area indicated on Fig. 3.6 has high plant endemism; perhaps one of the largest concentrations is around the area of the Pongo de Manseriche on the Marañón River (77° 36′W; 4° 24′S), for example, *Mouriri tessmannii* and *M. floribunda*. The endemic species of the Napo centre of endemism are quite distinct from those of the East Peru-Acre

centre to the south east, as has been shown for example for Peruvian Rubiaceae (Simpson 1972). Napo has several palms in common with the Chocó centre (see above). Another disjunct species is *Licania arborea* found here and also common in Central America and northern Colombia.

14, 15. *São Paulo de Olivença and Tefé*. There is a small area of high plant endemism in the vicinity of São Paulo de Olivença (68°W, 3°24′S). Many species from this region are related to those of the more northern centres, for example *Mouriri micradenia* and *M. monopora*. This centre, which was first proposed by Prance (1973), seems necessary for plants although not for animals. Likewise, farther east the Tefé centre of plant endemism has not been identified by many zoologists, but agrees with that of Brown (1977b) for butterflies. This area, which must have been near to the margin of the inland sea at the time of maximum embayment, has many endemics. Some extend south into other areas. For example the rare *Couepia edulis* occurs here and in the upper Purus region near to the East Peru–Acre centre.

16. *Manaus*. The Manaus area was first proposed by Prance (1973) and later Brown (1977b) recognized a centre of endemism for butterflies extending from Manaus to the south of Guiana. Both areas are centres of plant endemism and there are certainly some species well distributed throughout the range of Brown's centre, for example *Hirtella mucronata* and *Licania sprucei* both of the Chrysobalanaceae. Figure 3.7, however, shows clearly that there are a large number of plant species confined either to the Guianas or to the vicinity of Manaus. This is supported, for example, by the two vicariant species pairs of *Mouriri* cited by Morley (1975), *M. crassifolia-M. ficoides* and *M. dumetosa-M. densifoliata*. In each case one species occurs in the West Guiana centre and the other in the Manaus centre.

Gentry (1979) considered the Manaus centre of endemism controversial since it has not been recognized by many zoogeographers and since ten species of Bignoniaceae once thought to be Manaus endemics have since been collected elsewhere. For example, *Tabebuia incana*, first known from Manaus, was collected for the second time on the Río Ucayali in Peru. This is perhaps a disjunction with the East Peru–Acre centre or the result of the wind-dispersal mechanism of the Bignoniaceae. Gentry himself pointed out the importance of the consideration of dispersal mechanism of any plant under biogeographical study and noted that canopy lianas with light wind-dispersed seeds are perhaps too easily dispersed to have retained present day distribution patterns which can be correlated with centres of endemism.

Manaus is botanically one of the best collected regions of Amazonia and extreme caution is needed in order not to confuse well-collected areas with centres of endemism. However, even allowing for this, there is strong evidence of a high degree of plant endemism near to Manaus and that it should be regarded as an important centre.

17. *Rio Trombetas*. An area of interesting plant endemism occurs in the basin of the Rio Trombetas which has been comparatively well collected by Adolpho Ducke and his associates. This area is not rich in endemic Chrysobalanaceae and was not proposed as a refuge by Prance (1973). However, in many other plant families there is some endemism, with different species from the Manaus and Guiana areas. Brown (1977b) almost included this area in his irregularly shaped Guiana-Manaus centre. Since various species of trees of this region are different from either the Guiana or the Manaus refugia, I prefer to separate the Trombetas region as a distinct centre of endemism.

18, 19. *Belém, Tapajós*. The Belém area has been recognized by all workers as a centre of endemism and possible refuge. Only the size and extent of it has varied. The area directly south of Belém, which was made accessible by the Belém–Brasilia highway, is well-collected botanically and contains many local species, for example *Votomita monadelpha* and *V. pleurocarpa*. Prance (1973) indicated rather a large centre, extending west to the Rio Xingu. Limited knowledge made it difficult to be any more precise at that time. Recent field work by the author along the Trans-amazon Highway and also the publications of RADAMBRASIL (for example Projeto RADAM 1974) have helped to clarify the botany of this region. It can now be seen that the Belém centre extends only slightly west of the Tocantins river and the area between the Tocantins and Xingu has much liana forest, an impoverished type with very little endemism (Fig. 2.1, pp. 20–1). There is some plant endemism in the Serra dos Carajás region but this is edaphic and altitudinal speciation on this iron-rich range of hills. However, farther west in the Tapajós river basin there is very much plant endemism. The Belém centre is consequently now reduced in size and a separate centre is proposed near to the Rio Tapajós, south of Itaituba, extending east to the Rio Curuá in the areas where liana forest does not occur.

20. *Rondônia–Aripuanã*. This area on the eastern boundary of Rondônia and extending east to the Rio Aripuanã is another clear centre of plant endemism, but one which is hard to map with any accuracy because of limited collections. Nearby in Rondônia the Serra does Parecis has many montane endemics. All families studied are rich in lowland rain forest endemics from this region, for example *Cariniana kuhlmannii*, *Hirtella juruensis*, and *Licania bellingtonii*.

21. *East Peru–Acre*. In western Acre and adjacent Peru there are many forest endemics which do not extend northwards into the Napo centre (e.g. Rubiaceae, Simpson 1972). The clear-cut separation indicates that this region east of the Andes was broken into at least two refugia, although some plants such as *Stephanopodium peruvianum* (Fig. 3.9) extend into both, which is to be expected considering their present-day climatic similarity. Brown (1976*a*) subdivided this region into at least three refugia based on butterflies. Plant collecting from this area is too poor to distinguish any more than two centres of endemism at present. Moore (1973) pointed out the disjunction of the palms *Wettinia* and *Phytelephas* between here and Chocó and Darién/ Rio Magdalena-Cauca respectively.

22. *Beni*. This is an area of lower montane forest, which is quite distinct and contains species definitely related to those in other Amazonian centres. It was not included in Prance (1973) because of doubts about its validity as a refuge at this higher altitude. However, species from this area are clearly related to other lowland species. Figure 3.7 shows the distribution of *Hirtella lightioides* and *Licania boliviensis*, endemics which occur in or near this centre and Fig. 3.8 shows *Trigonia echiteifolia* and *T. floccosa*, both confined here. The Beni centre was recognized by Brown (1976*a*) and called Yungas. In Brown and Ab'Sáber (1979) Yungas was subdivided into two, La Paz and Cochabamba. Bolivia remains botanically one of the poorest collected areas in South America, and it is impossible to localize endemism centres with greater precision. There are not enough data to indicate whether the La Paz and Cochabamba areas are separate.

23. *Pernambuco*, 24. *Bahia*, 25. *Rio-Espírito-Santo*. These three centres of endemism lie in the Atlantic coast rain forest, well known to be rich in plant endemics, many of them highly distinctive. Soderstrom and Calderón (1974) indicated this area as the source of various primitive bamboos. They studied the tribes of Bambusoid grasses Olyreae and Parianeae, especially the genera *Diandrolyra* and *Piresia* of the former tribe, and found that the primitive species occur in the forests of eastern Brazil particularly in Bahia and north of the Rio Doce in Espírito Santo. They hypothesized that eastern Brazil, particularly Bahia, represents a centre of the primitive elements of these genera and that migration occurred south along Serra do Mar and northwest into Amazonia. The forest areas from Bahia north to Paraiba is considered to have been a refuge for at least some primitive herbaceous bambusoid grasses. *Piresia* has four species in Bahia, one in East Peru– Acre and two more which are widespread in northeastern Amazonia and sympatric in the Guiana centre.

There are clear relationships between the Atlantic coast rain forests and those of Amazonia. Andrade-Lima (1953) listed fifty-eight species in common which are now isolated by the dry area of caatinga to the north and the cerrado of central Brazil to the west. The endemic species in the Atlantic coast rain forests delimit these three centres (see Fig. 3.7 for Chrysobalanaceae). To the north there are a group of species in the forests of Pernambuco, for example *Couepia impressa* and *C. pernambucensis*. There are eleven species in Bahia and northern Espírito Santo and eight more south of the Rio Doce region in Espírito Santo and Rio de Janeiro. There are other more widespread species which extend between both areas. Figure 3.8 shows the range of *Stephanopodium blanchetianum* in Bahia, with other species in the more southern centre. The great species diversity of the Atlantic coast forests, and their high amount of endemism indicate the area has been relatively stable throughout the Pleistocene climatic changes. The area of Atlantic coast rain forest is small compared with Amazonia during the Quaternary. The link was more likely to have been in the late Tertiary which means that these have been isolated for longer than the refugia of lowland Amazonia and, like the Chocó refuge, those of the Atlantic coastal region have contributed fewer species during the coalescence of the Amazonian refugia. Any contribution which these areas have made to the re-establishment of continuous forest in Amazonia must have been through the gallery forests of central Brazil. There are various species which show this connection today, for example *Hirtella burchellii* (Fig. 3.10) common in Acre and Rio de Janeiro and with a more or less continuous range through the gallery forests of the cerrado on the planalto.

Fig. 3.10. The distribution of *Hirtella burchelli* (Chrysobalanaceae), a species of southwestern Amazonia that enters well into the cerrado region through its occurrence in the gallery forests of central Brazil.

26. *Araguaia*. There are a number of endemics northwest of Brasília in the Araguaia/Tocantins watershed which occur in gallery forest and rain-forest patches in the cerrado, for example *Licania* *araneosa*. This centre was first proposed by Vanzolini (1970) and discussed by Brown (1977*b*).

The gallery forest and the rain-forest patches or islands, which are often quite extensive, are a reminder that during the Pleistocene dry periods rain forest persisted outside the refugia in just such pockets. This was discussed further in Chapter 2.

Phytogeographical regions

The eight phytogeographical regions of Amazonia first proposed by Ducke and Black (1953, 1954) and slightly modified by Prance (1977) are shown in Fig. 2.3 (p. 26). These zones, based on their content of plant species, show the major division of the Amazon rain forest, rather than just the centres of endemism. The different centres of endemism within each phytogeographical zone can be considered to be more closely related to each other than to those of other phytogeographical zones. For example, the Imataca, East and West Guiana, and Belém centres all fall within the Atlantic coastal phytogeographical region which indicates a closer relationship than with those which fall in the southeast Amazonian region.

PRESENT-DAY REFUGIA

Dry climate refugia

At the present day the climate is almost as wet and warm as at any time in the Pleistocene and rain forests are at or near their maximal extent. Conversely, areas of low rainfall and seasonally dry climates are at or near their least. Dry forests have contracted to isolated patches. Xerophytes have much more restricted ranges than at some periods in the past and are today restricted to these xeric refugia.

Bignoniaceae is a family well adapted to the drier parts of the tropics. They have recently been studied by Gentry (1979). Many species are today found restricted to dry forest patches in Central America and scattered around the fringes of Amazonia, especially in the inter-Andean valleys of Colombia and Peru. Some species, for example *Tabebuia impetiginosa*, show little differentiation between the various isolated populations, but others, such as *T. ochracea*, which has been divided into various taxonomic subspecies by Gentry, appear to be actively differentiating at present. The several subspecies occur in the different contemporary dry-forest refugia. Gentry cited further examples of variability in dry-forest species.

Brejos in the caatinga

Andrade-Lima (1982) studied the plants of humid forest refugia within the dry caatinga region of northeastern Brazil. Here, several small mountains are high enough to attract cloud cover and rainfall and to have a dense forest locally known as *brejo*. The brejos are isolated from each other, and also from other large forest areas. They contain many species of Amazonia, for example such lowland rain forest species as *Gallezia gorarema*, *Huberia ovalifolia*, *Manilkara rufula*, *Myrocarpus fastigiatus*, and *Phyllostyllon brasiliensis*. Some brejos which are higher and face towards the cooler winds also serve as refugia of more southern species characteristic today of areas of cooler climate, such as *Podocarpus lambertii* and *P. sellowii*. Brejos demonstrate the potential of even small hills to act as moist-forest refugia during an arid climatic phase.

BOTANICAL OPPOSITION TO REFUGE THEORY

Only one botanist, Morley (1975), has produced a detailed critique of the refuge theory. It is based on the distribution of species of the tribe Memecyleae of the Melastomataceae. I have discussed Morley's work in some detail elsewhere (Prance 1982c). He argued that present-day areas of dry climate could account for the distribution of all species of the group except the Guiana–Western Amazonian disjunct, *Mouriri oligantha*. He analysed the distribution of thirty-one species considered to be local endemics and marked their distribution with the proposed refugia of Prance (1973). Seventeen species fitted into the refugia and fourteen did not. In fact, eleven of these fourteen occur near to those refugia. The fourteen species which formed the basis for his argument are either local endemics for edaphic

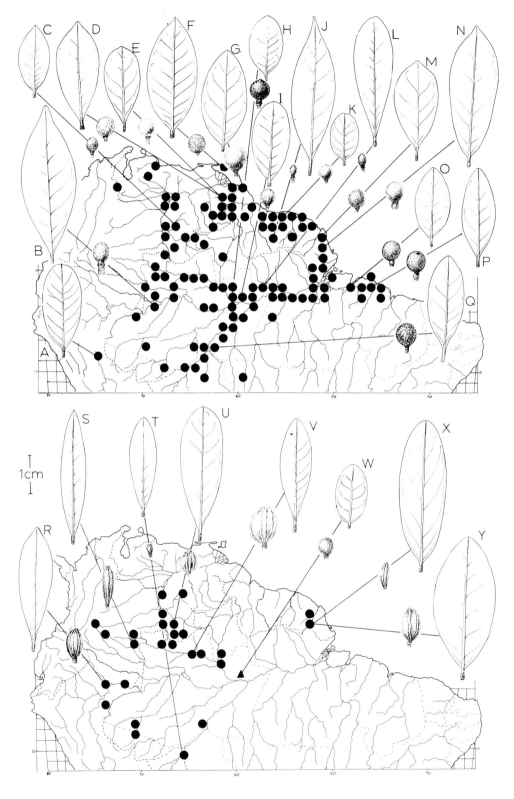

reasons (e.g. the white-sand species, *M. ambiconvexa* and the black-water species, *M. eugeniifolia*), or they fit into centres of endemism as redefined in this chapter. The data of Morley and other more recent data from other plant families have helped to define centres of endemism and modify them slightly, rather than disprove the refuge theory.

DIFFERENTIATION IN REFUGIA

The above discussion has shown that plant distributions support the existence of areas of terra firme rain forest with high endemism which are interpreted to have remained stable through drier climatic periods. The level of differentiation of plants within these refugia is less clear. Except for the hybrid zone data which exist for butterflies little is in fact known about differentiation within refugia. Refugia undoubtedly allowed a certain amount of speciation in newly allopatric populations of formerly coherent species. Subsequent redistribution explains the frequent existence today of several closely related species in the same forest. For example, the twelve species of *Eschweilera* (Lecythidaceae) in a single hectare of forest near to Manaus discussed in Prance *et al.* (1976) could have become sympatric from such an origin. The comparatively short time span in which populations were isolated in refugia represents rather few generations for tree species. Many canopy species under optimal circumstances require about thirty years to reach reproductive maturity and there is a long period during which individuals of different generations overlap. This means that differentiation is likely to be much slower than in birds or in insects with several generations each year and which overlap less in time.

In most genera which I have studied, and in many other large genera which have been monographed, there are always a few widespread, polymorphic, ochlospecies. These species have a large amount of morphological variation which is hard to classify in formal terms much of which could have occurred in refugia. Examples are *Licania apetala* and *L. heteromorpha* (Fig. 3.11), *Tapura guianensis*, *Caryocar glabrum*, *Couratari guianensis*, *Mouriri grandiflora*, and *M. vernicosa*.

The type of differentiation which actually has occurred in woody plants in refugia will only be known after a lot more experimental work has been done.

Fig. 3.11. The distribution and morphological variation of a polymorphic ochlospecies, *Licania heteromorpha* (Chrysobalanaceae). Note the variation in leaf shape and size, and in fruit shape.

4

BIOGEOGRAPHY AND EVOLUTION OF NEOTROPICAL BUTTERFLIES

SUMMARY

Butterflies have long attracted the attention of naturalists and biogeographers in the neotropics. Some groups offer abundant material for the analysis of biogeographical and evolutionary patterns, though care must be taken to obtain adequate biosystematic information, geographical coverage away from major rivers, and ecological and genetic data (Table 4.1). Butterflies with aposematic (warning) coloration are especially suitable for biogeographical analysis. They are common, easily sampled, and show conspicuous regional variation in adult wing pattern which can be related to ecology and major genes. Two such groups, the Heliconiini and the Ithomiinae, are chosen here for detailed examination. For these, classic museum data have been supplemented by intensive collecting, concentrating on blank areas, and zones of intergradation (Fig. 4.2). Studies of behaviour, ecology, and genetics have also been undertaken.

The available palaeontological evidence indicates that most butterfly speciation occurred up to the early Pleistocene (Fig. 4.1), with later environmental fluctuations leading mainly to infraspecific differentiation and range alterations. At both the species and the subspecies levels a number of consistent patterns of geographical differentiation occur in the two groups chosen for study (Figs. 4.3-4.10) and in other butterflies too (Fig. 4.11). The four main biogeographical regions seen in neotropical forest butterfly species, including over 80 per cent of the species examined, are the same as those seen in many other forest organisms: Transandean, Andean, Hylaean (Amazonian), and Atlantic (Fig. 4.12). Much more detail is seen by examination of the ranges of recognized geographical subspecies. When these are mapped and overlain forty-four strongly marked centres of subspecies endemism can be recognized of which sixteen are shown in Fig. 4.13. The area of each centre is usually 50 000-200 000 km^2, and usually includes 15-25 differentiated endemic subspecies out of the 20-40 species present and analysed. All forty-four centres of subspecies endemism are shown on Fig. 4.15, corrected to allow for subspecies mixing. Although the two butterfly groups (Heliconiini and Ithomiinae) have very different ecological preferences, their centres of endemism closely coincide (Fig. 4.16); this is believed to reflect a similar evolutionary history for subspecies differentiation. Each of the endemic centres shows strong affinities with some others, often but not always with those immediately adjacent or of similar environment (Fig. 4.17). It seems that different sorts of continuity have existed between populations in neighbouring regions during periods of evolutionary divergence. Some of the more surprising links, such as those across the Brazilian Shield and between eastern Ecuador and southwestern Brazil, are found also in other groups, for example plants and snakes.

The species limits and zones of contact between the subspecies often lie away from present-day physiographical and ecological barriers. This suggests a spreading out from smaller areas in the past and that the genesis of present patterns must involve historical as well as ecological factors.

The number of endemic subspecies associated with the different centres varies from five to over forty. The richest centres are those with the most complex mosaic of different habitat types, reflected in topography, soils, and vegetation.

Anomalous distributions have been found (Fig. 4.6) and are believed to have resulted from major changes in vegetation distribution near savanna areas in the past.

Some other butterflies are today confined to isolated patches of open vegetation scattered through the forest. Detailed study of these would be interesting, to elucidate processes of adaptation and differentiation such as must also have occurred when the forest was restricted to isolated patches. There are yet but few critical studies of these butterflies and of those of the similarly isolated montane habitats (Fig. 4.18). The latter show evidence of past closer connections between the northern Andes and isolated peaks (Pantepui, Guiana) and between the south central Andes and southeastern Brazil. Both examples are

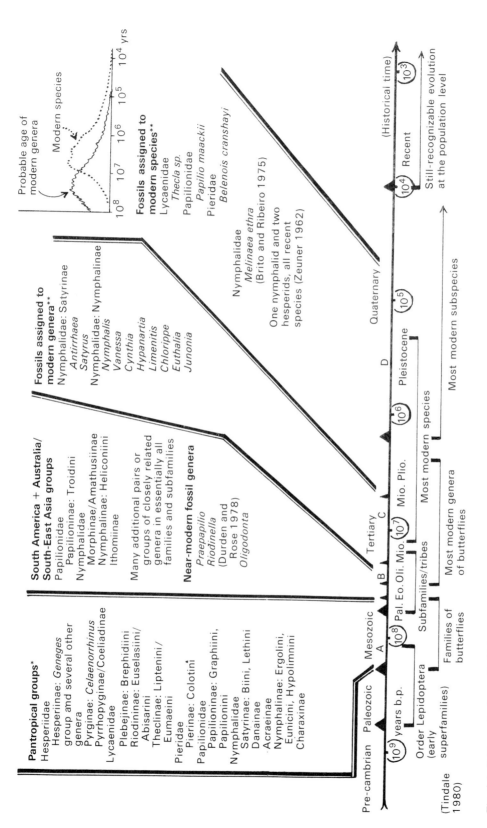

Fig. 4.1. Evidence for butterfly higher evolution through time. *Many pantropical groups have limited representation on northern continents. **Lepidoptera represent but a minute part (0.05 per cent) of known fossil insects. The other orders, especially Coleoptera, show the same temporal patterns of generic and specific evolution far more clearly (Coope 1978).

TABLE 4.1. Evaluation of recently revised neotropical butterfly groups for biogeographical analysis.

Butterfly Group (at least 5 species)	Aposematic?	Batesian mimic?	STUDY Biosystematic (field)	STUDY Zoogeographical	STUDY Experimental genetics	GENETICS Reproductive strategy	GENETICS Chromosome complement†	APTITUDE Degree of differentiation	APTITUDE Polymorphism	APTITUDE Material available	Usual habitat	ECOLOGY Dispersal capacity	ECOLOGY Food-plant specificity	ECOLOGY Population size (usual)	ECOLOGY Altitude preference	TOTAL‡	RATING**	REFERENCES Taxonomic revisions (whole group)	REFERENCES Other broad comparative biology
(Ideal group)*†	+	–	4	4	4	K	(known)	3	1–=	3	Restricted	1–=	3	1–?	?	20+	A	many	many
HESPERIIDAE																			
Pyrrhopyginae†	?	+	2	3	1	r	Ins	2	2	2	Forest, marsh	3	?	3	1–3	7	C	a, b	b
Pyrginae: *Phanus*	–	±	1	2	1	?	?	2	1	2	Forest	2	?	2	1	7	C	c, d	e
Aguna	–	–	1	2	1	?	?	1	1	2	Forest, scrub	3	?	3	1–3	6	D	e	e
Other Pyrginae	–	±	1	2	1	?	St	2	2	2	All types	3	3	2	1–3	5	D	b, c	b
Hesperiinae	–	+	1	2	1	?	St	2	2	2	All types	3	2	2	1–3	5	D	b, f	b
LYCAENIDAE																			
Theclinae: *Arcas*	–	–	1	2	1	K	?	2	1	1	Forest	2	?	1	1	7	C	g	h
Symbiopsis	–	–	1	2	1	K	?	2	1	1	Forest	2	?	1	1	7	C	i	h
Panthiades	–	–	1	2	1	K	?	2	1	1	Forest, scrub	3	?	1	1	6	D	j	h
Magnastigma	–	–	1	2	1	K	?	1	1	1	Forest, field	2	?	1	1	6	D	k	h
Parrhasius	–	–	2	2	1	K	?	2	2	2	Forest	2	?	1	1	9	C	l	h
Calycopis, Calys-tryma	–	–	1	2	1	r	?	1	2	2	Forest, open	3	?	3	1	4	D	m	h
Riodininae: *Lasia*	–	–	2	2	1	?	?	1	2	2	Scrub	3	?	2	1	7	C	n	
Nymphidium group†	–	+	2	2	1	K	Ins, St	2	2	2	Forest	1	?	1	1	9	C	o	
PIERIDAE																			
Dismorphiinae†	–	+	2	3	2	K	Ins	3	2	3	Forest	1	?	2	1–3	13	B	p, q	h
Pierinae: *Catasticta*	–	±	2	3	3	K	?	3	1	3	Cloud forest	1	?	2	2	14	B	r	h
Tatocheila-Phulia†	–	–	3	3	2	K	?	2	1	2	Paramo	2	?	2	3	12	B	s	t
PAPILIONIDAE																			
Troidini†	+	–	3	3	2	K	St	3	2	3	Forest	1	3	1	1	14+	B	u, v	v, w
NYMPHALIDAE																			
Morphinae*	–	–	2	3	2	K	St	3	2	3	Forest	2	2	2	1–3	11	B	x	y
Satyrinae																			
Brassolini*†		?	3	3	2	K	St	2	1	2	Forest, edges	2	2	1	1	12	B	z	
Haeterini			2	2	2	K	St	2	1	2	Forest	1	1	1	1	11	B	a′, c′, b′	b′
Biini†			2	3	1	K	Ins	2	2	2	Forest	1	?	2	1	12	B	c′, d′	b′
Euptychiini*			3	2	2	r	Ins	2	2	2	Forest, open	2	?	3	1–3	8	C	c′, e′	e′
Pronophilini†			3	1	2	K	Ins	3	1	3	Cloud forest, paramo	2	?	2	2–3	13	B	c′, f′	g′, f′
Danainae†	+	–	3	3	3	r	St	1	2	3	Open, forest	3	3	3	1–2	10+	B	h′, f′	h′
Ithomiinae*†	+	–	3	3	2	K, r	Ins	3	2	3	Deep forest	1	3	1	1–2	14+	B	i′, j′	k′, l′
Acraeinae†	+	–	2	3	2	r	Ins	2	2	3	Open, forest	2	3	3	1–2	10+	B	m′	

															j', n'	A
															j', l'	
															o'	C
															p'	B
															q'	C
															r'	B
															s'	C
															t'	B

Nymphalinae																
Heliconiini*	+	−	4	4	3	K	St	3	1	3	Mostly forest	1	3	1	19+	A
Melitaeini	−	+	2	2	1	r	St	3	2	3	Forest, open	?	2	1	9	C
Callicorini	−	?	3	3	1	K	St	3	1	3	Forest, open	3	2	1	10	B
Adelpha*	−	±	1	2	2	K	Ins	2	1	2	Forest	2	3	1–2	8	C
Nessaea*	−	−	1	2	2	K	Ins	2	1	2	Forest	1	3	1	11	B
Charaxinae																
Agrias/Prepona**	−	?	2	3	2	K	Ins	2	3	3	Forest	3	1	1	8	C
Anaea* (s.l.)	−	±	3	3	2	K	Ins	2	2	3	Mostly forest	3	2	1	10	B

*In these groups, the recent revisions cited included an analysis of the biogeography and evolution of regional subspecies.

+ These groups are presently undergoing intensive biogeographical and evolutionary analysis, in relation to palaeoecology.

**The 'rating' is related to the sum of the data available and favourable characteristics for biogeographical analysis at a regional level (usually subspecies), and is no reflection upon the *quality* of the taxonomic revision or the biological information.

‡ The sum includes a subtraction for the difference between the value assigned and the optimum value of 1 in the two columns marked $\underline{1}$, and does not include the final three figures.

†Those listed as 'Stable' are more likely to undergo classical differentiation in allopatry (Bush 1975): those indicated as 'Instable' can undergo parapatric divergence easily.

Assigned values: first three columns, 1 = Little, 2 = Fair, 3 = Good, 4 = Excellent; other columns, 1 = Low, 2 = Medium, 3 = High (for Altitude, 1 = tropical lowland forests, 2 = cloud forests, 3 = paramo)

Groups which have not been recently revised, whose size and biological characters suggest that they would be favourable for analysis (at least in part), include:

LYCAENIDAE: Riodininae: *Euselasia, Mesosemia, Mesene, Eurybia, Helicopis, Audre* (open vegetation), *Symmachia, Emesis.*

PIERIDAE: *Eurema* (s.l.), *Hesperocharis–Archonias–Charonias–Pereute* group, *Perrhybris.*

NYMPHALIDAE: Satyrinae: many genera of Pronophilini and Euptychiini.
Nymphalinae: *Marpesia, Dynamine, Myscelia, Epiphile, Pyrrhogyra, Eunica, Catonephele, Doxocopa, Asterope, Perisama* (cloud forests).

References

a Evans (1951); b O. H. H. Mielke (unpublished); c Evans (1952, 1953); d Miller (1965); e Mielke (1971); f Evans (1955); g Nicolay (1971a); h Robbins and Small (1981); i Nicolay (1971b); j Nicolay (1976); k Nicolay (1977); l Nicolay (1979); m Field (1967a, b); n Clench (1972); o C. J. Callaghan (unpublished); p Lamas (1979a); q G. Lamas (unpublished); r Reissinger (1972); s Herrera and Field (1959), Ackery (1975), Field and Herrera (1977); t Shapiro (1978a, b, c, d, 1979); u Rothschild and Jordan (1906), D'Almeida (1966); v K. S. Brown (unpublished); Brown, Damman, and Feeny (1981); w Moss (1919); x Lemoult and Real (1962); y Otero (1971); Young and Muyshondt (1972), Young (1973); z Blandin (1978); Bristow (1981); a' Brown (1942a, 1948); D'Almeida (1951a); b' Masters (1970); c' Forster (1964); Miller (1968); d' R. I. Vane-Wright (unpublished); e' M. C. Singer and P. R. Ehrlich (unpublished); f' Adams and Bernard (1977, 1979, 1981), Brown (1942b); g' Adams (1973, 1977, unpublished); h' G. Lamas, P. R. Ackery, and R. I. Vane-Wright (unpublished); i' Lamas (1973, unpublished); i' Papageorgis (1975); D'Almeida (1978), Mielke and D'Almeida (1978); j' Brown (1979); k' Drummond (1976), Haber (1978), Brown and Vasconcellos-Neto (1976); l' Papageorgis (1975); m' D'Almeida (1935), Potts (1943), F. Fernandez Yepez and G. Lamas (unpublished); n' Brown (1976a, 1981); o' Hall (1928–30), Higgins (1981); p' Dillon (1948); q' Hall (1938), C. Samson (unpublished); r' Vane-Wright (1979); s' Descimon (1977a, unpublished); t' Comstock (1961).

Note added in press (mid-1986): Neotropical butterflies are continually more used in systematic, biological and ecological studies; a landmark book has appeared bringing the literature up to 1983 (*The Biology of Butterflies*, edited by R. I. Vane-Wright and P. R. Ackery, Academic Press, 1984). Reviews have now appeared covering *Myscelia, Catonephele* and *Epiphile* (D. Jenkins, *Bull. Allyn Museum* 87 (1984): 92 (1985) and 101 (1986)). S. S. Nicolay (see g, i–l) has revised another Thecline genus, *Olynthus* (*Bull. Allyn Museum* 74 (1982)), and C. J. Callaghan has published part of his revision of *Nymphidium* (ref. o) (*Bull. Allyn Museum* 98 (1985) and 100 (1986)). Reference b has been partly published in various papers and a thesis; Shapiro (ref. t) has many new papers on *Tatochila* biology. Bristow (ref. z) has further works on Brassolini (also revised in a thesis by M. Casagrande), and Otero and Brown (ref. v) has many new papers on Troidine biology (*Atala* 10–12, 2–16 (1986). The group 'Bimi' (ref. d') has been split between Morphinae (for *Antirrhaea* and *Caerois*) and probably Brassolini (P. DeVries, I. R. Kitching and R. I. Vane-Wright, *Systm. Ent.* 10, 11–32 (1985); unpublished observations of Otero and Brown). An important book on Danaine biology (ref. h') has appeared (P. R. Ackery and R. I. Vane-Wright, *Milkweed Butterflies: their Biology and Cladistics*, British Museum, 1984). The Ithomiinae (j', k') have been completely revised (see Brown, *Nature, Lond.* 309, 707–9 (1984) and *Rev. bras. Biol.* 44, 435–60 (1985), others in preparation), and the American Acraeinae have been placed in much better order (Penz, Francini, Lamas, in preparation). An important work has appeared on *Adelpha* (q') biology (A. Aiello, *Psyche* 91, 1–45 (1984)), and many new papers are available on Charaxinae systematics and biology (s', t'). All of these works have brought further groups of Neotropical butterflies to the condition of utility for biogeographical analysis.

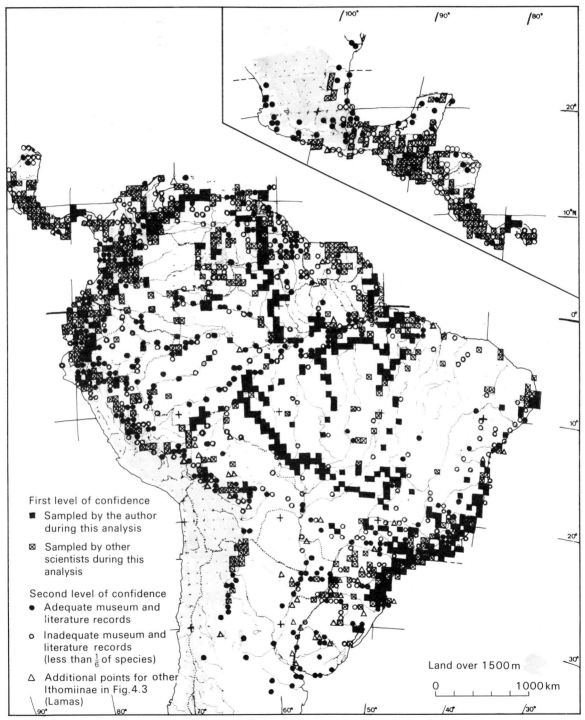

First level of confidence

■ Sampled by the author during this analysis

⊠ Sampled by other scientists during this analysis

Second level of confidence

● Adequate museum and literature records

○ Inadequate museum and literature records (less than $\frac{1}{6}$ of species)

△ Additional points for other Ithomiinae in Fig.4.3 (Lamas)

Land over 1500 m

0   1000 km

Fig. 4.2. Sampling points for Heliconiini and Ithomiinae in the Neotropics, grouped as half-degree quadrants of latitude and longitude.

further witnesses to past climatic change and evolutionary processes (Fig. 4.19).

Finally, a remarkably close correlation is shown to exist (Fig. 4.20) between the centres of subspecies endemism of heliconiine and ithomiine butterflies (Fig. 4.15) and the geoscientific model for forest continuity in the last, Wisconsin–Würm, ice age (Fig. 2.8). Thirty-two of the forty-four endemic centres enclose forest refugia, the rest show a reasonable fit (except Ilha de Marajó which has since been inundated). Correlation does not prove causation but the fit between the biogeographical patterns and the geohistorical model based on very different data sources is so good that causation is probable. Patterns of species diversity are poorly correlated with endemism and do not form recognizable centres (Fig. 4.14) and show if anything a negative relationship to the palaeoecological model, but are closely tied with environmental microheterogeneity in the present, as mentioned above for the different centres. Again it is emphasized that to be meaningful such correlations must be based on adequate sampling (not biased simply to river sides) of organisms whose biology, ecology, and interbreeding patterns are known.

INTRODUCTION

Butterflies have long attracted the attention of naturalists, especially those who have visited the tropics. In the famous writings about his experiences in the Amazon River basin, Henry Walter Bates gave a complete biological rationale for his prediction that 'the study of butterflies — creatures selected as the types of airiness and frivolity — instead of being despised, will some day be valued as one of the most important branches of Biological science' (1864, pp. 412–13). He also laid the foundations for the modern biogeographical analysis of these organisms, in his comments on mimicry rings, composed of superficially convergent species representing widely divergent families, which 'were found to change in the uniform country of the Upper Amazon from one locality to another not further removed than one hundred to two hundred miles. [These] species, however, have often the characters of local varieties, some of them indeed showing the connecting link' (Bates 1862, pp. 73–4).

Further data on the parallel geographical variations in neotropical mimetic butterflies were accumulated by Kaye (1907) and Moulton (1909). With modern biosystematics, it has become possible to confirm that these mimicry rings are in fact usually composed of essentially the same set of widespread species, which change their patterns in harmony in different parts of the Amazon Basin and elsewhere in tropical America. Broader studies have also revealed that the parallel variation is not confined to the 'tiger-patterned' mimetic species in the complexes observed by the early authors, but also appears consistently in less conspicuous forms, such as glasswings, morphos, owls, wood satyrs, swallowtails, leafwings, 88s, and many other groups of the neotropical forests.

Thus, it is not surprising that butterflies figured often among the organisms examined in early studies of biogeographical patterns in the neotropics. Michael's studies (1911*a*, 1911*b*, 1912, 1914–15) identified some of the best groups for analysis (Papilionidae, *Agrias*, *Heliconius*, *Morpho*) and their division patterns in the Amazon Basin. Krüger's paper (1933) focused on endemic regions for species (as then understood; many are now regarded as subspecies); his proposals, based on the family Papilionidae and the Nymphalid subfamily Morphinae and tribes Heliconiini and Haeterini, are still surprisingly coherent in light of today's much better knowledge. Krüger recognized discrete 'centres of evolution' in Mexico (to Nicaragua), western Panama and Costa Rica, the Pacific coast of Colombia and Ecuador, the Colombian cordilleras and interior valleys, northern Venezuela, the lower Amazon region, the northwestern Amazon (southeast Colombia and east Ecuador to Iquitos, Peru), the southwestern Amazon, and the Atlantic coast of Brazil.

Still finer biogeographical divisions, recognizable at the subspecies level in more obviously differentiated species, were discussed in early papers with specific and pioneering reference to their relationship to Pleistocene climatic fluctuations, by Fox (1949), Comstock (1961), Clench (1964), and Turner (1965). The groups covered by the first and last of these authors (mimetic ithomiines and heliconians; see Fig. 4.3–4.10) are precisely those which had incited wonder in Bates a hundred years earlier. They are ideally suited for a detailed study of the variation patterns throughout the neotropics; their biogeography and evolution are examined quantitatively in this chapter.

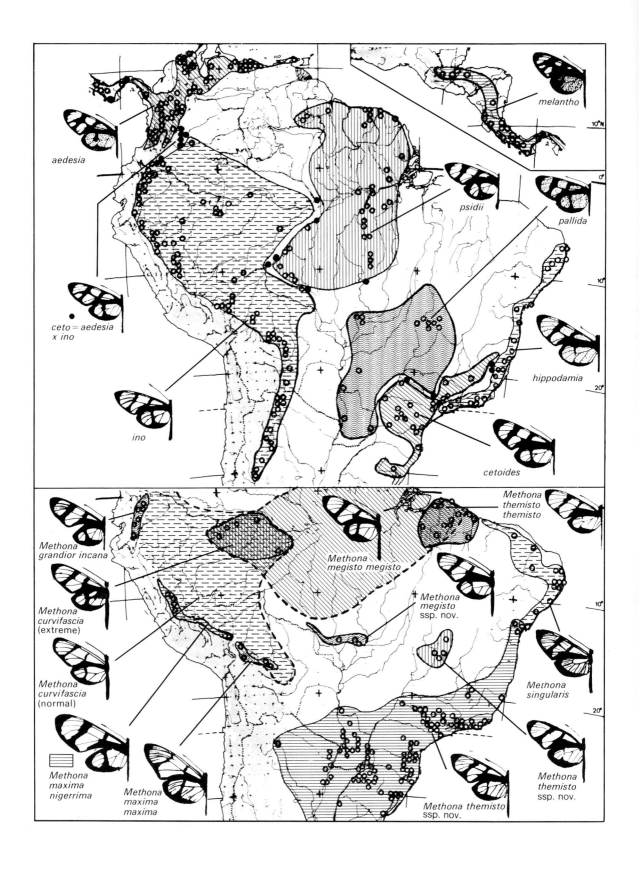

melantho

aedesia

psidii

pallida

ceto = aedesia
x ino

hippodamia

ino

cetoides

Methona
themisto
themisto

Methona
grandior incana

Methona
megisto megisto

Methona
curvifascia
(extreme)

Methona
megisto
ssp. nov.

Methona
curvifascia
(normal)

Methona
singularis

Methona
maxima
nigerrima

Methona
themisto
ssp. nov.

Methona
maxima
maxima

Methona themisto
ssp. nov.

HISTORICAL BIOGEOGRAPHY

If any analysis of biogeographical patterns in the light of evolutionary processes is to avoid some of the more serious errors inherent in non-testable narrative hypotheses (Ball 1976), it must be solidly built upon:

1. Biologically relevant taxonomy, preferably based upon characters whose significance has been investigated in natural systems, and whose rate and geographical scale of evolution can be estimated.

2. Thorough biogeographical coverage, revealing the patterns of variation in these characters through space and time and across environmental gradients, as well as any disharmonic variation patterns in other characters.

3. Reliable ecological data, including independent biological and geoscientific information on the history and present importance of as many biotic and abiotic factors as can be shown to have relevance to the life of the organisms.

4. Corroborating genetic, evolutionary, and population data (chromosomes, allozymes, radiation rates, test crosses, colonization strategies, dispersal rates, deme sizes, fossils), to permit a reasonable understanding of fundamental processes and velocities of differentiation and species formation.

Such information is not wholly available for any group of neotropical organisms known to me. Several fair-sized butterfly groups, however, are well enough studied to be strong candidates for analysis.

Biosystematics

Many neotropical butterflies are still poorly represented in collections and little studied in the field. Most groups are revised with the use of pattern or morphological characters whose relation to geographical ecology· and natural selection is still unknown. Speculative biogeographical and evolutionary schemes are erected upon limited numbers of museum specimens. The identification of true, recently conphyletic, units has made shambles of many such morphologically based schemes. Many supposedly different 'species' have been found to intergrade in hybridization zones, representing in fact only subspecies of single widespread polytypic species (Brown 1977*a*, *b*, 1979). Recent experience suggests that taxonomic or evolutionary proposals which do not include work with live organisms in natural environments are highly prone to misrepresentation. Fortunately, field work is once again becoming popular and is now possible in many well-placed stations in all parts of the neotropics. Fundamental revisions can confidently be predicted in the near future of most butterfly systematic groupings; they will be necessary before any serious biogeographical work can be undertaken, because analysis of vicariance or any other differentiation process in artificial or polyphyletic groups is quite meaningless.

Geographical coverage

The generalized locality labels on many early or commercially collected butterflies have plagued biogeographical work in the neotropics. 'Colombia' includes at least ten evolutionarily distinct biotas; recent specimens labelled 'Cali' represent five of these, all within a few hours' drive of that city. To add injury to insult, most heavily collected localities are in areas where distinct biotas meet and intergrade; typical examples are many lower Andean valleys, larger Amazonian and Guianan rivers, the Panama Canal Zone, and substantial sections of the east coast of Brazil. These are often very attractive areas to collectors, because of their high butterfly diversity (especially in unique hybrid phenotypes), abundance of certain easily commercialized species, and ease of access by river or road and to the more open forest interior.

The general pattern of rather haphazard sampling by naturalist-travellers and professional hunters has been broken by a very few exceptions, assiduous

Figs. 4.3–4.11 [Collective caption₄] Various patterns of geographical differentiation in neotropical forest butterflies. Open circles are locality records for essentially uniform phenotypes (as illustrated around the outside of the maps). Shading shows the probable limits of predominance of this phenotype. Solid circles show sampling localities with mixed phenotypes, usually recombinations of characters from adjacent subspecies. Small graphs (Figs. 4.4–4.10) show the preferred altitude range of each species mapped. (Figs. 4.4–4.10 after Brown 1979.)

Fig. 4.3. Ithomiinae: *Thyridia psidii* (upper), and (lower) *Methona maxima, M. grandior incana, M. megisto* ssp. nov., *M. singularis, M. themisto*, and *M. curvifascia* (note character-displaced phenotype where sympatric with *M. megisto megisto*; the ranges of these two entities are shown by dotted lines). (After Lamas 1973 updated by author.)

scientists who have kept careful records and made efforts to leave behind the coast, major rivers, and easily penetrated open forests and to explore the highlands, interfluvials, upper river valleys, and dense forests. The picture uncovered by them casts into disrepute the traditional body of knowledge, revealing its basis in zones of contact between vast and still largely unknown biotas. It is clear that intensive geographical sampling away from traditional collecting points must accompany any biogeographical analysis of neotropical organisms.

Ecology

Many genera of neotropical butterflies comprise populations and species with dramatically different ecological strategies and preferences. The single genus *Heliconius* includes in its ranks generalists and specialists along a multitude of environmental continua: forest and non-forest subspecies within a single species, forms with cryptic, flash-disruptive and warning coloration, vastly differing life-cycles, highland and lowland species, strong and weak flyers, migrants and residents, pollen-extractors and scavengers, as well as populations regulated and strongly selected by any combination of parasitism, predation, competition, social interaction, larval or adult resources, or other environmental variables which fluctuate through time and space (Brown 1981). This reduces, but does not eliminate, the possibilities for prediction and generalization of biogeographical data and hypotheses. Indeed, the fact that some general geographical patterns were discernible even to early authors argues for strong selective forces for coherence and integration over long periods of evolutionary history and geographical restructuring on a regional level, even though each population's characters seem to be under continual ecological pressure for massive divergence, dispersal, fragmentation, and reorganization in the present. Similar variation in ecology has been observed in most of the larger genera of neotropical Lepidoptera examined. Therefore, the simple inclusion of an unstudied taxon in an ecological habitat on the basis of close relationship may not be justified. Good and abundant ecological data for all members of any group of organisms analysed should be used in the evaluation of their biogeographical patterns.

Genetics

Only recently has genetic information become

available for a very few groups of neotropical lepidoptera. In the Heliconiini, this information has shown that gross morphological characters, and even compelling biosystematic data, may have little relationship to evolutionary reality (Brown and Benson 1974). Fortunately, in polytypic species it is often possible to observe qualitative interactions of presumed genes between subspecies and populations, in zones of rapid transition between them (actually, unit characters are observed, be they biochemical, morphological, or ethological, and these are hoped to bear a close relationship with major selected genes). Analysis of intergradation within broad zones of hybridization between subspecies has supported conspecificity for taxa long regarded as separate biological species, and sometimes even for taxa placed in separate species-groups or genera on the basis of classical morphological analysis. Genetic analysis, including cytogenetics, has also shattered many standard views and generalizations about rates, mechanisms, and options for evolutionary processes, especially the achievement of reproductive isolation (Bush 1975). At a higher level, the relationships among tribes and families of Lepidoptera are still in debate, and the most cherished ideas about selection, adaptive radiation, allopatric and parapatric speciation, genetic and chromosomal revolutions, must all clearly be tested and examined carefully in each individual species and group, preferably by experimental genetics (study of inheritance in hybrids and backcrosses) and the collection of abundant new field data, especially in natural zones of overlap.

Geology, palaeoecology, and butterfly evolution

Recent advances in the geosciences, especially the discoveries of plate tectonics and mountain-building, new discoveries of fossils, dramatic fluctuations in palaeoclimates and sea levels, and correlation of landforms with palaeoecology, have revolutionized the science of biogeography. It no longer seems strange that numerous genera and tribes of butterflies show disjunct pantropical distributions, since they could have been continuous as late as the mid-Tertiary (see Shields and Dvorak 1979 for a review). A critical look at plate tectonics and the still meagre fossil record (Forbes 1932; Shields 1976) permits an approximate representation of neotropical butterfly higher evolution through time (Fig. 4.1). Key information includes the existence of families or subfamilies which span all parts of an ancient Gondwanaland but are scarce or unknown from northern continents (period A; Shields and Dvorak 1979), of tribes which span South America and Australia–New Guinea but do not occur in Africa (period B), of modern genera represented by early Tertiary fossils (period C; Shields 1976), and of modern species represented by late Tertiary to early Pleistocene fossils (period D). A critical geological and palaeontological view of all insects (Coope 1970, 1978, 1979) has contrasted the stability of species during the two million years of the Pleistocene with the wide palaeoecological fluctuations which were occurring then. The case of *Heliconius hermathena vereatta* also suggests that a simple one-gene mutation frequent in other *Heliconius* species, which forms a favoured mimetic colour-pattern in this subspecies (Brown and Benson 1977) may have occurred in only one restricted locality within a vast range in which models and favourable selection have probably been present since after the last high sea-level episode, 5000 years ago (Fairbridge 1976). Thus it is acceptable that in *Heliconius* and possibly in most neotropical insects, climatic crises of the late Pleistocene may have led principally to changes in geographical distributions (Coope 1970, 1978, 1979) and population or subspecies differentiation, involving a few adaptive genes spread out over limited geographical regions. Similar conclusions emerged from a study of Antillean Lycaenid butterflies by Clench (1964).

Perspective

Neotropical biogeography is very complex and woefully immature, as are the data-bases which must support it. I know of no set of data about any important phenomenon, system, or group of organisms in the neotropics which can claim to have reached a clear asymptote on the gathering and interpretation of knowledge. Thus it is still impossible to know whether a given data-set and its derived hypotheses are 1 or 90 per cent along the road to meriting confidence and being used in predictions. In many cases, data which clearly show all previous interpretations to be deficient continue to appear regularly. With this caveat, I will attempt to examine here a few relatively clear biogeographical and evolutionary patterns in neotropical butterflies, which have incorporated large amounts of data and been used successfully for some predictions, and which suggest many lines for future research.

CHOICE OF GROUPS FOR THE ANALYSIS OF DIFFERENTIATION PATTERNS

An evaluation of the biosystematic, geographical, ecological, and genetic information available for a variety of butterfly groups which have been subjected to recent revisions is presented in Table 4.1. Coherent proposals for the evolutionary biogeography of many of these groups were included in the revisions; others are at present undergoing intensive study for the determination of their biogeographical differentiation patterns, and the relation of these to evolutionary trends and palaeoecology. Several groups of moths have also been analysed in recent years; these are discussed in Chapter 7.

Aposematic (warningly coloured) butterflies show some special advantages for biogeographical analysis, which favour the coherence of the patterns which may be seen. They often demonstrate clear, selection-mediated, regional differentiation in adult wing pattern, which can easily be related to ecology and major genes. They are usually abundant and easily sampled, and attract the attention of naturalists; they are thus amply represented in collections. Their biology is well known. They are often good ecological indicators, since food-plant specificity (Ehrlich and Raven 1965) and the incorporation of poisonous substances (Brower and Brower 1964) generate around them recognizable and tightly co-evolved food-webs (Gilbert 1977, 1980). It is important to note, however, that differentiation patterns very similar to (if often less fine-grained than) those observed in the aposematic Troidini, Danainae, Ithomiinae, Acraeinae and Heliconiini occur also in non-aposematic forest butterflies, but are more difficult to study owing to less obvious wing pattern differences and greater rarity. Thus, the aposematic butterflies do not seem to represent a special case, only easier patterns to see and understand.

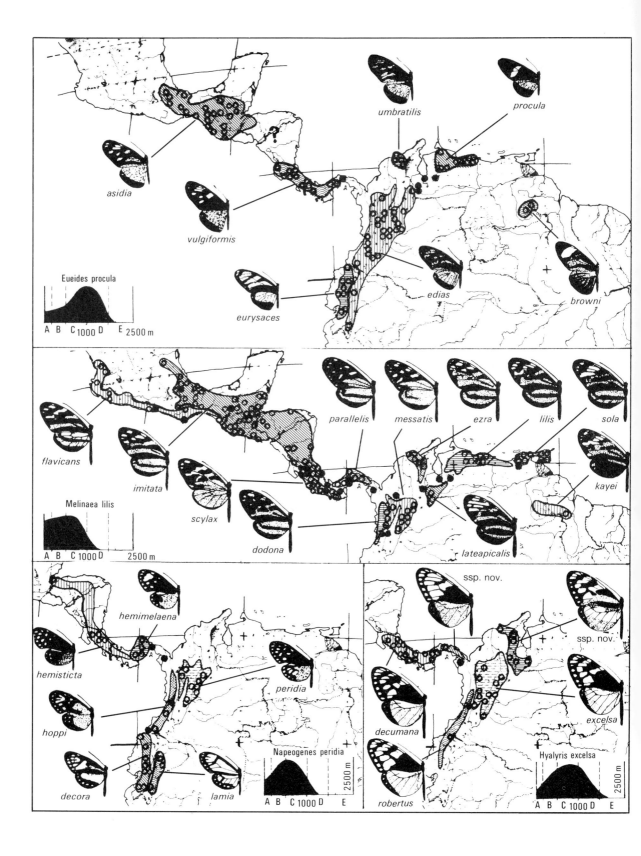

umbratilis

procula

asidia

vulgiformis

edias

browni

eurysaces

Eueides procula

A B C 1000 D E 2500 m

parallelis

messatis

ezra

lilis

sola

flavicans

imitata

scylax

dodona

lateapicalis

kayei

Melinaea lilis

A B C 1000 D 2500 m

hemimelaena

hemisticta

peridia

hoppi

decora

lamia

Napeogenes peridia

A B C 1000 D E 2500 m

ssp. nov.

ssp. nov.

decumana

excelsa

robertus

Hyalyris excelsa

A B C 1000 D E 2500 m

BASIC DATA FOR BIOGEOGRAPHICAL ANALYSIS OF HELICONIINI AND ITHOMIINAE

Before the biogeographical analysis of the Heliconiini was begun, their biosystematics were well developed, through careful field, laboratory, and museum work by a group supported by the New York Zoological Society and based in Simla, Trinidad (Crane 1955, 1957; Beebe, Crane, and Fleming 1960; Alexander 1961*a*, *b*; Turner and Crane 1963; Emsley 1963, 1964, 1965; Turner 1966, 1967, 1968, 1971). Later taxonomic revisions involved some separations and unions of species on biological grounds (Brown and Mielke 1972; Brown 1976*a*), and a revision of the especially difficult 'silvaniform' (tiger-patterned) group (Brown and Benson 1974; Brown 1976*b*). Many biological and ecological data, useful in biosystematic judgements, were collected in the field before the first overall analysis of biogeographical patterns in 1973. Many easily recognizable new subspecies were discovered in previously poorly sampled areas, and a few appeared in museum collections; but the general evolutionary and systematic arrangement of the tribe did not depart greatly from that of Emsley's revisions.

In contrast, the systematic arrangement of the Ithomiinae which pre-dated the biogeographical study was based on relatively restricted taxonomic viewpoints, dominated by monographs which incorporated little field data (Fox 1940, 1949, 1956, 1960*a*, *b*, 1965, 1967, 1968; Fox and Real 1971). Although Fox had an extraordinary 'feel' for the diversity and significance of the morphology of these butterflies, his ideas often contrasted with those of others who knew them in the field (Forbes 1948; D'Almeida 1951*b*, 1956, 1958, 1960, 1978; Brown and D'Almeida 1970; Lamas 1973). Biosystematic understanding of the ithomiines is still far behind that of the Heliconiini, but information is slowly accumulating to indicate that their biological and evolutionary patterns are dramatically different from those of most other nymphalids, often requiring systematic decisions in which classical concepts of morphology and species must be placed aside. The tidying up of the Ithomiinae has been a difficult and laborious project, and further field work will undoubtedly necessitate many additional changes in an arrangement which is already appreciably modified from that of Fox (Lamas 1973,

1981; Brown 1977*a*, *c*, 1979, 1980; Mielke and Brown 1979).

The known geographical distribution records for Heliconiini and Ithomiinae were sufficient in 1963 to permit the identification of broad differentiation patterns (Turner 1965; Emsley 1963, 1964, 1965; Fox 1967, 1968) and the general location of important hybridization zones. Additional sampling was then concentrated in these zones of subspecies intergradation, and in areas not yet covered. By mid-1979, very few major regions in the neotropical forests remained to be examined (Fig. 4.2). The actual sampling process for these butterflies was facilitated by the use of specific baits: red cloths and specific pollen sources for *Heliconius* (Brown and Mielke 1972; Brown 1973; Gilbert 1975), and sources of pyrrolizidine alkaloid decomposition products (such as *Eupatorium* flowers and drying *Heliotropium* plants) for male ithomiines (Moss 1947; Beebe 1955; Masters 1968; Pliske 1975). These techniques permit a reasonable analysis of the wing-pattern and morphological variation in most local populations in a single day.

Heliconians adapt easily to outdoor insectaries and glasshouses (Crane and Fleming 1953; Turner 1974), and thus are fit candidates for genetic and physiological experimentation (Sheppard 1963; Emsley 1964; Turner 1972; Brown and Benson 1974; Brown 1976*a*, *b*, 1979, 1981). This has permitted the accumulation of enough genetic information to give a considered understanding of the evolution of Müllerian mimicry in them (Turner 1978). Ithomiines are far more difficult to keep in captivity, and what little is suspected of their genetics comes only from examination of populations in hybrid zones between subspecies. Heliconiini mostly have invariable chromosome complements, usually at $n = 31$ or 21, except for a transitional group *Neruda*, the very advanced *sapho*-group and the aberrant *Laparus doris* which is polymorphic in colour and shows $n = 19$–33; many ithomiines, however, have highly variable chromosome complements between subspecies of a single species and even between individuals of a single population (de Lesse 1967, 1970*a*, *b*; de Lesse and Brown 1971; Wesley and Emmel 1975). This variation

---

Fig. 4.4. Transandean species. Heliconiini: *Eueides procula* (upper). Ithomiinae: *Melinaea lilis* (centre), *Napeogenes peridia* (lower left; note outlying subspecies in E. Ecuador), and *Hyalyris excelsa* (lower right).

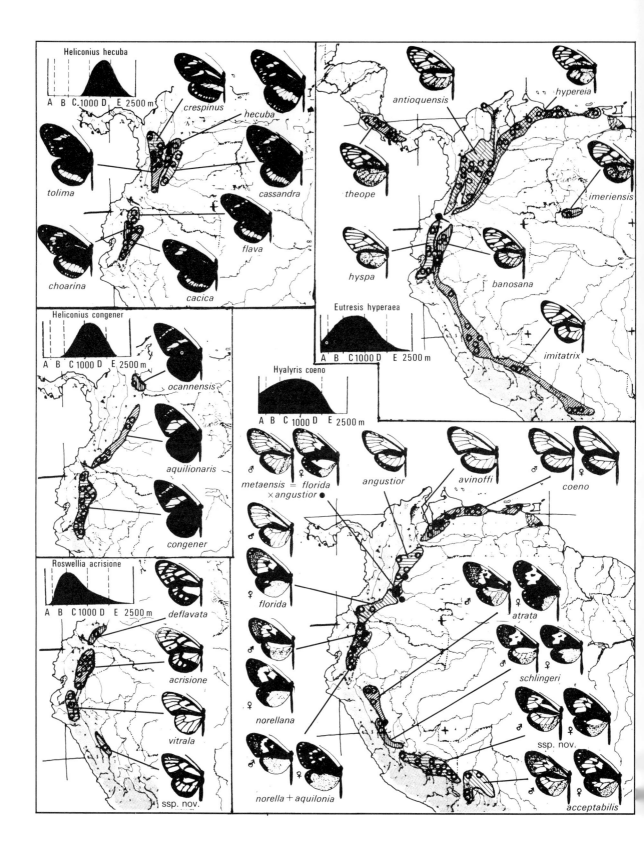

Heliconius hecuba

A B C 1000 D E 2500 m

crespinus
hecuba
tolima
cassandra
flava
choarina
cacica

Heliconius congener

A B C 1000 D E 2500 m

ocannensis
aquilionaris
congener

Roswellia acrisione

A B C 1000 D E 2500 m

deflavata
acrisione
vitrala
ssp. nov.

antioquensis
hypereia
theope
imeriensis
hyspa
banosana

Eutresis hyperaea

A B C 1000 D E 2500 m

imitatrix

Hyalyris coeno

A B C 1000 D E 2500 m

metaensis = florida
×angustior ●
angustior
avinoffi
coeno
florida
atrata
norellana
schlingeri
ssp. nov.
norella + aquilonia
acceptabilis

indicates that they may at present be undergoing rapid evolution. Species in some genera may be in the process of chromosome revolutions, leading to fuzzy or indiscriminate boundaries. Nevertheless, the patterns of regional geographical variation in most Ithomiinae are closely parallel to those of the Heliconiini, suggesting a preponderant action of environmental factors in their formation.

The majority of heliconians and ithomiines inhabit shady, humid forest, flying in the lower or middle storey and often showing a restricted 'home-range'. They have á limited range of specific larval food-plants (Passifloraceae and Solanaceae plus Apocynaceae–Echitoideae, respectively), and in many cases occur as relatively small and moderately stable populations. However, in strongly seasonal environments populations may fluctuate greatly and not always predictably through the year (Brown and Mielke 1972; Brown and Benson 1974; Brown and Vasconcellos-Neto 1976). Population regulation is often by predators and parasites on eggs and small larvae (Ehrlich and Gilbert 1973; Gilbert 1977); larger larvae and adults are sometimes taken by predatory spiders, pentatomid bugs, large ants and wasps, and rarely by birds and lizards (Brown and Vasconcellos-Neto 1976; Drummond 1976; Haber 1978). Pupae are cryptic, partly or wholly silvered and thus mirroring their surroundings, and probably rarely subject to mortality except by certain specific parasitoids. Competition for resources (space, larval and adult food, and mates) can be very keen and has probably determined many local, regional and species characters in these butter-flies (Gilbert 1969, 1975; Benson, Brown, and Gilbert 1976; Benson 1978; Smiley 1978). Density independent mortality factors include egg, larval, and adult death during violent tropical storms (Drummond 1976). Very few of the species extend into regions which experience annual frosts. The juvenile biology is very well known in the Heliconiini (Brown 1981) and becoming quite well studied in the Ithomiinae (Drummond 1976; Haber 1978), such that reasonable ecological statements are becoming possible for some populations in natural surroundings.

Adults are usually long-lived, up to six months in nature, and often show an extended reproductive period, seeking out unusual nitrogen sources to maintain egg production: pollen for *Heliconius* (Gilbert 1972, 1975), and carrion and bird droppings for the Ithomiinae (Gilbert 1969). In normal tropical localities, there are six to ten generations a year; in seasonal climates, late dry-season generations may be extended in the Ithomiinae (as adults in reproductive diapause) followed by wide dispersal at the beginning of the wet season.

Geographical variation in morphology in a single species may be as great as that in wing-colour, but some characters can usually be found (often among minor pattern elements, secondary genitalian characters, scent organs and scales, wing venation, leg segments, and body markings, though no one of these is always useful) which concord with the biological barriers perceived by the organisms themselves in their natural habitats.

BIOGEOGRAPHICAL PATTERNS IN FOREST BUTTERFLIES AT SPECIES AND SUBSPECIES LEVELS

After Emsley (1963, 1964, 1965) on Heliconiini, the first in-depth systematic, evolutionary, and biogeographical analysis of neotropical butterflies, involving wide travelling, essentially complete data from museum collections (including a list of 868 collecting localities with coordinates), and abundant biological and ecological data collected by a scientist resident in tropical America was presented in the doctoral thesis of Lamas (1973). In many ways, this study of Danainae and Ithomiinae has set a new standard for all subsequent work in these and other groups of neotropical butterflies; it is most unfortunate that it is still not publicly available. Based upon distributional analysis of nine polytypic species (a sample is shown in Fig. 4.3), Lamas suggested 27 centres for subspecies and species evolution (presented as 'refugia') in the

Fig. 4.5. Andean species. Heliconiini: *Heliconius hecuba* (upper left) and *H. congener* (centre left). Ithomiinae: *Roswellia acrisione* (lower left), *Eutresis hypereia* (upper right; note isolated subspecies in the Imerí massif), and *Hyalyris coeno* (lower right). (For species occupying both the Transandean and Andean regions, see Brown (1979) – *Olyras crathis*, a mimic of *E. hypereia*, Figs. 37, and *Tithorea tarricina*, Fig. 39.)

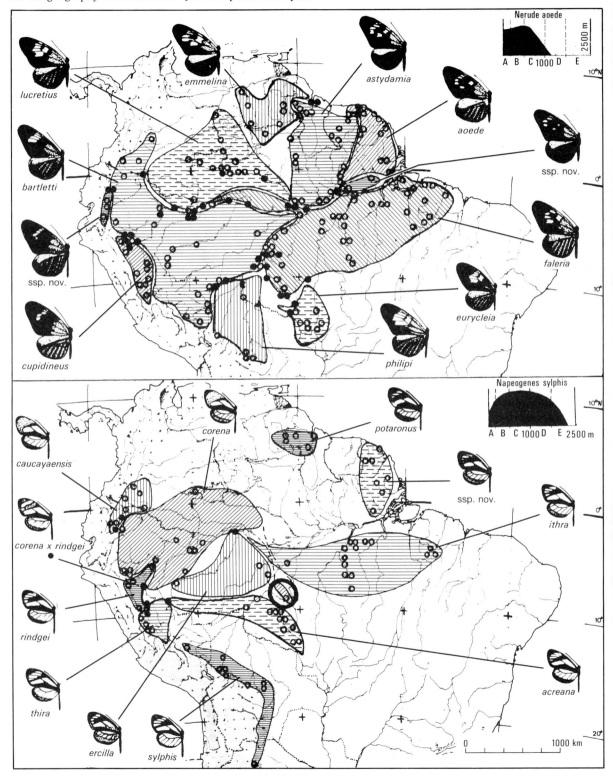

neotropical forests, and related these to the Quaternary refuges and centres for dispersal proposed by Haffer and Müller, respectively.

A more quantitative analysis, which was patterned after Lamas's approach and published in parts starting the year after his thesis (Brown, Sheppard, and Turner 1974) but completely only five years later (Brown 1979), covered five tribes of the Ithomiinae (92 species, including all those already revised by Fox and Lamas) and most of the Heliconiini (53 of the 65 species; the other 12 with minimal differentiation, high vagility, or non-forest habitat preferences were not included). It presented summary revisions of the groups, in order to identify monophyletic units (mostly considered as polytypic species), and a list of 3844 collecting localities with coordinates; species lists were prepared for 1519 of the 4646 half-degree quadrants in the neotropical forests. A selection of the distribution maps published in Brown (1979) is presented here, redrawn and organized by species regions, as Figs. 4.4–4.10.

For comparison, Fig. 4.11 illustrates presently known distributions of polytypic species in a very different and still much less studied butterfly group: the aposematic Troidine swallowtails *Parides*. Similar patterns may also be seen in the dusk-flying owl butterflies *Catoblepia* (Bristow 1981). These two groups show subspecies with differentiation patterns very similar to, though at times more complex than, the general patterns perceived in most Heliconiini and Ithomiinae (Figs. 4.4-4.10).

Species distributions and altitudinal preferences

The vast majority of neotropical forest butterfly species (more than 80 per cent; Brown 1982*b*) are found in one of four major biogeographical regions: Transandean, Andean, Hylaean (Amazonian), and Atlantic (Fig. 4.12). The rest are mostly widespread (Fig. 4.9); about 15 per cent are more restricted (Figs. 4.8 and 4.10). The Hylaea may be divided into upland (Guiana Shield) and lowland subregions, as well as further variants as shown in Figs. 4.6 and 4.7. A summary of the biogeographical regions observed in neotropical butterfly species is shown in Fig. 4.12, with lists and isolines for Heliconiini and Ithomiinae analyzed in Brown (1979, 1982*b*).

At a higher level, one genus of Heliconiini (*Podo-*

tricha) and six of Ithomiinae (*Roswellia*, *Athesis*, *Eutresis*, *Patricia*, *Elzunia*, and *Aremfoxia*, all regarded as primitive) are essentially restricted to the Andean region, which may represent the area for maximum diversification of the latter subfamily. Two further primitive ithomiine genera (*Athyrtis* and *Paititia*) occupy primarily the Andean foothills; two (*Epityches* and *Placidula*) occupy related cool habitats of the Atlantic region (*E. eupompe* may be congeneric with *Aremfoxia ferra*); and four (*Forbestra*, *Garsauritis*, *Rhodussa*, and *Sais*) are widespread over the lowlands of the Hylaea and its periphery. The lowland Hylaea is scrupulously avoided by one widespread ithomiine genus (*Hyalyris*, Figs. 4.4, 4.5, 4.8, and 4.10), which shows an essentially Andean/Atlantic (periamazonian) distribution, extending northwards to Costa Rica. The Transandean region (including Middle America) has no endemic genera in Heliconiini and Ithomiinae, possibly reflecting a southern origin for the two groups (Fig. 4.1). However, the appreciable species endemicity in the Transandean region (Fig. 4.12) makes it likely that widespread ancestral species occurred over both American continents by the mid or late Tertiary, and were then fragmented into northern and southern vicariants by tectonic processes. The results of this vicariance are commonly observed today as cis/trans-Andean species pairs, many of which show sympatric and character-displaced subspecies in a limited area of southwestern Venezuela and northeastern Colombia. Almost all genera in the four ithomiine tribes not analysed (Oleriini, Ithomiini, Dircennini, Godyridini) are either present in all the neotropics (sometimes without northern Central America, and usually poorly represented in the lowland Hylaea), or confined to the Andes (in two cases with outliers in the Atlantic region like *Hyalyris*) (Brown 1977*a*; D'Almeida 1978; Mielke and Brown 1979).

Most species show clear altitudinal preferences (shown as graphs within Figs. 4.4–4.10), which in a majority of lowland forest species (86 per cent of the Heliconiini and Ithomiinae analysed by Brown 1979) have a peak between 200 and 1200 m. Only 14 species of the 145 analysed in the two groups show preferences peaking at or below 200 m, where more than two-thirds of neotropical rain forests occur today. This reflects the great vegetational, topographic, and topoclimatic diversity which exists at

---

Fig. 4.6. Amazonian species. Heliconiini: *Neruda aoede* (upper). Ithomiinae: *Napeogenes sylphis* (lower; note unusually wide altitudinal range; encircled anomalous populations discussed on p. 91).

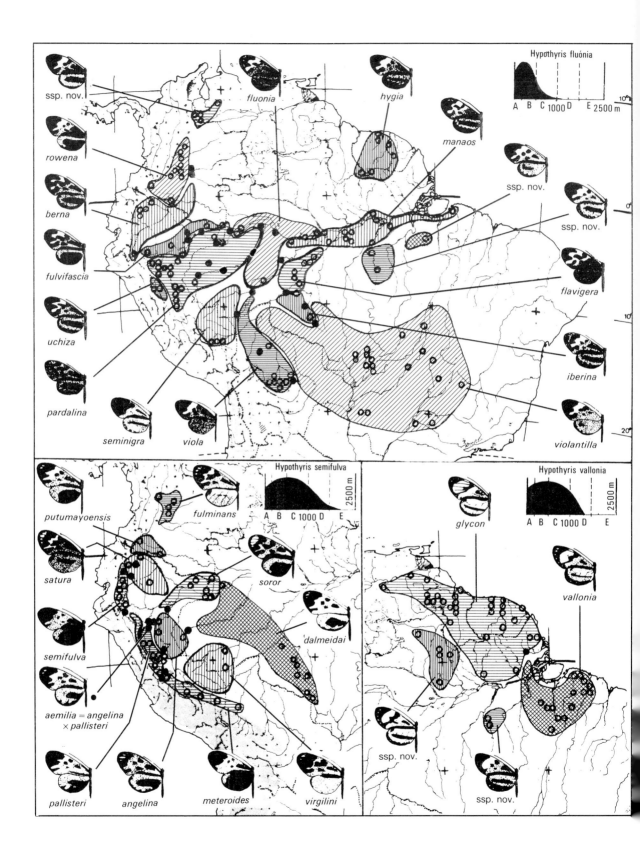

ssp. nov.

fluonia

hygia

manaos

Hypothyris fluónia

A B C 1000 D E 2500 m

rowena

ssp. nov.

ssp. nov.

berna

fulvifascia

flavigera

uchiza

pardalina

iberina

seminigra

viola

violantilla

putumayoensis

fulminans

Hypothyris semifulva

A B C 1000 D E 2500 m

glycon

Hypothyris vallonia

A B C 1000 D E 2500 m

satura

soror

vallonia

semifulva

dalmeidai

aemilia = angelina × pallisteri

ssp. nov.

pallisteri

angelina

meteroides

virgilini

ssp. nov.

moderate elevations in the neotropics, and perhaps also the fact that 95 per cent of the last two million years has been appreciably cooler than today — which may have led to deep-seated physiological and biochemical adaptation optima, for temperature regimes at present existing in the Andean foothills, where water also has probably always been abundant to maintain forest systems.

Some geographically restricted species show wide altitudinal ranges (Figs. 4.6, 4.8, and 4.10); these species are usually semimigratory, or occur in coastal mountain ranges or other peripheral situations, or demonstrate multiple mimetic associations with species of different ecological preferences. A few may represent lowland/upland semispecies complexes with introgressive capacity along a contact zone, despite extensive divergence and altitudinal displacement in the past.

Other very widespread species show restricted altitudinal ranges (Figs. 4.7 and 4.9); in these cases, the inconsistency between geographical and ecological amplitudes can usually be explained by strong adaptation to continual, expansive recolonization of favourable lowland habitats, or unavoidable competitive replacement of altitudinal sectors in closely related species (Benson 1978; Brown 1979, pp. 165-7).

It is easy to see that the species regions (Fig. 4.12) closely parallel the rough biogeographical patterns observed at this taxonomic level and above in many other groups of organisms. They certainly are not useful for finer biogeographical analysis, even though a limited number of good species (22 of the 145 analysed) show more restricted ranges, closer to those of the subspecies of others. It is at the regional subspecific level of differentiation that a more detailed picture of recent evolution can be observed and analysed in these organisms.

Subspecies distributions

When a sufficient number of polytypic species is mapped together, an 'overlay projection' appears (Lattin 1957; Müller 1972, 1973) which can be used to identify areas of high concentration of subspecies ranges. In the Heliconiini and Ithomiinae, an extraordinary coincidence of restricted geographi-

cal distributions for subspecies very soon became evident, even for widely different forest species of several mimetic complexes and different ecology. A total of 44 strongly marked and generalizable centres of subspecies endemism are now recognized (Brown 1979, 1982a); of these, 43 were evident by early 1976 (Brown 1977b), 40 by late 1974 (Brown 1976a), and 31 in the first analysis in early 1973 (Brown et al. 1974). The different endemic centres vary in area between 20 000 and 800 000 km², but most are between 50 000 and 200 000 km². Each contains at least 10, and most of them 15–25, differentiated and distinct endemic subspecies, out of the 20–40 heliconian and ithomiine species present and analysed. Exceptions were made for three areas, two with very poor soils and mixed vegetation (Ventuari and Marajó), and one an isolated mountain range with little tropical foothill forest (Santa Marta). In these three, as well as in most other centres, well over half the species present and analysed were represented by endemic subspecies. In addition, many other less important subcentres of endemism were recognized, which included fewer restricted subspecies or which shared well over half their subspecies with adjacent subcentres. In general, a very clear discontinuity in qualitative endemic character and composition could be observed between the few regions of species endemism (Fig. 4.12), the many centres of subspecies endemism, and the rather few additional subcentres with subspecies endemism.

After the recognition of the major endemic centres in the overlay, the next step was to assign each recognized subspecies of Heliconiini and Ithomiinae to its 'native' endemic centre, on the basis of the correspondence of its range limits with those of the set of subspecies which defined the centre. The relatively few subspecies which overlapped several centres were assigned to one of these by their mimetic association, abundance, ecological preferences, and population structure (a few of these assignments were necessarily arbitrary, and some underwent refinement during the later analysis). The subspecies lists for each half-degree quadrant (about 3000 km²) were then classified according to their representation of subspecies assigned to each endemic centre. Separate projection maps were also prepared for the range limits of all the subspecies assigned to each endemic centre; a sample of these is shown in Fig. 4.13. These maps revealed,

Fig. 4.7. Amazonian species of Ithomiinae, *Hypothyris*. Upper: *H. fluonia* (extends far SE in central Brazil); lower left: *H. semifulva* (W. Amazon); lower right: *H. vallonia* (NE Amazon).

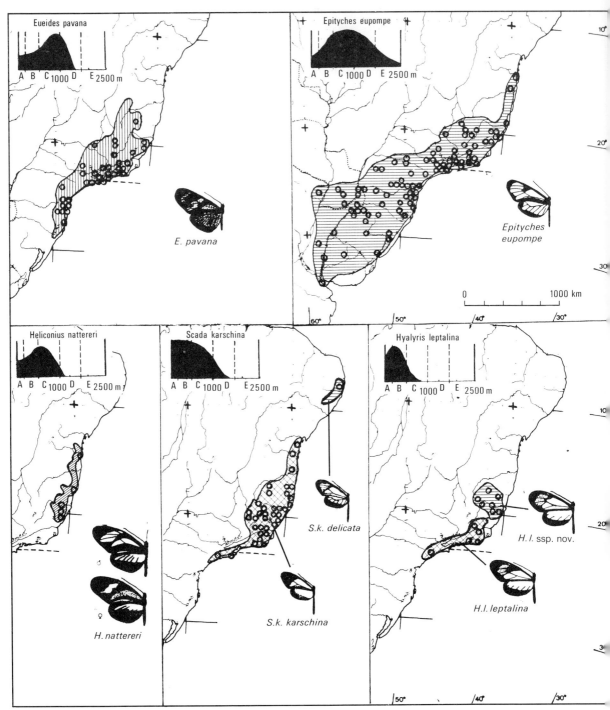

Fig. 4.8. Atlantic species. Heliconiini: *Eueides pavana* (upper left) and *Heliconius nattereri* (lower left). Ithomiinae: *Epityches eupompe* (upper right), *Scada karschina* (lower middle) and *Hyalyris leptalina* (lower right). (For species occupying both Amazonian and Atlantic regions, see Brown (1979) − good examples are *Heliconius numata*, Fig. 28, *Melinaea mnasias*, Fig. 41, *Napeogenes cyrianassa*, Fig. 59, and *Hypothyris ninonia*, Fig. 69.)

in general, a coherent pattern of coincident limits near major ecological barriers like the ocean, and near to or just beyond lesser ecological barriers such as broad rivers, mountain ranges, and white-sand strips; many subspecies range limits, however, did not reach or spread far beyond the latter type of barrier. In essentially isthmic endemic centres, for example narrow Andean valleys like Chanchamayo or Huallaga, or narrow strips of land like Darién, the subspecies ranges were mostly coincident along the sides of the centre, but showed little coincidence at the ends. Almost all the patterns were strongly suggestive of butterfly biotas which are either spreading out in a variable fashion from a smaller area occupied in the past, or else responding in a highly random, unpredictable, and disharmonic fashion to subtle ecological changes in their superficially non-dissected geographical centres (see Fig. 4.13). A few centres, like Marañón and Pantepui, showed the clear internal disharmony associated with the complex topography or ecology and the endemism subcentres recognized in them, and a few others, typically islands or strongly isolated valleys or mountain ranges, showed the clean external limits of regions thus surrounded by strong barriers.

When all the quadrant values for each recognized endemic centre were plotted on a map, adjacent centres usually revealed appreciable overlap, as shown also by the projection patterns (note also the overlapping of endemic centres for birds in Fig. 5.2, p. 107). The resulting quantitative maps of numbers of endemic subspecies per quadrant (sample in Fig. 4.14) also showed a most unexpected structure, with maxima often distributed in a broken ring around a lower centre — rather like a collapsed volcanic crater. If the centre was very narrow or bounded by strong ecological barriers, the structure was distorted towards higher values near the barriers, sometimes eliminating the central inflections; but the maxima for isthmic centres almost always appeared at the two ends of the isthmus. First suspected to be a problem of sampling, this 'peripheral diversity' phenomenon was later confirmed by a number of controlled sampling transects across endemic centres; its significance is discussed at the end of this chapter.

In order to obtain coherent isolines for 'peaks' of endemism and 'valleys' of hybridization, a correction was applied to the crude endemism value for each quadrant. This correction was initially established to produce a negative value for a quadrant if only a single but hybridized species was recorded, or more

hybridized than unhybridized species. Thus in the calculation of the corrected quadrant value for any given centre represented on its species list, a double subtraction (-2) was introduced for mixing or hybridization of any species. In this way, when over half of all the subspecies associated with a given endemic centre and found in the quadrant showed over 10 per cent hybridization, the corrected value fell below zero (Brown 1979). This correction produced even isoline representations for the 44 endemic centres (and their occasional subcentres), compensating for variations in degree of sampling in the contact zone (negative numbers occurred even in small samples, and changed little as the sample grew), and reducing most of the 'rim of the volcano' (high uncorrected values in a broken ring around the centre) to very low positive corrected values. A peak or plateau of corrected endemism was thereby produced in the geographical nucleus of each subspecies endemic centre. The isolines for Heliconiini and Ithomiinae subspecies endemism, and the quadrants in hybridization zones, are shown in Fig. 4.15; the transformation of six centres upon correction of quadrant values for hybridization is illustrated in Fig. 4.14.

As the Heliconiini and Ithomiinae have rather different ecological requirements (Table 4.1) and depend upon plants which prefer very different soil types (Passifloraceae, Cucurbitaceae, and Rubiaceae on sandier soils; Solanaceae, Apocynaceae-Echitoideae, and Compositae-Eupatorieae on fertile, deep, mixed-texture soils, respectively) it might be expected that the fine structure of their separate endemic centres should be quite different. Indeed, if intrinsic (genetic) or ecological factors were preponderant in their differentiation, it could be predicted that they should show very different centres for endemic concentrations. This they do not (Fig. 4.16), again suggesting the importance of strong environmental and historical factors in their regional differentiation. When the endemism isolines were plotted for the groups separately, however, they did reveal appreciable differences in the outlines of some centres, especially those in which the respective preferred soil types occur in large, contiguous and distinct areas (Fig. 4.16). The interpretation of this variation in terms of regional ecological factors, will be discussed more fully at the end of this chapter.

Most of the endemic centres showed clear affinities with some others, either by sharing certain subspecies, containing subspecies with very similar

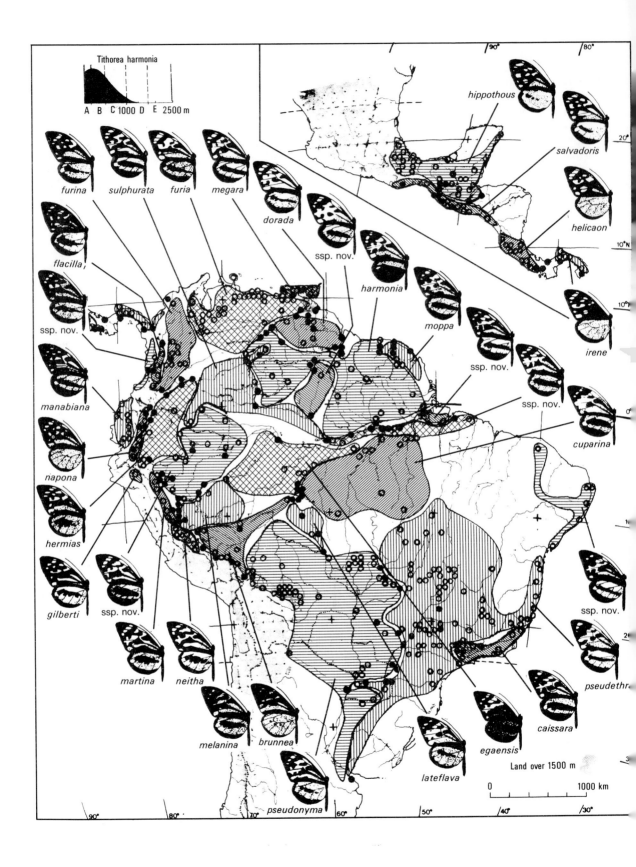

Tithorea harmonia

A B C 1000 D E 2500 m

furina sulphurata furia megara dorada

ssp. nov. harmonia moppa ssp. nov.

flacilla, ssp. nov. hippothous salvadoris helicaon

manabiana napona hermias irene

ssp. nov. cuparina

gilberti ssp. nov. ssp. nov.

martina neitha pseudethr

melanina brunnea caissara

egaensis pseudethr

lateflava pseudonyma

Land over 1500 m

0   1000 km

phenotypes, or both. Since centres with strong affinities do not necessarily show similar environments (Chapter 2), it seems that there has been continuity between populations during different periods of evolutionary divergence. Most of the strongest affinities (Fig. 4.17) are with the immediately adjacent centres, but surprising affinities are seen between widely separated areas, suggesting older biogeographic links. The three links **a–c** indicated in Fig. 4.17 across the Brazilian Shield (**a**, Belém to Pernambuco; **b**, Tapajós-Araguaia-Bahia; **c**, Yungas/Guaporé-Rio de Janeiro) are visible in the distribution of numerous organisms; in view of the belt of dry vegetation presently isolating the first members of these pairs from the second, the affinities should be very small. The first (**a**) in the northeast of Brazil, is strongest at the subspecies level, often even showing identical subspecies of forest-obligate species, and can be traced today through the interior of Maranhão, Ceará, southern Piauí, and western Pernambuco (not around the coast) as a series of presently widely separated patches of typical forest soils and dense humid forest (brejos) on steep northeasterly slopes (Figs. 2.1 and 2.2). This connection may have existed during the last glacial age (Brown and Ab'Sáber 1979) and persisted until after the climatic optimum 5000 years ago. The other two trans-Brazilian Shield affinities are strongest at the species level, often showing close but distinct species or very different subspecies, and can also be traced today across the northcentral (**b**) and southcentral (**c**) sectors of the Brazilian planalto, as disjunct humid forest patches in the background landscape of cerrado. These connections may have been more continuous in the earlier parts of the Pleistocene, but were probably interrupted during much of the Wisconsin-Würm ice age, according to geomorphological evidence.

Other extraordinary affinities (Fig. 4.17) include the strong association between Rondônia and the distant Napo-Sucúa-Huallaga axis **d**, also seen in snakes (W. Höge, personal communication); the Belém-Oyapock and Belém-Roraima links across the lower Amazon **e**, **f**; the Manaus/Guiana-Tapajós tie across the lower middle Amazon **g**; the Imerí-Yungas affinity spanning the breadth of the upper Amazon Basin **h**; and the unexpected links of Guatemala to Chocó and Rancho Grande across the intervening centres **i**, **j**, which show very low affinities to either of the latter. Most interesting are two pairs of strong affinities, each of which forms an 'X': Rancho Grande-Chocó crossing Darién-Nechí **k**, and Imataca-Roraima crossing Guiana-Pantepui **m**. These strong and sometimes crossed ties between distant centres cannot be explained by chance similarities in environment (which cannot be verified) or long-distance dispersal of whole segments of the respective biotas. It seems that they may represent ancient affinities, relics from times when these biotas were actually united. Major alterations in the landscape, slow during the Tertiary orogenies and more rapid in Quaternary climatic fluctuations, have placed first some and then other regions in effective contact, as sea levels and mountain ranges rose and fell, winds and rainfall patterns shifted, and mobile animals moved with their changing environments through evolutionary time and geographical space. Present-day biogeographical patterns are like a palimpsest (multiply-inscribed parchment; see Heyer and Maxson 1982) or a series of superimposed semitransparent layers, the result of this long turbulent history. Each previous period has left its traces. What can be seen of only the few most recent of these periods of major environmental rearrangements (Chapters 1 and 2) is but a meagre representative of the whole montage, with its multiple short- and long-lasting effects on genetic resources in the neotropics. Insect fossils have substantiated the massive movements of north-temperate biotas in late Pleistocene time (Coope 1970, 1978, 1979). Great reorganization of biogeographical patterns has certainly also taken place in the humid tropics. It is fortunate that. modern patterns, subtly influenced by indeterminable but still detectable historical factors through all the Tertiary tectonics and Quaternary glaciations and continually diffused by ecological forces, have none the less lent themselves to some analysis, with limited generalizations and predictions possible.

Barriers

Physiographical and ecological barriers exist in many forms in the neotropics, fragmenting the landscape

---

Fig. 4.9. A widespread single species, Ithomiinae: *Tithorea harmonia*. Other good examples are illustrated in Brown (1979) and include *Heliconius melpomene*, *erato*, and *sara* (Figs. 26–28), *Mechanitis polymnia* and *lysimnia* (Figs. 51–52) and *Hypothyris euclea* (Fig. 77; see also Brown 1980). There are also widespread species-complexes such as *Heliconius cydno* and its close relatives (Fig. 23 in Brown 1979).

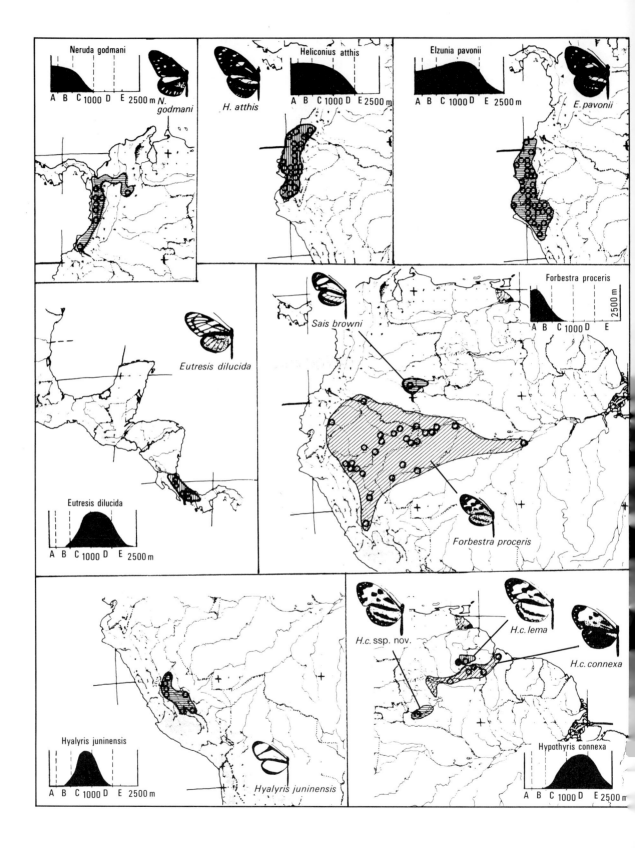

Neruda godmani

A B C 1000 D E 2500 m *N. godmani*

Heliconius atthis

H. atthis

A B C 1000 D E 2500 m

Elzunia pavonii

A B C 1000 D E 2500 m *E. pavonii*

Eutresis dilucida

Eutresis dilucida

A B C 1000 D E 2500 m

Sais browni

Forbestra proceris

A B C 1000 D E 2500 m

Forbestra proceris

Hyalyris juninensis

A B C 1000 D E 2500 m

Hyalyris juninensis

H.c. ssp. nov.

H.c. lema

H.c. connexa

Hypothyris connexa

A B C 1000 D E 2500 m

and biotas and reducing or impeding gene-flow between the resulting segments. Many zones of subspecies hybridization in the Heliconiini and Ithomiinae (Fig. 4.15) coincide roughly with these barriers. How important are such barriers in the genesis of species and subspecies endemism? At the species level (Fig. 4.12), an unexpected and contradictory picture emerges. The Transandean and Amazonian faunas, though separated in part by the high Andean cordillera, meet along a broad zone in southwestern Venezuela and northeastern Colombia, actually within the Hylaea though limited eastwards by the llanos grasslands. In this region some species form hybridization belts while others show character displacement indicating the fixation of reproductive isolation between Amazonian and Transandean species pairs. The cordillera itself is breached by many species, in northern and western Venezuela, the upper Magdalena valley of Colombia, northern Peru and even in Ecuador. The Andean endemic fauna extends down from the higher parts of the cordillera, mixing with the Hylaean fauna to the east and the Transandean fauna to the north and west. Some of its elements invade half of the lowland Hylaea, others are absent there but reappear on the Atlantic coast of southeastern Brazil (sometimes as closely related but separate species), and yet others extend northward through Central America to the very limits of the neotropics. Interchange between the Hylaea and the Atlantic region across the cerrado and caatinga domains of central Brazil is still strong today in some species, though weak in others and still not verified in many; it has surely been stronger in the recent past, as genetic analysis (Fig. 4.17) and, in some cases, experimentation can verify (Brown and Benson 1974). No general correlation can be observed between species limits and simple physiographical barriers.

At the subspecies level (Fig. 4.13), the contradictions are even greater. Some subspecies of some species stop at major rivers, but many do not. South of Manaus, almost all species cross the broad Rio Solimões but only a few cross the narrower Rio Negro, which just upriver is broken into a maze of narrow channels by forested islands (Brown and Mielke 1972). On closer examination, it is found that most subspecies actually do get across the Negro, but do not penetrate more than a few kilometres into competitor populations on the other side. Many homogeneous lowland subspecies span high Andean cordilleras. Most of the Marañón fauna seems to ignore both frigid altitudes and searing deserts in establishing a geographical range. Some sedentary butterflies of deep, humid forest shade occur in every forested pocket across central Brazil, and in both the Peninsula de Paria of Venezuela and Trinidad; other strong flyers, which frequent forest edges, are totally blocked by the cerrado mosaic, or the 15 km of sea separating the north Trinidad mountains from those on the mainland. Even where effective isolation has probably existed in subspecies of islands of the continental shelf (land-bridge islands) since the rise of the ocean after the Wisconsin–Würm glaciation (as in islands off Mexico, Honduras, Costa Rica, Panama, Colombia, and Venezuela), no effective differentiation can be seen in some resident heliconians and ithomiines. The careful examination of potential ecological barriers (marked transitions of temperature, rainfall, vegetation, substrate, mimicry complexes, abundance of potential predators, and competitors) has, in almost every case, resulted in their confirmation as conditioning agents for the velocity of gene-flow between some finely adapted populations on the two sides, but their disqualification as generators of biotic differentiation of the sort visible in the endemic centres (Fig. 4.15). As if to place the final nail in the coffin, the fact that the same species and even the same subspecies often exist on both sides and even on top of the physiographical barrier, and at times use the barrier itself for dispersal (Brown 1979, pp. 186–7) provides rich material for studies of physiological ecology and adaptation, but not for the elaboration of simple biogeographical explanations.

Indeed, the only type of barrier which appears to be completely effective in stopping the inexorable spread of the butterfly subspecies examined is another, locally better adapted, conspecific race. Since the biogeographical analysis of subspecies is based upon a genetic approach, however, the fact that the effective barriers observed are of a genetic nature is mere tautology. That such genetic barriers often exist

---

Fig. 4.10. Restricted species. Heliconiini: *Neruda godmani* (upper left: Chocó) and *Heliconius atthis* (upper middle; Chimborazo). Ithomiinae: *Elzunia pavonii* (upper right; Marañón + Chimborazo), *Eutresis dilucida* (centre left; Chiriquí), *Forbestra proceris* (Napo) and *Sais browni* (Imerí) (centre right), *Hyalyris juninensis* (lower left; Chanchamayo), and *Hypothyris connexa* (lower right, Pantepuí subregion).

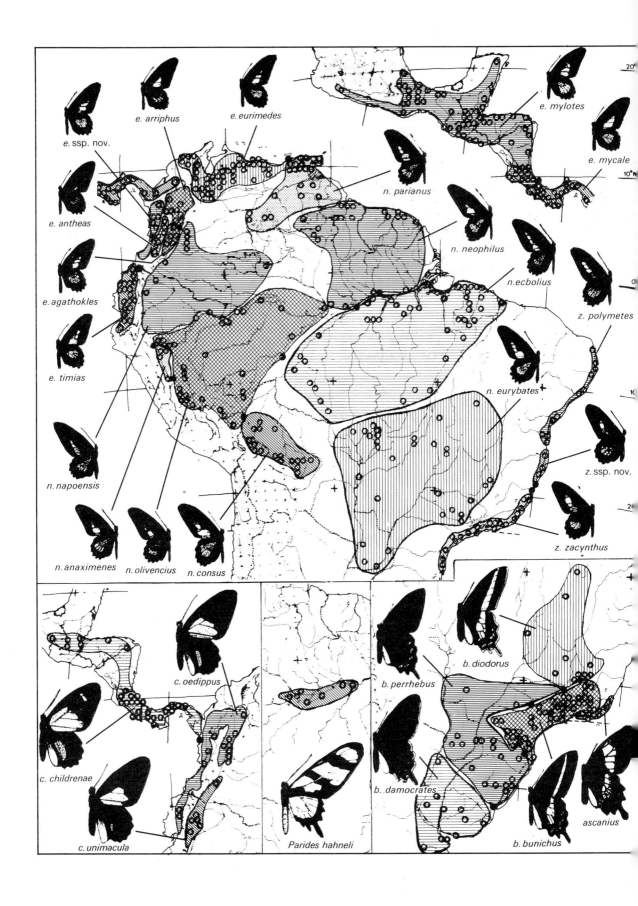

e. ssp. nov.

e. arriphus

e. eurimedes

e. mylotes

e. antheas

n. parianus

e. mycale

e. agathokles

n. neophilus

n.ecbolius

e. timias

n. eurybates

z. polymetes

n. napoensis

z. ssp. nov.

n. anaximenes   n. olivencius   n. consus

z. zacynthus

c. oedippus

b. diodorus

b. perrhebus

c. childrenae

b. damocrates

c. unimacula

Parides hahneli

b. bunichus

ascanius

20°

10° N

10°

20°

in regions where no corresponding ecological or physical barriers can be verified, is much more interesting. The population characteristics such as dispersal capacity, mimetic association, population size, and adaptability of the numerous species examined in the quantitative analysis of endemism should make them respond in predictable manners to recognized barriers. Instead, they seem to form zones of contact in unpredictable places, often far away from or beyond such barriers but sufficiently grouped to form regular patterns (Figs. 4.13 and 4.15) — once again suggesting a spreading out from smaller areas in the past. It is evident that the genesis of these patterns can have no simple explanation; detailed research is demanded, into both modern and past historical factors.

Richness of endemic centres

The number of endemic heliconian and ithomiine subspecies associated with the different centres varies from only five to over forty. Similar spreads are seen in other butterfly groups, such as from one to eleven in troidine swallowtails. Attempts to 'normalize' these numbers to a narrower scale, using such corrective factors as amount of study, minimum size, isolation, altitude range, and rainfall were only partly successful (Brown 1977*a*, *b*); many still remained outside of 50 per cent above and below the mean. More complete ecological study of the centres has suggested that other factors, especially fertility and mosaics of soil, degree of internal coherence (formation of subcentres), the integrity of the humid forests, and topography irrespective of altitude (rolling country being able to maintain more varieties of topoclimate), may be more important in determining the quantitative value (if not necessarily the qualitative character) of endemism. Even when all these factors are taken into account, certain centres seem to remain richer in endemic subspecies than predicted in many taxonomic groups. These centres, Chimborazo, Cauca, Villavicencio, Abitagua, Ucayali, Chanchamayo, and Yungas are all on humid, medium-elevation slopes of the Andes, where many species reach their greatest abundance and maximum tendency to accumulate and persist. A few other centres, Chocó, Catatumbo, Apure, Imataca, and Bahia are

poorer in endemic subspecies than expected. All these have great deviations in present climate, which may put heavy strains on insect populations (especially juvenile individuals) and make species-persistence less probable. All of these areas need more research to discover the major ecological processes which have helped or hindered the development of their endemic biotas. It is significant that the factors identified so far which seem to help predict the richness of the biota, can all be evaluated by studies of present-day ecological conditions; this is in accord with the major conclusions of this study (see the end of this chapter).

Biogeographical anomalies

In the course of recent intensive geographical sampling, a few striking anomalies have appeared. One, shown as a circle in Fig. 4.6, represents a completely disjunct occurrence of an entire mimetic complex (transparent, orange-tip) in an isolated low chain of mountains on the Amazonas/Mato Grosso border, 1000 km southeast of the nearest regular occurrence of the same complex and totally surrounded by other, non-orange-tipped subspecies of all the same species. This could be attributed to a recrudescence of ancestral genes in this locale (a phenomenon often observed in more widespread, migratory or plastic species than those in this complex; Brown 1973, 1977*c*, 1980), producing the 'polytopic subspecies' which are so difficult to evaluate in evolutionary biogeography and taxonomy. Alternatively, it could represent a very ancient link which has been broken by recent restructuring of the upper Amazonian landscape. It seems significant that the low chain of mountains which sports this anomaly occurs alongside a large and even more anomalous relict Amazonian savanna, lying in the middle of the humid north Rondônia-west Tapajós forests (Fig. 2.4) as a testimony to the complex vegetational history of this part of the Amazon Basin. The observation acquires greater significance because all the other major anomalies observed occur beside other major Amazonian savannas. These anomalies may represent a tantalizing witness to the importance of major historical restructuring of the landscapes, in the lowland Amazon rain forest which is today nearly continuous. Perhaps such populations became 'stranded' during some past

---

Fig. 4.11. Papilionidae, Troidini, *Parides* (aposematic *Aristolochia* feeding swallowtails): *P. eurimedes*, *P. neophilus*, and *P. zacynthus* (upper), *P. childrenae* (lower left), *P. hahneli* (lower middle; Maués subcentre of Tapajós; a mimic of *Methona*, Fig. 4.3) and *P. bunichus* and *P. ascanius* (lower right; a closely related species is endemic to SE Ecuador).

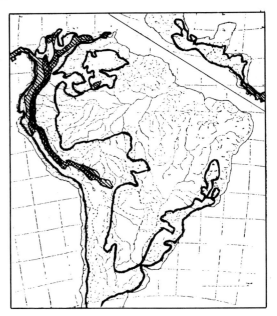
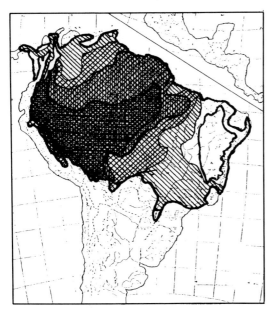
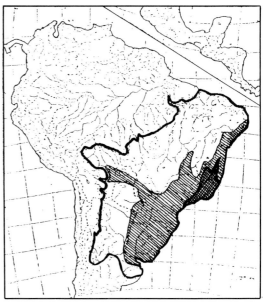

Limits for all endemic species examined
Limits for all endemic species analysed
Five or more of species analysed
Ten or more of species analysed
Twenty or more of species analysed
Thirty or more of species analysed

vegetation reorganization in their complex environment and have retained an ancestral, but well-adapted colour-pattern even in the face of potential gene introgression from surrounding differentiated subspecies.

NON-FOREST BUTTERFLY FAUNAS

A number of butterfly species and genera seem to be restricted to open vegetation formations. Where such vegetation predominates (as in the cerrado and caatinga), these organisms are widespread and common; where it is fragmented into small relicts (as in much of the Amazon Basin), they are often found as widely disjunct isolates, recognizably differentiated from each other (Brown and Benson 1977). This situation parallels exactly that of plants (Chapter 3) and birds (Chapter 5), and may be detected also in the distribution of some savanna-adapted groups of man (Chapter 6). Indeed, all of these elements and many others show tightly co-adapted relationships within the large and small savanna systems associated with and bordering the neotropical rain forests, revealing the age, importance, and possible past coalescence of these systems.

Studies on savanna butterflies are at a very early stage. Many Amazonian relict savannas have not yet been sampled at all, perhaps because their overall fauna is not very rich. Typical endemic residents include a wide variety of Satyrinae (Euptychiini), mostly still unclassified. This group would surely be the best to examine in depth, to determine the biogeography and evolutionary history of savanna organisms, since local endemism is already known at the subspecies, species, and generic levels. Endemic species of the nymphalid genera *Heliconius*, *Phyciodes*, *Eunica*, *Callicore*, and *Anaea* (s.l.) are among the few other larger resident and differentiated butterflies in neotropical savannas. Lycaenids are common; large numbers of theclines and riodinines (especially *Audre*), many still undescribed, make up an important fraction of the endemic fauna. A few pierids (especially *Eurema*) and papilionids have adapted to savanna vegetation, along with many resident species and genera of Hesperiidae — usually not well differentiated in the separate savanna regions.

Fig. 4.12. Principal regions of species endemism in neotropical butterflies, with isolines given for Heliconiini and Ithomiinae. (After Brown 1982*b*). Outer line (white), limits for all butterflies considered endemic to the region; lightest shading (isoline for 1) limits for all Heliconiini (Hel) and Ithomiinae (Ith) analysed (isol = total number of species used for isolines, as listed below; Troi = number of endemic species of Troidini; GEN = number of endemic genera in all three groups). Progressively heavier shading indicates isolines for 5, 10, 20, and 30 species. The species considered are.

TRANSANDEAN: *Eueides procula*, *E. lineata/emsleyi*/libitina**, *Heliconius ismenius*, *H. cydno*, *H. hecalesia*, *H. hortense*/clysonymus*, *H. sapho*, *H. eleuchia*, *Olyras insignis*, *Melinaea lilis*, *Mechanitis menapis*, *Napeogenes peridia*, *N. tolosa*, *N. cranto*, *Hypothyris lycaste*, and *Hyalyris excelsa* (Total 19).
ANDEAN: *Heliconius hecuba**, *H. hierax*, *H. telesiphe*, *H. congener*, *Roswellia acrisione*, *Patricia dercyllidas*, *P. oligyrtis*, *Eutresis hypereia*, *Elzunia humboldt*, *Napeogenes harbona*, *N. glycera*, *N. omissa**, *N. apulia*, *Hypothyris meterus*, *H. mansuetus*, *Hyalyris coeno*, *H. latilimbata*, *H. praxilla**, *H. oulita*, and *H. frater* (Total 20).
AMAZONIAN: *Eueides lampeto*, *E. eanes**, *E. tales*, *Neruda metharme*, *N. aoede*, *Heliconius xanthocles*, *H. wallacei*, *H. burneyi*, *H. egeria*, *H. astraea**, *H. pardalinus*, *H. elevatus*, *H. antiochus*, *H. demeter*, *Athyrtis mechanitis**, *Melinaea maenius*, *M. marsaeus*, *M. menophilus*, *Sais rosalia*, *Forbestra equicola*, *F. proceris**, *F. olivencia*, *Mechanitis mazaeus*, *Napeogenes achaea**, *N. aethra**, *N. sylphis*, *N. pheranthes*, *Garsauritis xanthostola**, *Rhodussa cantobrica**, *Hypothyris fluonia*, *H. moebiusi**, *H. thea*, *H. semifulva*, *H. vallonia**, *H. mamercus*, *H. daphnis**, and *H. anastasia* (Total 37).
ATLANTIC: *Eueides pavana*, *Heliconius besckei*, *H. nattereri**, *Scada karschina*, *Epityches eupompe*, *Placidula euryanassa*, *Hyalyris leptalina**, and *H. fiammetta** (Total 8).
HIGHLY RESTRICTED RANGES; TRANSANDEAN. *Neruda godmani*, *Heliconius pachinus*, *H. atthis*, *H. hewitsoni*, *Elzunia pavonii*, *Eutresis dilucida*, *Scanda zemira*, *Scada kusa*, (Total 8). ANDEAN. *Heliconius heurippa*, *H. timareta*, *Hyalyris antea*, *H. juninensis* (Total 4); AMAZONIAN. *Sais browni*.
PANTEPUI (ANDEAN/AMAZONIAN, highlands): *Heliconius luciana*, *Hypothyris connexa*, *H. gemella*.
TRANSANDEAN + ANDEAN: *Olyras crathis*, *Tithorea tarricina*.
TRANSANDEAN + HYLAEAN: *Eueides lybia*, *Heliconius hecale*, *Scada zibia*[+], *Napeogenes stella*[+].
HYLAEAN + ATLANTIC: *Heliconius numata*, *H. ethilla*, *Melinaea mnasias*, *M. ethra*[+], *Scada reckia*, *Napeogenes cyrianassa*, *N. inachia*[+], *Hypothyris ninonia*, *H. leprieuri*.
WIDESPREAD: *Heliconius melpomene*, *H. erato*, *H. sara*, *Tithorea harmonia*, *Melinaea ludovica*[+], *Mechanitis polymnia*, *M. lysimnia*[+], *Hypothyris euclia*.
* = Restricted to less than half of region.
[+] = Will probably be divided into regional allospecies when more biological data available.

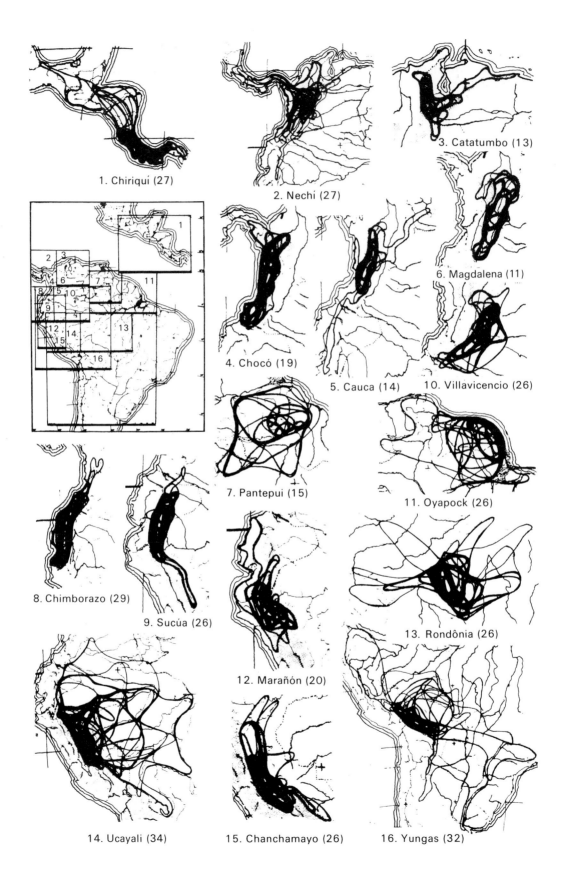

1. Chiriquí (27)

2. Nechí (27)

3. Catatumbo (13)

4. Chocó (19)

5. Cauca (14)

6. Magdalena (11)

10. Villavicencio (26)

7. Pantepui (15)

11. Oyapock (26)

8. Chimborazo (29)

9. Sucúa (26)

12. Marañón (20)

13. Rondônia (26)

14. Ucayali (34)

15. Chanchamayo (26)

16. Yungas (32)

The fauna is completed by migrants and transients, some of whom seek only nectar in savanna plants, and by cosmopolitan species which occur in all non-forest habitats including clearings and man-made systems.

The study of disjunct open-vegetation faunas promises to permit direct modern observation of the adaptation and differentiation processes which are hypothesized to have operated in similarly·disjunct forest faunas in the past, when the savannas were probably more continuous. For this reason, it is hoped that more biologists will dedicate an increasing effort to the classification and comparison of savanna butterflies, especially in the relict patches within the Amazonian forest.

HIGHLAND BUTTERFLY FAUNAS

Many butterfly species and genera are restricted to highland locations in tropical America. Nearctic elements, outside the scope of this book, extend down the Sierra Madre at progressively higher elevations as far south as Guatemala and Honduras and very rarely Costa Rica; some are closely related to South American groups. Their congeners or equivalents which occupy Andean forest, subpáramo, páramo, and puna vegetation patches in the high Andes, and pseudopáramo (a name suggested for similar, wind-conditioned vegetation at lower elevations) in the Guiana shield and southeastern Brazil, also show a few ancient ties to groups in the Himalayan or Australasian mountains, revealing the multiple origins and affinities of South American temperate animals, which may have moved far northward, coupled to cool environments or tectonic mountain-building, after the separation of continents (see Simpson-Vuillemier 1971; Shields and Dvorak 1979).

The butterflies of the higher Andes (above 2000 m; see Fig. 1.1, p. 2) include some tropical species and genera, but the fauna is dominated by endemic taxa which do not normally occur in tropical or even subtropical forests. The most typical groups of Andean forest to páramo are Satyrinae (especially Pronophilini), a few Nymphalinae (especially Argynnini and *Vanessa*, and *Dione glycera* in the Heliconiini), some transparent Ithomiinae (mostly *Hyalenna* and *Greta*), a few hardy Acraeinae (*Altinote*), rare Lycaenidae (mostly Theclinae), Pieridae (especially Pierinae and *Colias*), a few *Papilio*, and several groups of skippers, including a few Pyrrhopyginae. Very few of these groups have been well enough sampled for critical analysis, but an older paper on the pronophiline satyr genus *Lymanopoda* is available (Brown 1942*b*), and recently revisions have been published for a group of pierine genera (the Tatocheilae–Phulia complex; Herrera and Field 1959; Ackery 1975; Field and Herrera 1977). These indicate that separate and differentiated biotas may exist at high elevations in the Sierra Perijá (Valledupar + Motilones), the Venezuelan Andes (two subunits), the Colombian and Ecuadorian páramos as far south as extreme northern Peru (before the great Marañón discontinuity), the central Peruvian cordilleras, the eastern Bolivian chain (often differentiated from others in the Titicaca puna, the Arequipa region and at the Bolivia/Chile border), and the Patagonian Andes (with at least five subdivisions on both sides of the Andes and south to Tierra del Fuego), in addition to many evident islands like the Sierra Nevada de Santa Marta in Colombia, and other isolated mountain tops (Fig. 4.18). The *Tatochila* group and pronophiline satyrs, like a surprisingly large number of other south-Andean elements, have representatives in southeastern Brazil, becoming increasingly more local at ever higher altitudes and sometimes with subspecies differentiation, north to the Serra dos Órgaos (Rio de Janeiro). A few 'alpine' Andean species penetrate north as far as central Minas Gerais and Espírito Santo, especially in the Serra do Caparaó which reaches almost 3000 m and has a large expanse of treeless pseudopáramo above 2400 m.

The fact that many south-Andean genera reach southern Brazil leaves no doubt that links of suitable vegetation and humid temperate climate existed across northern Argentina in the distant past, through which these species could spread freely. More interesting yet are the mid-elevation (not páramo) elements which invade the summits of the table mountains in the Pantepui region of Venezuela from

Fig. 4.13. The ranges of all studied Heliconiini and Ithomiinae subspecies associated with 16 of the 44 endemic centres recognized in the Neotropics. (After Brown 1979.) The name of each centre and the total number of subspecies mapped is given in each case. Locations shown at left.

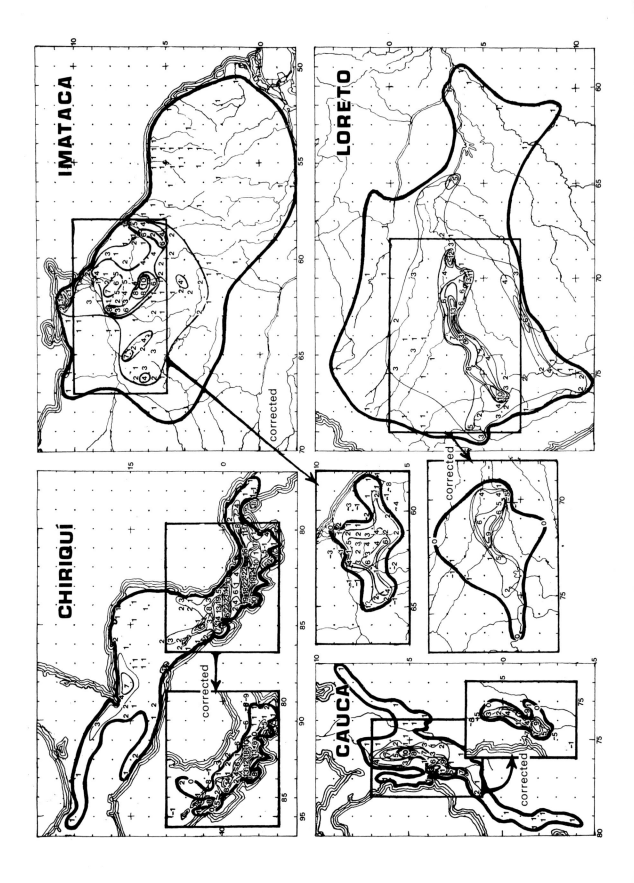

IMATACA

LORETO

CHIRIQUÍ

CAUCA

corrected

corrected

corrected

corrected

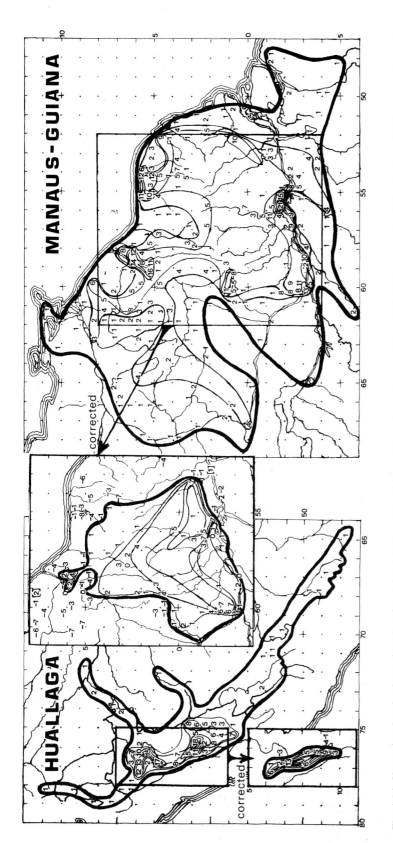

Fig. 4.14. Uncorrected endemism values (total associated Heliconiini and Ithomiinae subspecies present in each half degree quadrant) for the six endemic centres Chiriquí, Cauca, Huallaga, Loreto, Imataca, and Manaus/Guiana, and the corresponding corrected endemism maps and isolines. Note that higher values do not occur in coherent nuclei, but in scattered points arranged roughly in a ring around the centre as defined by correcting isolines (smaller box).

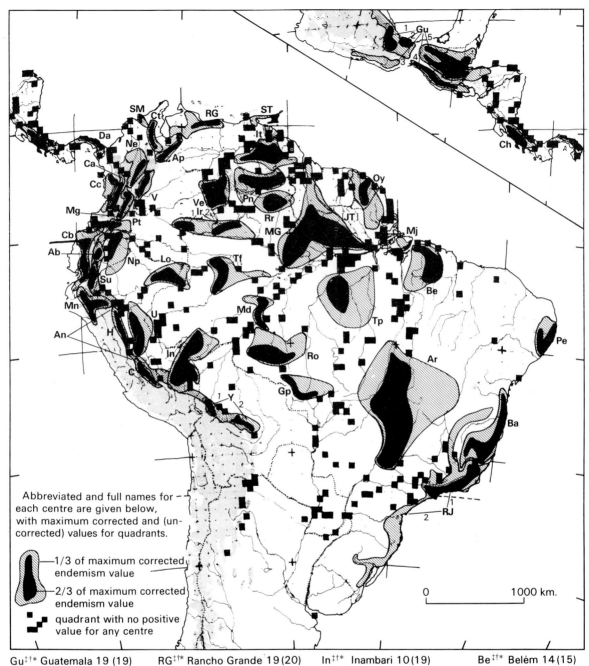

Abbreviated and full names for each centre are given below, with maximum corrected and (uncorrected) values for quadrants.

— 1/3 of maximum corrected endemism value

— 2/3 of maximum corrected endemism value

■ quadrant with no positive value for any centre

0   1000 km.

Gu‡†* Guatemala 19 (19)
  1   Veracruz 18
  2 †* Tuxtlas 14
  3 †* Guerrero 8
  4   Pacifico 14
  5   Verapaz 19
Ch‡†* Chiriquí 18 (24)
Da †* Darién 10 (11)
Ne‡†* Nechí 13 (23)
Cc‡†* Chocó 12 (18)
Cb‡†* Chimborazo 24 (24)
Ca‡ * Cauca 9 (13)
Mg‡* Magdalena 4 (11)
SM * Santa Marta 6 (6)
Ct †* Catatumbo 7 (9)

RG‡†* Rancho Grande 19 (20)
ST‡ * Sucre/Trinidad 10 (10)
Ap‡* Apure 10 (10)
V ‡†* Villavicencio 17 (25)
Pt * Putumayo 10 (20)
Np‡†* Napo 7 (21)
An   Andes (8)
Ab‡†* Abitagua 12 (22)
Su * Sucúa 14 (18)
Mn‡†* Marañón 9 (10)
H ‡†* Huallaga 13 (18)
U ‡†* Ucayali 14 (26)
C ‡†* Chanchamayo 11 (19)

In‡†* Inambari 10 (19)
Y ‡†* Yungas 19 (25)
  1   La Paz 19
  2   Cochabamba 19
Gp †* Guaporé 6 (10)
It * Imataca 5 (9)
Pn‡ Pantepui 8 (8)
Rr * Roraima 6 (8)
Ve * Ventuari 3 (3)
Ir ‡* Imeri 11 (13)
  1   Vaupés 8
  2   Neblina 11
MG‡†*Manaus/Guiana 8 (21)
[JT]‡* Jari-Trombetas 8
Oy ‡* Oyapock 18 (22)

Be‡†* Belém 14 (15)
Mj * Marajó 4 (4)
Tp‡†* Tapajós 22 (26)
Ro †* Rondônia 16 (25)
Md * Madeira 5 (7)
Tf   Tefé 5 (12)
Lo †* Loreto 9 (13)
Pe‡†* Pernambuco 6 (6)
Ar‡†* Araguaia 7 (7)
Ba‡† Bahia 8 (10)
RJ‡†* Rio de Janeiro 12 (1

the western Pico da Neblina (3045 m, with a large altiplano of pseudoparamo around it) to the north-eastern Mount Roraima (2810 m). These include typical Andean genera and species like *Eutresis hypereia* (Brown 1977a; Fig. 4.5), *Catasticta*, and *Pedaliodes*, and cold-adapted members of (sub) tropical genera like *Dismorphia, Leptophobia, Eueides* (*procula*, Fig. 4.4), *Greta, Siproeta* (*trayja*), *Thespeius*, and many other hesperids and lycaenids. An eastern link between the Andes and Pantepui can be traced across the lower Orinoco and the Altiplanicie de Nuria in Venezuela; it seems to be very ancient. A more continuous western link, possibly active in gene exchange during the Pleistocene, can be seen from the Serranía La Macarena (eastern Colombian Andes) across the Guaviare, Guainía and Vaupés highlands, crossing the Rio Negro at São Gabriel da Cachoeira and entering the Serra Imerí, the south-western tip of the Pantepui complex (Fig. 4.18).

Many high-altitude butterflies may be expected to be undergoing divergent evolution today, isolated on their highland 'islands' by surrounding forests (see Fig. 3.3). This has been verified as a long-term process in the Sierra Perijá and especially the Sierra Nevada de Santa Marta (Adams 1973, 1977; Adams and Bernard 1977, 1979, 1981; Ackery 1975). Its effects can be seen in almost any species or genus examined on a north–south transect. Adams (1977) has suggested that repeated warm–cool climatic cycles could lead to accumulation of closely related species in altitude bands of isolated high-elevation areas. The mechanism is shown in Fig. 4.19. More study is needed to determine the boundaries of the centres of high-altitude neotropical species and subspecies endemism, from Nicaragua to Patagonia, the Paria Peninsula of Venezuela and eastern Brazil (see also Lamas 1982).

Correlation of biogeography with present and past ecological factors

Many suggestions have already been made about the possible historical roots of the patterns observed in

neotropical butterfly biogeography, at the species and subspecies levels, as illustrated and discussed here for the Heliconiini and Ithomiinae. The detailed palaeoecological information incorporated into the forest stability model for the late Wisconsin–Würm ice age (Fig. 2.8, p. 44), and the broad data available on present-day ecological features (Figs. 1.1–1.4, 2.1, and 2.4; Brown 1979), invite geographical comparison with the quantitative centres of subspecies endemism (Fig. 4.15) and the locations of high species diversity (Fig. 4.14), respectively.

Indeed, the correlations are readily seen to favour exactly these ties. Superimposition of forest endemic centres for subspecies of Lepidoptera (Fig. 4.15) and high-probability forest refuges (Fig. 2.8) gives an almost complete correspondence (Fig. 4.20). A total of 32 endemic centres and four additional subcentres show complete enclosure of regions of high probability for forest continuity. All of the remaining endemic centres except one (Marajó) either correspond to and enclose more than one forest refuge or partially overlap a single region of palaeoecological forest stability. As the Ilha de Marajó was almost totally underwater during the climatic optimum 5000 years ago, its evident but limited endemic biota (four heliconians, two or three ithomiines, and one *Parides* swallowtail recognized to date; endemic plants and vertebrates are also known) may be of more recent origin. Ten of the forest refugia fall mostly between the endemic centres for butterflies; the seven of these which have been sampled are rich areas of subspecies mixing in Heliconiini and Ithomiinae (representing two or more nearby endemic centres), but show recognizable endemism in other tropical butterflies, plants, and diverse sedentary organisms examined.

Correlation does not prove causation, but the fit on Fig. 4.20 is so good that causation is probable. The model for forest stability was based wholly on geoscientific evidence for different climate in the late Pleistocene (vegetation structure is taken as a physical factor, dissociated from taxonomy and diversity) (Chapters 1 and 2). The biogeographical patterns in heliconian and ithomiine butterflies described in this chapter closely match the areas of forest stability

Fig. 4.15. Endemic centres for Heliconiini and Ithomiinae subspecies in the Neotropics, after applying the double correction for subspecies mixing to the crude endemism values (Fig. 4.14) for each quadrant in each centre, see text. Maximum corrected quadrant value given after the name of each centre; isolines for 2/3 and 1/3 of this shown by double and single hatching. Quadrants without any positive values for any nearby centre (in transition or overlap zones of endemic subspecies sets) are shown as solid squares. Maximum uncorrected endemism values for each centre are given in parentheses. Centres marked † appeared in Lamas's original analysis (1973; Fig. 4.3) or his later paper on Peru (1976); those marked * are seen in *Parides* swallowtails at the present level of sampling (Fig. 4.11); those marked ‡ are evident in brassolines (Blandin 1978; Bristow 1980).

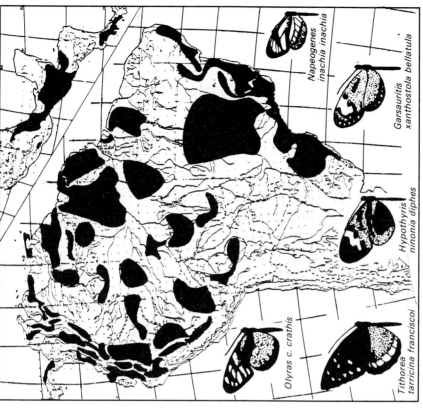

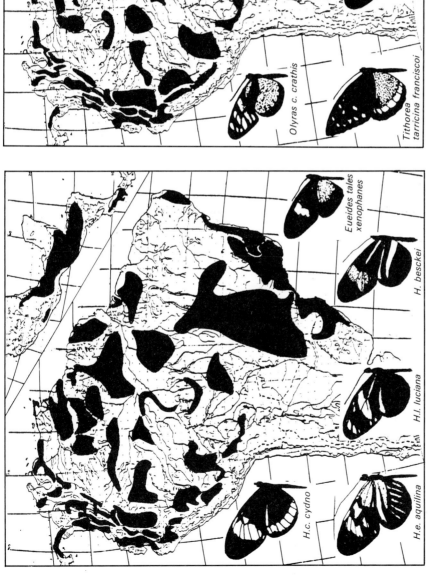

Fig. 4.16. Endemic centres seen in only subspecies of Heliconiini (left) and Ithomiinae (right), isoline for one-third of maximum corrected endemism value. Note the appreciable variation in shape, but not in position, of the centres in the two groups. This probably reflects different ecological preferences but a similar evolutionary history. The centres seen in Troidini (see Fig. 4.11) are also very similar both in number and in location (Brown 1982b, Fig. 3).

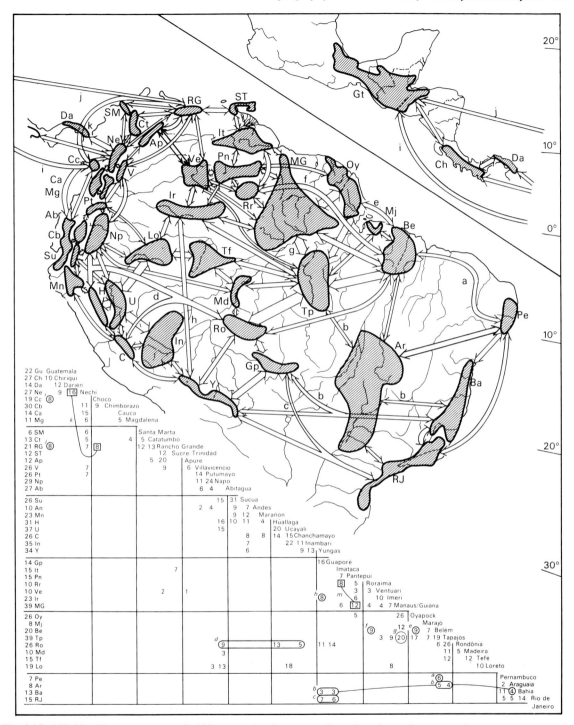

Fig. 4.17. Affinities (shown as arrows of width proportional to the total number of shared or very similar subspecies, as shown in Table) among the various endemic centres (represented as the 50 per cent isoline of the maximum corrected value) for Heliconiini and Ithomiinae subspecies. Centres as on Fig. 4.15. Exceptional affinities are indicated by letters above the arrows, and by circles in the Table (see text).

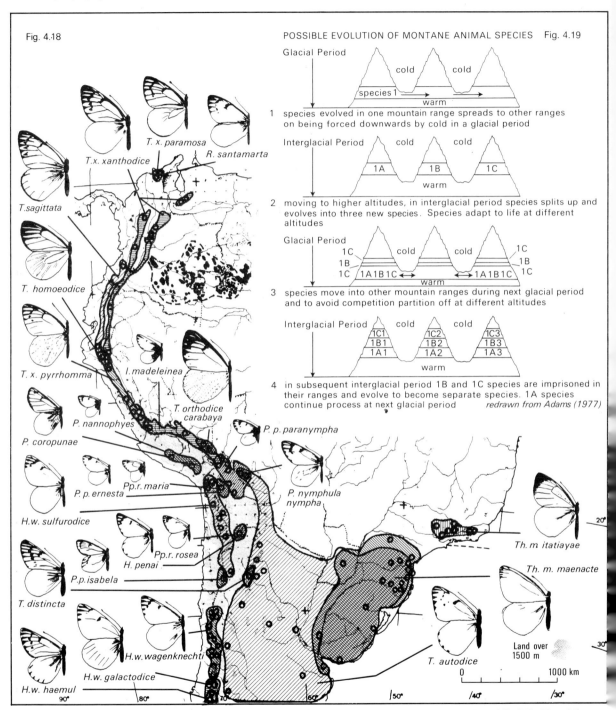

Fig. 4.18

POSSIBLE EVOLUTION OF MONTANE ANIMAL SPECIES   Fig. 4.19

Glacial Period

1   species evolved in one mountain range spreads to other ranges on being forced downwards by cold in a glacial period

Interglacial Period

2   moving to higher altitudes, in interglacial period species splits up and evolves into three new species. Species adapt to life at different altitudes

Glacial Period

3   species move into other mountain ranges during next glacial period and to avoid competition partition off at different altitudes

Interglacial Period

4   in subsequent interglacial period 1B and 1C species are imprisoned in their ranges and evolve to become separate species. 1A species continue process at next glacial period   *redrawn from Adams (1977)*

Fig. 4.18. Distributions of some high-Andean butterflies in the 'Tatocheilae–Phulia' complex (Genera *Tatochila* (T.), *Theochila* (Th.), *Hyposchila* (H.) *Phulia* (P.), *Pierphulia* (Pp.), and *Infraphulia* (I.)). Note very few isolated entities in the N, many in the S including across temperate Argentina and Uruguay and N in the SE Brazilian mountains, suggesting a spreading from S to N in the group. Many subspecies and species are restricted to a single mountain top. (After Field and Herrera (1977), Ackery (1975), and Herrera and Field (1959).) Western link from Andes to Pantepui shown as black spots representing higher hills across E. Colombia (mostly over 500 m, some over 1000 m), based on radar images.

('palaeoecological forest refuges') but are based on a wide diversity of genetical, ecological, and systematic data on present-day populations. No other model can explain the existence of marked endemism in places which are ecologically unfavourable, nor can it explain the zones of transition in ecologically favoured regions today showing high species diversity yet not correlated with clear ecological barriers. There are further lines of circumstantial evidence for biotas which occupied smaller areas and adapted to cooler temperatures in the past. The correlation between palaeoecology and patterns of endemism is strongly suggestive of a preponderant influence on some aspects of regional evolution of forest butterflies, of fundamental re-organization of the neotropical landscapes, promoted by very different climatic regimes in the late Pleistocene.

These butterflies do not show recognizable centres of species diversity. Instead, high species richness appears in scattered localities, mostly corresponding to low corrected values for endemism except at the limits of tropical forest (Fig. 4.14). Indeed, high butterfly diversity appears in small areas of neotropical forest which are characterized by interdigitating ecological systems in complex topography (Ab'Sáber's 'zones of ecological tension'). Local values for interacting species in neotropical systems are also far below regional or even half degree quadrant species lists which are those usually quoted in papers on 'diversity' (Gilbert 1977; Benson 1978; Brown 1978, 1979). Diversity values both locally and regionally fall off with regular frosts, fragmentation of forest systems (due to extensive penetration by very poor soils, or a highly unfavourable rainfall regime), or unusually high pressure by parasites and predators. They may be substantially increased locally by unpredictable mild disturbance, leading to environmental microheterogeneity (Brown 1978; Connell 1978; Fox 1979). The 'diversity' values quoted for quadrants or larger regions are largely determined by the fineness of the packing of taxonomically different but ecologically subequivalent component systems, favoured by small-scale complexity of soil, topography, microclimate, and vegetation type (including successional series). Thus, the species richness values on the uncorrected endemism maps (Fig. 4.14) as well as for the endemic centres themselves, are strongly correlated with identifiable modern ecological factors promoting environmental heterogeneity on progressively smaller scales of time and space.

Values for species diversity correlate poorly with endemism and with the model for ice-age forest stability (Figs. 2.8 and 4.20). It is clear that local diversity is an ecological phenomenon, related only distantly to recent evolutionary processes. Any particular place becomes saturated at values far below the regional potential for species richness, but the species–area curve rises far faster in areas of environmental microheterogeneity. A rich and complex physical environment and vegetation structure (such as characterizes many zones of hybridization in butterflies) will promote a closer packing of different component systems, each invaded at random by only a few of the potential ecological equivalents in the region.

These conclusions tend to make irrelevant many of the 'refuge' analyses and commentaries which are based on regional and usually scanty data about species diversity. Indeed, given the deficient biosystematic work and geographical sampling in many groups of neotropical organisms (especially away from the high-diversity hybridization zones along major Amazonian rivers), high species richness may often turn out to be merely phenotype and phenological multiplication resulting from invasion and mixing of subspecies and semispecies from various adjacent areas, in a region of rapid and complex change or environmental heterogeneity. The present analysis (Fig. 2.8) indicates that palaeoecological forest refuges correspond to little-explored interfluvial regions of relative ecological and taxonomic homogeneity, while local diversity or species richness (Fig. 4.14) characterize well-explored riverside or Andean-foothill regions of relative ecological and taxonomic heterogeneity. On detailed examination of adequately known organisms, as in this chapter, it is found that only the interfluvial regions correlate well with the model for forest stability during a recent major restructuring of neotropical environments and with high corrected values for subspecies endemism.

Other models which have been advanced to explain the observed biogeographical patterns in neotropical organisms will be discussed in Chapter 7.

(Top right, p. 102)
Fig. 4.19. A mechanism proposed by Adams (1977) for species multiplication in isolated mountain ranges during Quaternary climatic cycles. See also Fig. 5.26.

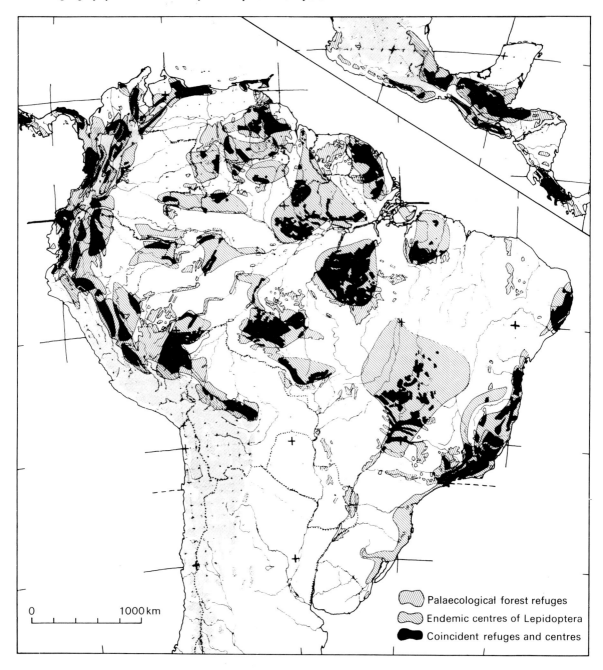

Fig. 4.20. Superimposition of endemic centres for subspecies of forest Heliconiini and Ithomiinae (isolines for one-third of maximum corrected value; Fig. 4.15) and the model for forest stability during the end of the Wisconsin–Würm glaciation (60 per cent probability; Fig. 2.8). Overlapping areas shown in black.

5

BIOGEOGRAPHY OF NEOTROPICAL BIRDS

SUMMARY

The forest avifauna of Amazonia is richest in species near the Andes. The total number of species decreases gradually toward the east in lower Amazonia. Among the 600 to 650 Amazon forest birds approximately 400 species are endemic. These include 60 endemic genera 40 of which are monotypic. Besides numerous taxonomically isolated (independent) bird species without close relatives there are in Amazonia and in the Neotropical Region generally many species which form superspecies. Up to 70–80 per cent of the species of several bird families are allospecies of superspecies.

Distribution patterns of avian species in Amazonia vary conspicuously despite the continuity and wide expanse of the forests. Many species occupy all or most of Amazonia while others are restricted to western, central, eastern, northern, or southern Amazonia. Species with geographically restricted ranges cluster in certain areas forming six 'centres of species endemism' (Fig. 5.2) each characterized by 10 to 50 species.

Large Amazonian rivers delimit, at least for some distance, the ranges of many bird species, especially the Amazon River itself and the lower portions of the Rio Negro, Rio Madeira, Rio Tapajós, and Rio Tocantins. However, in many cases interspecific competition rather than the inability to cross the watercourses or to circumvent rivers in the headwater region probably causes coincidence of borders of species ranges with a river.

Numerous endemic species, including many monotypic genera, render the rain forests of the Atlantic coast of southeast Brazil an important centre of endemism in tropical South America. However, the large number of bird species shared with the Amazonian forest indicates close relations.

Similarly close relations exist between the forest bird faunas of Amazonia and of the Pacific coast rain forests into Middle America. Numerous species are shared by both these faunas. Nearly 200 species of the trans-Andean avifauna are endemic, although many of them are merely the geographical represen-tatives of Amazonian species and have barely reached species status. Thirty species are members of highly distinct trans-Andean genera 15 or 16 of which are monotypic. Clusters of 7–32 endemic species occur in centres A–E of species endemism of Fig. 5.2. The total number of forest bird species decreases gradually northwards along the Middle American isthmus from Panama to Mexico.

Contact zones between geographical representatives are clustered in several areas of the forested neotropical lowlands forming faunal 'suture zones' (Fig. 5.22). Contact of the representatives leads to (a) hybridization (subspecies), (b) geographical exclusion due to ecological competition without hybridization (parapatric species), or (c) range overlap ('good' species).

Numerous endemic species and genera occur in the non forest region of the Brazilian tableland and eastern Bolivia south to northern Argentina. Two centres of species endemism are found in this region, northeastern Brazil and Paraguay to northern Argentina (Figs. 5.24 and 5.25). Contact zones of representative species and subspecies are clustered in the intervening area. A third centre exists in the campos cerrados region of central Brazil.

The basic ornithogeographical pattern observed in forest and non-forest regions of tropical America is therefore seen to be characterized by three features. Firstly, there is geographical clustering of endemic species and well defined subspecies with restricted ranges (centres of species endemism). Secondly there are widely disjunct populations of species with poor dispersal capabilities, and thirdly there are conspicuous contact zones many of which are clustered as faunal suture zones between centres of species endemism. Coterminous distribution patterns have been mapped in diverse bird families which differ widely in their feeding preferences. Examples are omnivorous toucans and trumpeters, insectivorous puffbirds, jacamars, woodpeckers, antbirds, fly-catchers, icterids, frugivorous guans, parrots, co-tingas, manakins, and tanagers.

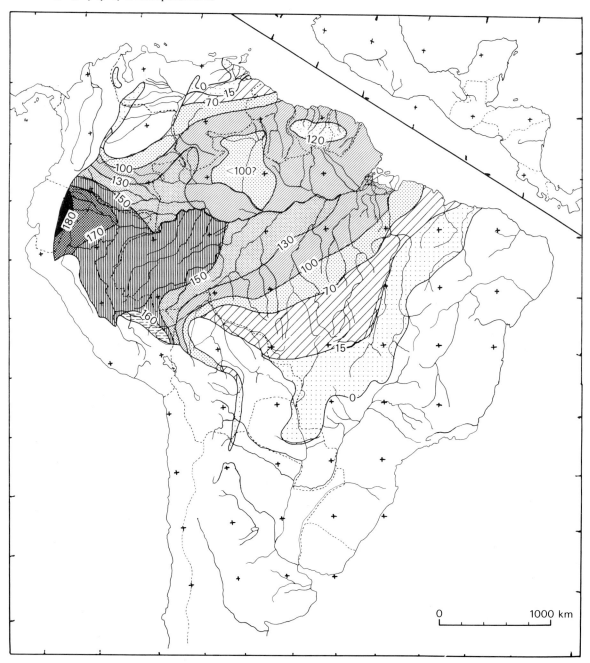

Fig. 5.1. The distribution of 360 Amazonian forest and forest-edge bird species with restricted ranges. Notice centres of relatively high species numbers in E Ecuador, SE Peru, and the Guianas and low species numbers in the Rio Branco region of N Brazil. North of the Amazon in Colombia, a steep northeastward gradient of decreasing species numbers contrasts with a more gentle gradient south of the Amazon River in W Brazil thus causing a conspicuous eastward displacement of the 130 and 150 isolines on the southern river bank. There are also c. 200 wide-ranging species not included. This map and Fig. 5.5. are generalized because of insufficient knowledge of the outline of individual species ranges and the simplistic assumption of continuous (uniform) occurrence of all species within their ranges. Obviously this is not the case, many species occur in a rather patchy manner, presumably due to unevenly distributed resources and/or to interspecific ecological competition.

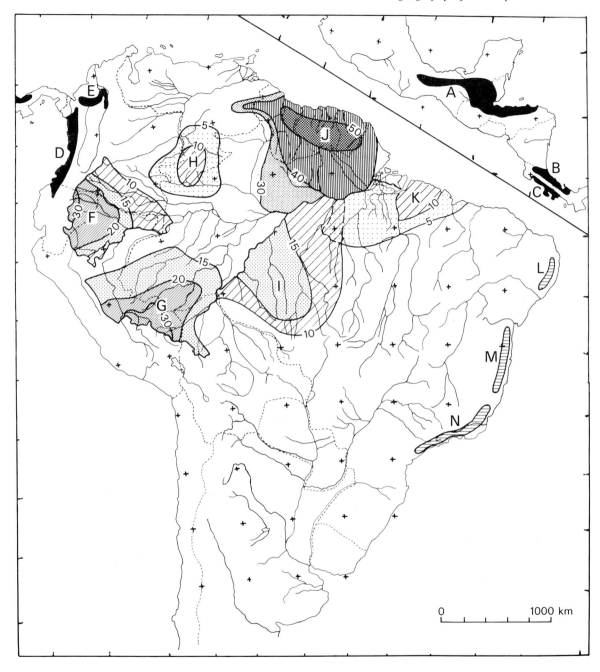

Fig. 5.2. Centres of species endemism in the neotropical lowland forest avifauna. Trans-Andean centres (solid): A Guatemala (7 spp), B Caribbean Costa Rica (14 spp), C Pacific Costa Rica (12 spp), D Chocó (32 spp), E Nechí (14 spp). East Brazilian centres (hatched): L Recife, M, N Serra do Mar. Amazonian centres (with numbers of endemic species shown by shading and simplified by omitting some displaced extensions across major rivers): F Napo, G Inambari, H Imerí, I Rondônia, J Guiana, K Belém. (Adapted from Haffer 1967a, 1974, 1975, 1978.)

The observed biogeographical patterns appear more likely to have been influenced by changing palaeogeographical and climatic-vegetational conditions of the Neotropical Region than by any other proposed causal mechanisms. The centres of species endemism may indicate in a general way the former existence of ecological refugia which formed during adverse climatic periods of the geological past, e.g. during the Pleistocene. However, the early history of many centres of endemism probably dates back to the pre-ceding Tertiary period. The contact zones between hybridizing subspecies and parapatric species — if interpreted as due to *secondary* contact of the respective animals involved — would indicate the former existence of ecological barriers which have since disappeared. Ecological aspects of the current environment and interspecific competition probably have been important in determining the details of distribution areas, especially the placement of distributional limits and of patterns of abundance.

INTRODUCTION

South America, the 'bird continent', harbours the richest avifauna of the world. There are almost 3000 species of birds including about 180 migrants from North America and the southern oceans. The number of species is increased to about 3300 if Middle America is included (Blake 1977; De Schauensee 1966; Eisenmann 1955). This compares with 1556 resident species in Africa south of the Sahara, 961 in the Oriental Region (India, southeast Asia to Java, Borneo, and Philippines) and 906 in Australasia (including New Guinea, the Moluccas, and Celebes) (Welty 1975). The vast lowland forests and the diverse habitats of the tropical mountains are among the chief ecological factors permitting the existence in the Neotropical Region of such highly varied bird life.

Among animals of tropical latitudes the distribution of birds and of certain groups of butterflies (e.g. Brassolinae, Ithomiinae, Heliconiini, and the genus *Agrias*, all of the family Nymphalidae) is comparatively well known because naturalists have accumulated a large amount of data from many remote areas over the last 150 years. Nevertheless, there are certain regions in Amazonia and along the slopes of the Andes mountains which remain poorly sampled today leaving gaps to be filled by future field workers. New species of birds are still being described from some of these areas (review by Haffer 1983). Despite this enough detail has been published to work out the basic distribution of the majority of neotropical birds. But little is known of their nesting habits, behaviour, interspecific relations, population density, seasonality, and migration, although many important data have been gathered over the last few decades.

From a zoogeographical point of view the most useful groups of neotropical birds are those families which are restricted to South and Middle America not even reaching the West Indies or the Antillean Islands. These families probably originated in South America to which region they are either restricted or represented by the greatest species number. The main families are tinamous (Tinamidae, 40 species), trumpeters (Psophiidae, 3), potoos (Nyctibiidae, 5), jacamars (Galbulidae, 17), puffbirds (Bucconidae, 30), toucans (Ramphastidae, 33), woodcreepers (Dendrocolaptidae, 52), spinetails and horneros (Furnariidae, 218), antbirds (Formicariidae, 238), and manakins (Pipridae, 59). There are several other small families in this group. More widely distributed families which occur in South and North America are the hummingbirds (Trochilidae), tyrant-flycatchers (Tyrannidae), tanagers (Thraupidae), and New World orioles (Icteridae). The curassows and guans (Cracidae) are a mainly neotropical group today but fossil members of this family have been found in North America. Most species of wrens (Troglodytidae), thrashers (Mimidae), vireos (Vireonidae), wood warblers (Parulidae) and motmots (Momotidae) are found in southern North America and Central America. Among bird families which are represented in all tropical regions of the world we find in South America parrots (Psittacidae), trogons (Trogonidae), barbets (Capitonidae), pigeons (Columbidae), pheasants (Phasianidae), cuckoos (Cuculidae), crows (Corvidae), thrushes (Turdidae), and a few others. Wide-ranging taxa mostly with worldwide ranges in the neotropical avifauna include barn-owls (Tytonidae), owls (Strigidae), hawks (Accipitridae), falcons (Falconidae), nightjars (Caprimulgidae), swifts (Apodidae), woodpeckers (Picidae), and swallows (Hirundinidae).

This review is concerned mainly with the avifauna

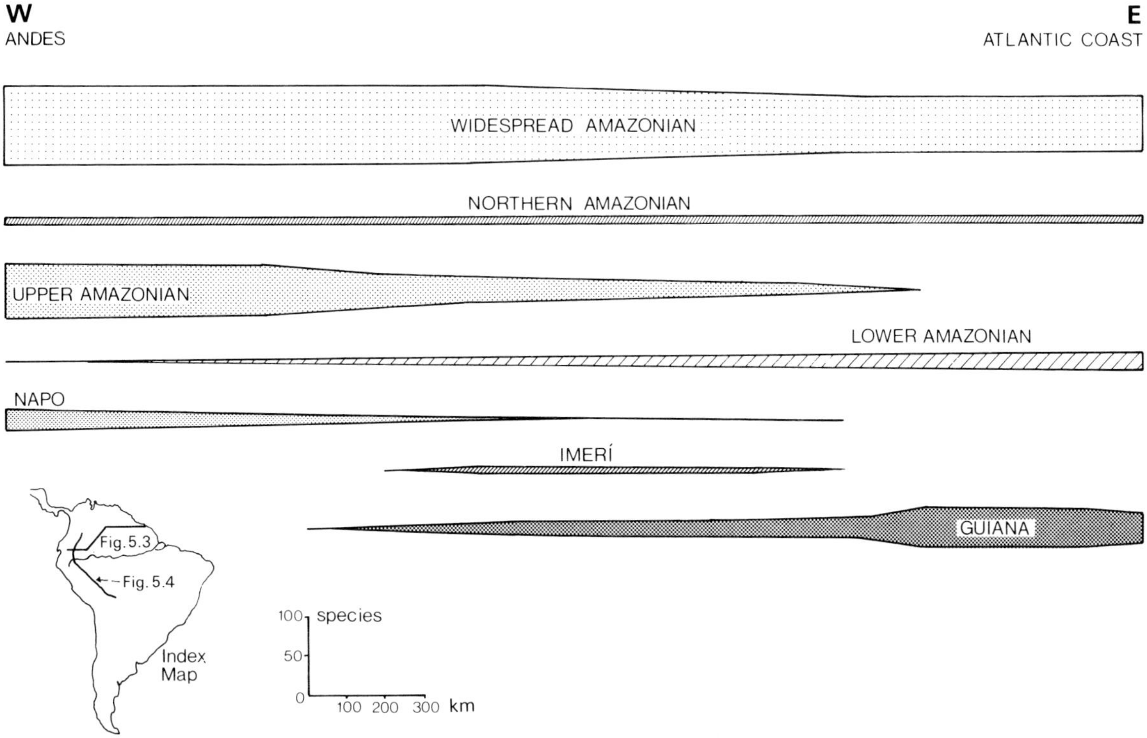

W
ANDES

E
ATLANTIC COAST

Fig. 5.3. Transect across northern Amazonia to show overlap of different species groups. Analysis based on a sample of c. 320 species.

of the tropical lowlands, especially that living in the humid forests which cover vast portions of the continent. I discuss briefly the avifauna of open non-forested lowland regions but mention only few aspects of the bird faunas of the Andes and of other mountain ranges in tropical America. I emphasize the description and illustration of the various distribution patterns among birds of tropical America in order to facilitate a comparison with the other groups of organisms treated in this volume. The majority of the distribution maps of species and superspecies have been compiled during the preparation of this article. The legends are fairly lengthy, as the text mentions only few details illustrated by the maps. The bio-geographical interpretation of the data is kept brief for reasons of space. More detailed discussions have been included in previous publications (Haffer 1974, 1979, 1981, 1982, 1983, 1985; Simpson and Haffer 1978).

DISTRIBUTION PATTERNS OF NEOTROPICAL LOWLAND BIRDS

The basis for zoogeographical analyses consists of detailed distributional and faunistic data including information on abundance in space and time as well as on the geographical variation and ecological requirements of the species inhabiting a given region. Sufficient data on these lines are available for many groups of neotropical birds.

A basic distinction is made between the widely different *'forest'* and *'non-forest'* avifaunas, each comprising numerous characteristic species. Forest birds are here defined to include species of the forest interior as well as birds inhabiting forest edges, woodland, riparian vegetation in forested areas and species that are generally restricted to forested regions and disappear from an area once the forest has been cut. 'Non-forest' birds live in regions covered with grassland, savannas, open dry woodland, cactus wastes, thorn scrub, etc. The separation of such 'open' and 'forested' areas in some cases is rather artificial, particularly in transitional regions of wooded savannas to light deciduous forests and in areas where man has cut the forest recently.

There are only a few bird species which range through both open and forested areas in South America. These are ecological generalists whose distributions tend to be wide and with little or no relations to major habitat boundaries. An example is certain groups of tyrant-flycatchers (Tyranninae and frugivorous Elaeniinae) which occupy the upper strata of both scrub and forest; in effect, these birds treat the forest canopy as an elevated scrub (Fitzpatrick 1980). In each of the forested and open regions species may be widely distributed or restricted to small areas and either common or rare. By comparing or superimposing the range outline maps of forest and non forest birds we establish groups of species with similar distributions and attempt to recognize regional distribution patterns characteristic for major sections of the avifauna.

I exclude from this review water birds such as grebes, herons, ducks, geese, rails, gulls, plovers, and others. Fjeldså (1985) and Reichholf (1975) discussed the relevant issues for the Andes and the South American lowlands, respectively. Reichholf (l.c.) compared geographical trends in the fish-eating herons and detritus-eating ducks. Neotropical species of herons (Ardeidae) increase in number of species and individuals toward the lower latitudes in accordance with the general pattern. By contrast, ducks (Anatidae) decrease in both abundance and number of species toward the tropics possibly because they are in competition for food with the rich tropical fish fauna. Slud (1976) also pointed out that the ratio of water bird species to land bird species drastically decreases with decreasing latitude being lowest in the humid tropics. Slud also discussed geographical patterns of the neotropical avifauna from another viewpoint. He analysed and mapped relative proportions of the major taxonomic components of the avifauna (passerines, non-passerines, suboscines, oscines) which correlate with climate and vegetation and showed that the suboscine proportion is highest in the Amazon forests while the non-passerine proportion decreases with increasing elevation and rainfall.

Besides numerous taxonomically isolated (independent) bird species without close relatives within or outside their respective genera there are in the neotropics many species which form superspecies (e.g. Mayr 1963). The member species (allospecies) of a superspecies replace one another geographically and are either allopatric or in contact (parapatric). Allospecies exclude one another geographically (or presumably do so) because they are (still) strong ecological competitors. Absence or near-absence of phenotypic indications for hybridization in the populations along the contact zone is taken as a criterion for species status of parapatric forms. Seventy-five per cent of the species of curassows and guans (Cracidae) are allospecies, 75 per cent of the jacamars (Galbulidae), 85 per cent of the toucans (Ramphastidae), and 75 per cent of the manakins (Pipridae); these figures are derived from the taxonomic studies of Vuilleumier (1965), Haffer (1974), and Snow (1975). The geographical range of many superspecies resembles a large-scale mosaic composed of neatly interlocking patches formed by the ranges of its component allospecies. Some of these assemblages consist of four to six (or more) allospecies of which two and two or two and three species (etc.) are more closely related among themselves than to the allospecies of the other cluster. Such assemblages should be considered as parapatric superspecies, even though none of the various allospecies have as yet achieved ecological compatibility. Haffer (1986) combined two or more parapatric first-order superspecies in a second-order superspecies or megasuperspecies. This emphasizes a conclusion of Mayr and Short (1970) that 'newly evolved species frequently maintain a parapatric distribution pattern long after the geographic speciation is complete' and, I would add, even after several successive periods of speciation within a group have occurred. Many of the avian allospecies and independent species vary geographically, especially when they occupy extensive areas in South America, and numerous often clinally related subspecies have been described. I do not consider this lowest level of taxonomically recognized regional differentiation in this chapter on birds.*

---

*In biogeographical studies the unqualified use of 'subspecies' is inadvisable where large continuous populations are concerned such as those of many South American forest animals. In these cases subspecies names may refer to clinal forms based on varying subjective criteria or to widespread uniform populations or they may designate highly variable

In zoogeographical analyses the superspecies and their ranges are more instructive than the individual distribution areas of their component species. Together with independent species not belonging to a superspecies they have been designated as 'zoogeographical species' by Mayr and Short (1970). Consideration of superspecies puts the number of taxonomic species into proper zoogeographical perspective. Thus, Haffer (1974) lists only 14 zoogeographical species in the toucans (out of a total of 33 taxonomic species) and eight zoogeographical species in the jacamars (out of a total of 17 taxonomic species). No systematic analysis of the entire neotropical avifauna based on the superspecies concept has been carried out; it would be of considerable zoogeographical and evolutionary significance.

Forest birds

Three major lowland tropical rain forest regions or blocks of rain forest in South America are Amazonia comprising the central portion of the continent, the southeastern Brazilian forest region along the Atlantic coast and the trans-Andean forest region including the Pacific lowlands of northwestern Ecuador and Colombia, portions of the Caribbean lowlands of Colombia, as well as the lowlands of Middle America (mainly along the Caribbean slope). A number of widely distributed forest bird species and superspecies occur in all three forest regions from subtropical South America north to northern Middle America, attesting to the close relationship between these widely separated bird faunas of the Neotropical Region. On the other hand, there are many bird species restricted to one or two of these three forest regions or to portions of them.

Amazonia

Amazonia is here understood to comprise the Amazon basin as well as the Guianas and southern to eastern Venezuela (southern part of the Orinoco drainage). These latter areas form part of the continuous forest region of central South America. About 650 bird species inhabit forest and forest edges in Amazonia. A more conservative count excluding birds restricted to fairly open forest edges probably would reduce this number to approximately 600 species. About 400 species or allospecies are restricted (endemic) to Amazonia. Many of them are the Amazonian representatives (allospecies) belonging to widespread neotropical superspecies, having other allospecies in southeastern Brazil and/or in the forest region west of the Andes into Middle America. Other endemic Amazonian species have no close taxonomic allies outside central South America. These form 60 endemic Amazonian genera (40 of which are monotypic; Haffer 1974, p. 56). Some examples are (number of species or allospecies in parentheses) the trumpeters (*Psophia*, 3), the parrots *Pionites* (2) and *Deroptyus* (1), the jacamars *Galbalcyrhynchus* (2), the antbirds *Frederickena* (2), *Myrmoborus* (4), *Hypocnemis* (2), *Pithys* (2), *Rhegmatorhina* (5), several genera of cotingas, manakins and tyrant-flycatchers.

The Amazonian forest bird fauna is richest in species per square of 1 degree longitude by 1 degree latitude near the Andes where Haffer (1978) estimated the totals to be around 370–380 species for Ecuador and 360 species for southeast Peru* (Fig. 5.1) decreasing eastward to 320 species in the interior Guianas and to 300 species near the mouth of the Amazon River. More important than absolute numbers (whose count will differ somewhat between authors because of disagreement about the delimitation of the group of 'forest birds') is the general pattern of an eastward decrease in total number of species by about 20 per cent. This decrease may be related in part to an overall eastward decrease in annual precipitation and greater seasonality of climate. Ecological factors associated with higher rainfall in upper Amazonia such as, for example, more luxuriant forest growth and increased plant species diversity (i.e. increased structural complexity) may be correlated with the maintenance of westwardly somewhat

---

hybrid populations. On the other hand, 'subspecies' are of more direct use in island or montane species where geographic variation is discontinuous and isolated populations are fairly uniform. In any case, analyses of population structure of the species involved should precede or accompany biogeographic studies which use 'subspecies' to establish faunal differences between areas and to interpret the history of their faunas.

*This estimate agrees closely with the results of recent counts made over extended periods of time at two localities in southeastern Peru by Terborgh, Fitzpatrick, and Emmons (1984; 5 km² in the Manu National Park: about 350 species of resident forest birds (broadly defined) plus 160 species of migrants and non-forest birds including those of large clearings as well as swallows, waders, gulls, etc.) and by Donohue, Parker, and Sorrie (Ms; 100 km² in the Tambopata Nature Reserve: about 350 species of resident forest birds plus 166 species of non-forest birds and migrants).

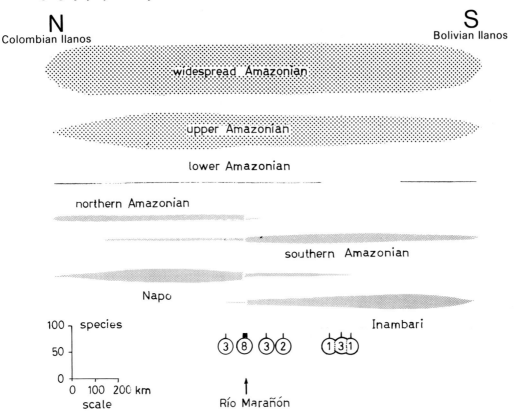

Fig. 5.4. Transect through upper Amazonia to show overlap of different species groups. Analysis based on a sample of c. 280 species. Position of contact zones shown by circled figures which also indicate the number of pairs of parapatric species and hybridizing subspecies. See Fig. 5.3 for location of transect.

increased numbers of bird species. Historical factors (as discussed below) may also have contributed to this situation. Rather low total numbers of bird species have been mapped in the Rio Branco region of north-central Amazonia (north of the lower Rio Negro east to include the surroundings of the town of Manaus, see Fig. 5.1). This situation is unlikely to be an artefact of insufficient sampling.

Avian distribution patterns in Amazonia vary conspicuously despite the continuity and wide expanse of the forests. There are widespread species occupying all of Amazonia, whereas others are geographically more restricted and inhabit only a part of this large region. A number of these more restricted species are in fact surprisingly localized. In order to establish general patterns underlying the immense variety of avian species ranges in Amazonia, the species considered have been grouped into several

assemblages based on the geographical similarity or congruence of their ranges. In spite of this procedure the resulting species groups are not sharply delimited and assignment of certain transitional cases to a particular group remains subjective. All species of a given distributional group occur together in the variously extensive central portion of the 'group area'. The number of regionally sympatric species in each group decreases in directions away from the central region resulting in gentle to steep gradients of the species totals in several directions.

There are six clusters of these geographically restricted bird species with largely congruent ranges, forming what I call 'centres of species endemism'. Each centre is characterized by 10–50 species (Figs. 5.2–5.4). Lists of species composing these distributional groups of forest and forest edge birds are not repeated here as they may be obtained from a

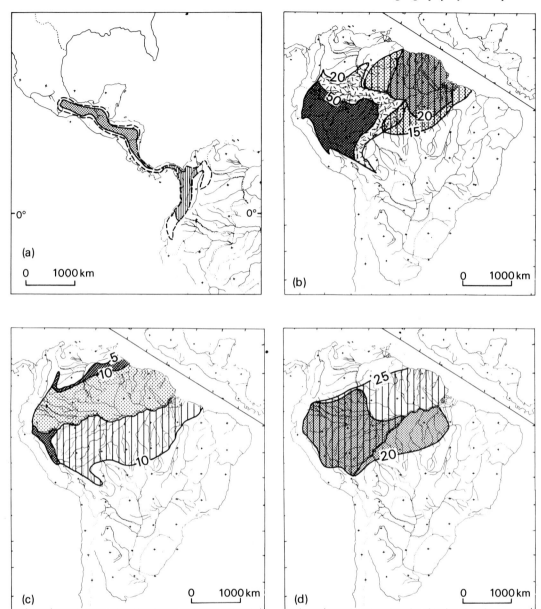

Fig. 5.5. Ranges of some major distributional groups of (a) trans-Andean and (b–d) Amazonian forest birds. Areas in Amazonia are outlined by contours indicating the number of sympatric species in the respective groups. (After Haffer (1975, 1978).) See legend to Fig. 5.1. (a) Trans-Andean forest birds: shaded. – Middle American species; hatched – NW Colombian–Panamanian species; dashed outline – species of the entire trans-Andean forest region; (b) upper Amazonian (west) and lower Amazonian birds (east, hatched), notice the barrier effect of the lower Rio Negro and of the lower Rio Madeira in C Amazonia; (c) northern and southern Amazonian birds; notice that, near the Andes, some northern Amazonian species occur south of the Rio Marañón; conversely, a few southern Amazonian species occur north of this river and advance to the base of the Colombian Andes (not shown on this map; but see Fig. 5.9); (d) two groups of wide-ranging Amazonian birds which are missing from SE (hatched) and NE Amazonia, respectively.

previous publication (Haffer 1978*); see also the lists compiled for the Amazonian areas of endemism by Müller (1973) and especially Cracraft (1985*a*). The Napo and Inambari centres in upper Amazonia and the Guiana centre in northeastern Amazonia are weakly discernible on the composite map Fig. 5.1.

On the following pages I briefly discuss the distributional groups of Amazon forest and forest edge birds and illustrate the distribution areas of selected species or superspecies. Additional detailed distribution maps have been published among recent authors by Willis (1968, *Gymnopithys*; 1969, *Rhegmatorhina*), Short (1972, woodpeckers of the *Celeus elegans* superspecies), Delacour and Amadon (1973, Cracidae, based on Vaurie's taxonomic studies), Fitzpatrick (1976, *Todirostrum* flycatchers), Novaes (1981, *Pionites* parrots), Haffer (1967*a*, 1969, 1970*a*, 1974, 1975, 1977*a, b*, 1978, parrots, trogons, ground-cuckoos, jacamars, toucans, antbirds, cotingas, manakins, tanagers, and others), and Snow (1982, cotingas).

Widespread Amazonian birds. A large number of species inhabit all of Amazonia, where they occur at many localities. For many others, however, only a few widely scattered locality records are available and it remains unknown whether they are found throughout Amazonia. Two smaller groups of birds (not all of them endemic to Amazonia) are also wide ranging but are missing either from the northeastern portion of Amazonia (26 species) or from the southeastern portion (28 species) (see Fig. 5.5(d)). In some of the species lacking in northeastern Amazonia a close ally (allospecies) in the Guianan region probably prevents, through ecological competition, the wide-ranging representative from occupying northeastern Amazonia or vice versa, e.g. *Celeus grammicus/C. undatus* (Haffer 1974, p. 97), *Veniliornis affinis/V. cassini* (Fig. 5.13(b)), *Xiphorhynchus ocellatus/X. pardalotus*, *Myrmotherula hauxwelli/M. guttata* (Fig. 5.14(b)), *Iodopleura isabellae/I. fusca* (Fig. 5.14(d)), *Tyranneutes stolzmanni/T. virescens*, *Euphonia rufiventris/E. cayennensis* (Haffer 1970*a*, p. 316). Potentially competing allies in the other group of wide-ranging species which are missing from

portions of the area south of the lower Amazon can be identified in only two cases, *Capito niger/C. dayi* and *Xiphorhynchus guttatus/X. eytoni*.

Northern and southern Amazonian birds. Fourteen species occur in northern Amazonia from the Andes to the Atlantic coast, the Amazon River often forming the southern range boundary. These species are missing from all or most of southern Amazonia (Figs. 5.5(c), 5.6, and 5.7). Some of the northern Amazonian birds have crossed the Río Marañón near the Andes and/or the lower Amazon near its mouth in this way occupying small areas south of this river (Fig. 5.8). The opposite is true for several southern Amazonian bird species. Lists of these distributional groups have been prepared by Haffer (1978).

The ranges of 21 southern Amazonian species are essentially complementary to those of the above birds, the Amazon River representing the northern range limit in most cases (Figs. 5.5(c) and 5.9). Several species are the southern representatives (allospecies) of the northern forms such as *Leucopternis melanops/L. kuhli* (Fig. 5.6), *Pionites melanocephalus/P. leucogaster* (Haffer 1977*b*, p. 270; Novaes 1981), *Galbula albirostris/G. cyanicollis* (Haffer 1974, p. 325), *Hypocnemoides melanopogon/H. maculicauda* (Fig. 5.7), *Pipra erythrocephala/P. rubrocapilla* (Haffer 1970*a*, p. 307), *Lanio fulvus/L. versicolor* (Haffer 1977*b*, p. 275). A few of the southern Amazonian species occur across the Río Marañón and inhabit portions of upper Amazonia near the base of the Andes in eastern Ecuador (or even reaching southeastern Colombia), e.g. *Trogon curucui*, *Electron platyrrhynchum*, *Malacoptila rufa* (Fig. 5.9(a)), *Xiphorhynchus spixii*, *Myrmotherula ornata*, *Myrmeciza hemimelaena* (Fig. 5.9(b)), *Phlegopsis nigromaculata* (Fig. 5.9(c)), *Campylorhynchus turdinus*. Competition between these species and other more distantly related birds of the same or another genus may prevent further range expansion in northern Amazonia where no close allies exist that might be considered as the geographical representatives of these southern Amazonian species.

Western Amazonian birds. This large group comprises widespread species inhabiting most or all of upper Amazonia and two groups of species restricted to northwestern and southwestern Amazonia, respectively. Seventy upper Amazonian species occur near the Andes both north and south of the Marañón River ranging for varying distances eastward into

---

*Six species which Haffer (1978) included in these lists should be deleted as in Amazonia they inhabit second growth and wooded savanna (*Ortalis motmot*, *Cyanocorax heilprini*), campina and Amazonian caatinga (*Xenopipo atronitens*, *Neopelma chrysocephalum*), or second growth and dry scrub (*Sakesphorus melanothorax*, *S. luctuosus*).

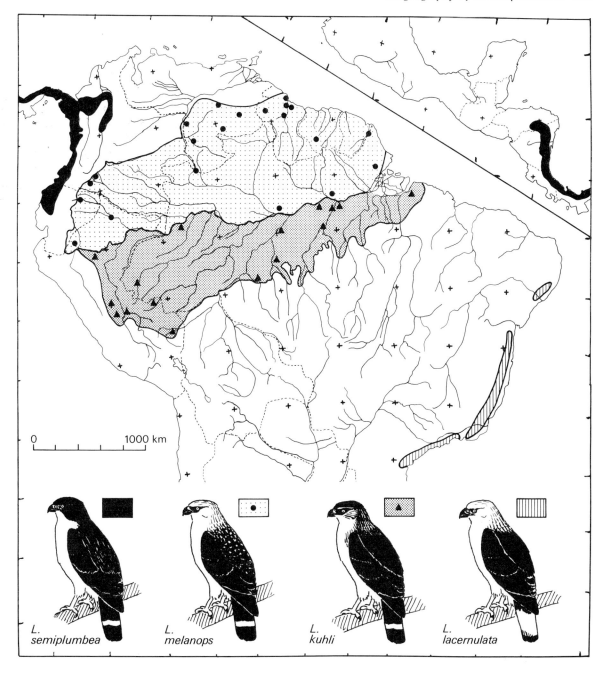

Fig. 5.6. Distribution of four species of forest hawks, *Leucopternis*. The three cis-Andean forms are considered conspecific by some authors. Unpublished records shown on the map include the Peruvian localities of *L. melanops* (boca Rio Curaray and boca Rio Santiago), (AMNH), and *L. kuhli* (boca Rio Urubamba) (AMNH) and for the latter São Paulo de Olivença, W Brazil (CM) also. These are based on museum collections as shown. A female specimen of *L. melanops* supposedly from 'Tauary, lower Rio Tapajós' (AMNH), within the range of *L. kuhli*, probably has been incorrectly labelled and is not shown.

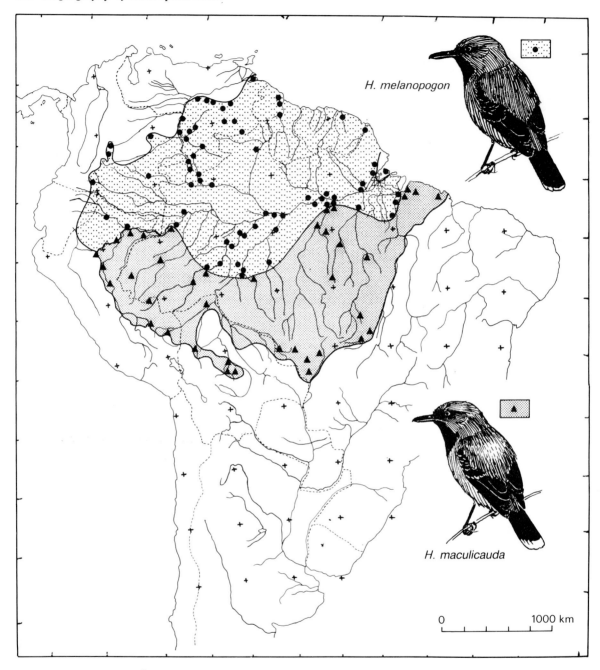

Fig. 5.7. Distribution of the *Hypocnemoides melanopogon* superspecies (antbirds, Formicariidae). These small birds inhabit tangled thickets along forest streams and the undergrowth of swamp forest. *H. maculicauda* has the black throat more extensive, the white fringes of wing coverts and rectrices are broader and a large white interscapular patch is present. Specimens of both species have been collected at São Paulo de Olivença (CM), Cachoeira, Rio Purús (Snethlage) and along the lower Rio Tapajós (AMNH, CM) indicating the location of the contact zone; no hybrids are known. Three specimens of *H. melanopogon* supposedly from 'Lagarto, boca Rio Urubamba' (AMNH) in E Peru (within the range of *maculicauda*) probably have been labelled incorrectly and are not shown.

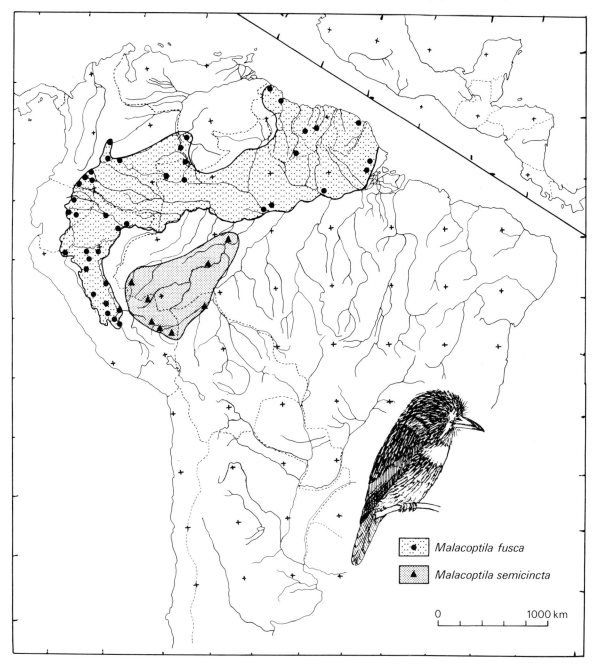

Fig. 5.8. Distribution of the *Malacoptila fusca* superspecies (puffbirds, Bucconidae): These closely similar taxa differ mainly in the presence (*M. semicincta*) or absence of a rufous half-collar; they inhabit the understory of the forest interior where they perch singly to hunt insect prey. *M. fusca* extends south of the Rio Marañón along the base of the Andes. In large areas of S Amazonia, *Malacoptila rufa* represents *M. fusca/semicincta* ecologically and is sympatric with these in E Ecuador and W Brazil (see Fig. 5.9(a)).

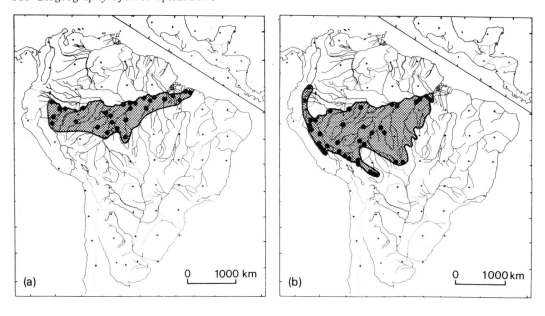

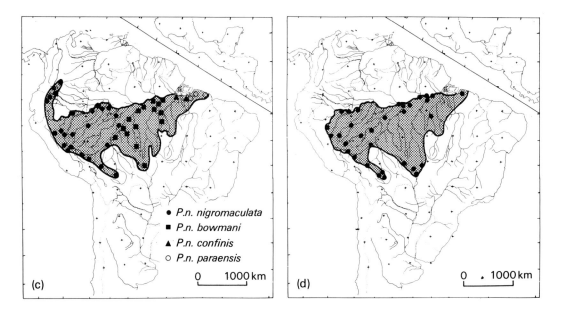

Fig. 5.9. Distribution of four southern Amazonian bird species. Notice that all ranges include small areas north of the Amazon. No close relatives are known that might prevent further range expansion of the southern species in northern Amazonia except perhaps in *Malacoptila rufa* (a), an inhabitant of the low understory of the forest interior: *M. fusca/semicincta* (Fig. 5.8) may prevent its extension to upper Amazonia. (b) *Myrmeciza hemimelaena* inhabits thickets in the forest. (c) *Phlegopsis nigro-maculata* is found at low levels inside the forest where it commonly follows army ants; four subspecies distinguished on the

central Amazonia. Fewer than half of these birds reach the upper Rio Negro and the Rio Madeira in central Amazonia (Fig. 5.5(b)). Rivers, acting as partial barriers to dispersal, have modified the steep but fairly smooth eastward gradient of decreasing total number of species within this group. Thus, on the southern bank of the Rio Solimões, many species range for a greater distance eastward than on the northern bank. This is shown by the eastward displacement of the contour lines on the southern bank of the upper Amazon (Fig. 5.1). Other discontinuities in the eastward gradient are caused by the lower Rio Negro and by the Rio Madeira.

Several of the upper Amazonian species are taxonomically isolated and are placed in monotypic genera such as the curassow *Nothocrax urumutum*, the hummingbird *Polyplancta aurescens*, the Plushcrown *Metopothrix aurantiacus*, the hookbill *Ancistrops strigilatus*,* the antbirds *Megastictus margaritatus*, *Dichrozona cincta*, the tapaculo *Liosceles thoracicus*, the cotinga *Porphyrolaema porphyrolaema*, the oropendola *Clypicterus oseryi*. A number of the upper Amazonian species have representative populations in the forested lowlands west of the Andes and/or in eastern Amazonia (Figs. 5.10–5.15). In central Amazonia, many of the western and eastern Amazonian allies are in contact. Here they either hybridize extensively (Fig. 5.10(d)), exclude each other geographically (Figs. 5.11 and 5.12) or show overlaps in their ranges (Fig. 5.11, Guyana and southern Venezuela). The ranges of several representative species as mapped on the basis of currently available locality records do not abut, indicating a real or potential geographical gap between these allies (Fig. 5.14). The ranges of the western ground-sparrow *Arremon aurantiirostris* and of the eastern species *A. taciturnus* approach each other along the eastern base of the Andes in southeastern Colombia and in eastern Peru (Fig. 5.16). However, both species are widely separated in the intervening areas because *A. taciturnus* (which is quite common in the areas it occupies) avoids the entire upper Amazonian lowlands. We have no explanation for its absence from western Amazonia.

*Included in the genus *Philydor* by Vaurie (1980).

The two clusters of restricted upper Amazonian birds occur in northwestern Amazonia (centring in eastern Ecuador and the Río Napo region; *Napo centre*) and in southwestern Amazonia (centred in southeastern Peru; Rio Inambari region north to the hills in the headwater region of the Rio Purús; *Inambari centre*) (see Figs. 5.2–5.4). I have assigned 38 and 47 species and well defined subspecies respectively to these centres of endemism. Conspicuous gradients of decreasing total numbers of species within each group occur in a northeastern direction parallel to the main rivers in these regions. On the other hand, the Rio Marañón–Solimões delimits the ranges of many Napo and Inambari birds truncating several contour lines on the composite map of these species groups (Fig. 5.2). As an illustration of clustered and geographically restricted ranges in Amazon forest birds, Figs. 5.17–5.19 show a series of distribution maps of selected southeastern Peruvian species, the ranges of some of which extend for varying distances into western Brazil. Similar sets of maps could be prepared for the clusters of Napo, Imerí, and Guiana birds; see also the discussions and maps for upper Amazonian centres of endemism by Müller (1973) and Cracraft (1985a).

A number of the Napo and Inambari birds are geographical representatives which are either in contact or separated by a distributional gap. Examples are *Pyrrhura albipectus/P. rupicola* (Andean foothills, Fig. 5.18(a)), *Heliodoxa gularis/H. branickii* (Andean foothills, Fig. 5.19(b)), *Phlogophilus hemileucurus/P. harterti* (foothills, Fig. 5.19(a)), *Galbula tombacea/G. cyanescens* (Fig. 5.10(b)), *Galbalcyrhynchus leucotis/G. purusianus* (Haffer 1974, p. 315), *Pteroglossus f. flavirostris/P. f. mariae* (Haffer 1974, p. 220), *Selenidera r. reinwardtii/S. r. langsdorffii* (Haffer 1974, p. 240), *Myrmeciza melanoceps/M. goeldii* (Fig. 5.18(b)), *Grallaria dignissima/G. eludens* (Fig. 5.18(c)), *Todirostrum c. calopterum/T. c. pulchellum* and *Poecilotriccus capitale/P. albifacies/P. tricolor* (Fig. 5.18(d)). In other cases one of the allies inhabits not only upper Amazonia but also parts or all of lower Amazonia such as in the species pairs *Pyrrhura picta/P. melanura* (Fig. 5.13(c)) and *Malacoptila fusca/M. semicincta* (Fig. 5.8). Additional Napo (N) and Inambari (I) species without close allies

basis of plumage characters are separated by southern Amazonian rivers, a record of *P. n. paraensis* from Macapá on the north bank of the lower Amazon near its mouth(Novaes 1957) may be based on a specimen that was incorrectly labelled, the locality needs verification. (d) *Attila bolivianus* is a flycatcher of the subcanopy. Unpublished locality records indicated on the map include Rio Pichana, Dep. Loreto, NE Peru (FM), São Paulo de Olivença and Tonantins (CM), Caviana and Manacapurú (CM), Arimã and Huytanahan (CM).

are the following: *Nonnula brunnea* (N), *Thamno-philus praecox* (N, known only from the female type specimen), *Phlegopsis barringeri* (N, known only from the type specimen which Willis (1979) believes may be hybrid *P. erythroptera* × *P. nigromaculata*), *Neoctan-tes niger* (N), *Cacicus sclateri* (N), *Formicarius rufifrons* (I), *Conioptilon mcilhennyi* (I), *Cacicus koepckeae* (I).

Central Amazonian birds. Comparatively few species are restricted to the central portion of Amazonia. Four of them are widespread. Fifteen and twenty species have restricted ranges in northcentral Amazonia (Imerí region) and in southcentral Amazonia (Rondônia region), respectively (Fig. 5.2). Representative Imerí and Rondônia species are *Selenidera nattereri/S. gouldii* (Haffer 1974, p. 240) and *Hetero-cercus flavivertex/H. linteatus* (Fig. 5.20). The parakeet *Pyrrhura rhodogaster* inhabits southcentral Amazonia (Fig. 5.13(d)); its eastern allospecies is *P. perlata.** Other central Amazonian species have geographical representatives in upper and/or lower Amazonia such as the species of *Mitu* (Fig. 5.13(a)), *Rhegmatorhina* (Willis 1969) and *Pipra* (Haffer 1970*a*, p. 312).

Eastern Amazonian birds. About 80 species are restricted to eastern (lower) Amazonia. Among them are widespread and more localized birds; the latter comprise a large group of Guianan species and a small group of Belém species (Fig. 5.2). All of the wide-spread species occur to the north and south of the lower Amazon River their ranges extending for varying distances westward into central or even upper Amazonia (Fig. 5.5(b)). The barrier effect of the lower Rio Negro and the Rio Madeira results in the development of stepped westward gradients in the decreasing total number of species in this group the steps diminishing or disappearing in the headwater regions where the rivers cease to function as barriers.

*The two type specimens of *P. perlata perlata* (Spix 1824) without precise locality data represent the immature plumage of *P. rhodogaster* (Sclater 1870) as determined by Arndt (Spixiana [München], Suppl. 9, 425–428. 1983). If confirmed this may require the following nomenclatural changes: The name *rhodogaster* becomes a synonym of *P. perlata*. The eastern representative inhabiting the area south of the mouth of the Amazon River and east of the lower Rio Xingu, previously known as *P. perlata*, henceforth may have to be designated *P. lepida* (Wagler 1832) with the subspecies *lepida, coerulescens,* and *anerythra.*

The large cluster of 51 endemic *Guianan species* renders the avifauna of northeastern Amazonia highly distinctive. The lower Rio Amazonas and Rio Negro form the range boundaries of many of these species. Some of them occur across the lower Rio Amazonas and occupy a small area south of the mouth of this river. Many Guianan species are the local representatives (allospecies) belonging to widespread Amazonian superspecies: *Crax alector* (Delacour and Amadon 1973), *Pionopsitta caica* (Haffer 1970*a*, p. 292), *Galbula galbula* (Fig. 5.10(b)), *Pteroglossus viridis, Selenidera culik, Celeus undatus, Veniliornis cassini* (Fig. 5.13(b)), *Xiphorhynchus pardalotus, Frederickena viridis* (Fig. 5.15), *Cyanocorax cayanus, Euphonia cayennensis, Gymnopithys rufigula.*

Several other Guianan species are taxonomically isolated. They are placed in monotypic genera and have no close relatives and probably represent derivatives of old (Tertiary) isolates in this region. These species are *Haematoderus militaris, Perissocephalus tricolor, Rupicola rupicola* (with one Andean allospecies, Fig. 5.21), *Periporphyrus erythromelas* and *Cyanicterus cyanicterus.*

Three localized species whose ranges, however, are insufficiently known may indicate the existence of a separate small centre of endemism in the eastern Guianas. The nightjar *Caprimulgus maculosus*, the hummingbirds *Threnetes niger* and *Phaethornis malaris* occur in French Guiana, the latter two species extend into extreme northeastern Brazil. Mees (1977) encountered *P. malaris* also in eastern Surinam.

There is a small group of Belém species which characterize the avifauna south of the lower Amazon River. This group includes *Pyrrhura perlata* (Fig. 5.13(d)), *Aratinga guarouba* (see Oren and Willis 1981), *Pionopsitta vulturina* (see Haffer 1970*a*, p. 292) *Xipholena lamellipennis* (see Haffer 1970*a*, (p. 298), and *Pipra iris* (see Haffer 1974, p. 100). Three additional species are known only from the region between the Rio Tapajós and the Rio Tocantins. Among these, *Rhegmatorhina gymnops* is fairly widespread, whereas *Pipra vilasboasi* and *Hemitriccus* ('*Idioptilon*') *aenigma* so far have been collected at only one and two localities, respectively.

Several bird species inhabit the humid lowlands south of the mouth of the Amazon and occur south-westward in peripheral portions of Amazonia into eastern Bolivia. These species then range northward along the Andes mountains through Peru and Ecuador into Colombia inhabiting lower montane levels. Examples are the antbirds *Thamnophilus palliatus* and *Pyriglena leuconota*. The significance of this distribution pattern is not yet known. We may speculate that competing allies prevented the entry of these species into Amazonia from the south. The southern species, however, managed to expand their ranges northward in the lower montane levels of the Andes (possibly because there the competitive pressure was less severe). The Amazonian species *Thamnophilus murinus* has a range complementary to that of *T. palliatus* and both species might compete ecologically where they are in contact in the lowland of southeastern Peru and along the eastern base of the Andes of Peru and Ecuador. Additional examples of birds which inhabit tropical lowlands along the southern margin of Amazonia and ascend into montane levels of the northern Andes include the tinamou *Crypturellus obsoletus* and the tanager *Tangara cyanicollis*. The latter species has restricted lowland populations in central Brazil. The trogon *Trogon collaris* and the tanager *Tangara gyrola* are widespread in Amazonia but range in montane levels along the Colombian Andes and Middle American mountains north to Mexico and Costa Rica, respectively. The distribution areas of the allospecies *Cephalopterus ornatus/penduliger/glabricollis* (Fig. 5.21 (a)) as well as *Rupicola rupicola/peruviana* (Fig. 5.21(b)) also illustrate relations between the Amazonian and the Andean forest avifaunas.

Significance of Amazonian rivers. Large Amazonian rivers delimit, at least for some distance, the ranges of many bird species as can be seen on several of the maps. Notice in this respect the eastward displacement of contour lines along the southern bank of the Solimões–Amazon River on the composite map Fig 5.1. In northcentral Amazonia, the barrier effect of the lower Rio Negro is noticeable and among southern tributaries of the Amazon the lower Rios Madeira, Tapajós, and Tocantins are especially important. Numbers of species ranges delimited for some distance by the Amazon River itself increase in the sample of about 360 species from about 10 near the Andes to over 150 along the wide lower Rio Amazonas which seems to be a formidable barrier zone for many birds (Haffer 1978, p. 71, 1985).

At least three explanations of the barrier effect of Amazonian rivers may be offered, depending upon whether the rivers act as absolute or partial barriers to dispersal for bird species (Haffer 1978, with examples). Rivers may delimit the ranges of species which are unable to cross or to circumvent them. Or

secondly, rivers, acting as partial barriers to dispersal, may have stabilized the equilibrium between competing species or between hybridizing subspecies (in the latter case by drastically reducing gene flow). In other words, the birds under consideration probably would cross the river or circumvent it in the headwater region but for the existence there or across the river of a competing or hybridizing ally (subspecies or allospecies). Or thirdly, competition, with more distantly related species and/or somewhat different ecological conditions on the opposite river bank may prevent the spread of some species despite the fact that, in some cases, a 'bridge head' exists on the opposite river bank (Figs. 5.9 and 5.13(a)). Generally speaking, interspecific competition rather than the inability to cross the watercourses probably determines that many species range borders coincide with a river. Direct evidence to support or disprove this statement will be difficult to furnish. Detailed behavioural and ecological studies of differing species populations on opposite river banks in areas where the rivers are narrow or where islands might facilitate river crossings would be of great interest.

Even though Amazonian rivers, or portions of them, form range boundaries for many species, the cumulative total of boundaries located away from rivers is much larger (Haffer 1978). The composite distribution maps of species groups and the transect profiles published by Haffer (1978) illustrate the varying number of species range boundaries per unit distance in the Amazonian lowlands, i.e. the decrease of numbers of endemic species around centres of endemism. These gradients are certainly independent of the rivers, although displaced or locally modified by the water courses acting as partial barriers.

Many bird species are adapted to riparian swamp forest and thickets on flood plains and avoid the extensive terra firme forest away from major rivers (Remsen and Parker 1983). Examples of species restricted to the immediate vicinity of the Amazon and the lower parts of some of its tributaries include *Myrmoborus lugubris* (Fig. 5.21(c)), *Mymochanes hemileucus*, *Conirostrum margaritae*, *Cranioleuca muelleri*, and *Myrmotherula klagesi*. The eastern allospecies of the antshrikes *Thamnophilus cryptoleucus/ T. nigrocinereus* reaches the upper Rio Orinoco drainage and eastern Colombia (Fig. 5.21(d)).

About 100 species (15 per cent) of the nonaquatic avifauna of the Amazon basin are restricted to habitats created by rivers such as beaches and sandbars, sandbar scrub, river-edge forest, varzea forest,

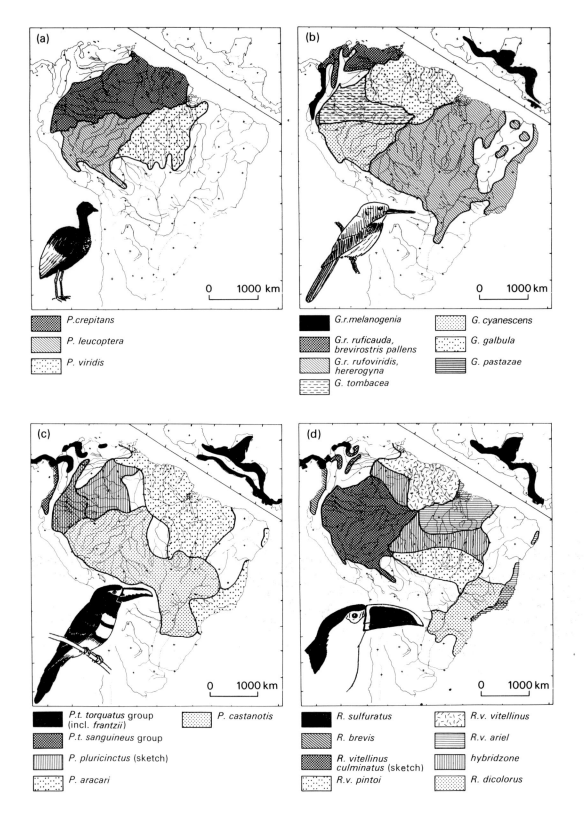

(a)

0 ____ 1000 km

▨	*P.crepitans*
▨	*P. leucoptera*
▨	*P. viridis*

(b)

0 ____ 1000 km

■	*G.r.melanogenia*	▨	*G. cyanescens*
▨	*G.r. ruficauda, brevirostris pallens*	▨	*G. galbula*
▨	*G.r. rufoviridis, hererogyna*	▨	*G. pastazae*
▨	*G. tombacea*		

(c)

0 ____ 1000 km

■	*P.t. torquatus* group (incl. *frantzii*)	▨	*P. castanotis*
▨	*P.t. sanguineus* group		
▨	*P. pluricinctus* (sketch)		
▨	*P. aracari*		

(d)

0 ____ 1000 km

■	*R. sulfuratus*	▨	*R.v. vitellinus*
▨	*R. brevis*	▨	*R.v. ariel*
▨	*R. vitellinus culminatus* (sketch)	▨	hybridzone
▨	*R.v. pintoi*	▨	*R. dicolorus*

transitional forest, and water-edge. Neither the Congo or Mississippi basin avifaunas show such a high per-cent of species restricted to river-created habitats (Remsen and Parker 1983).

Southeastern Brazil

Numerous bird species (*c*. 160) are restricted to the tropical rain forests of the Serra do Mar region in southeastern Brazil and thus make this area an important centre of endemism in tropical South America. Many of the endemic species are so distinct that they are placed in monotypic genera, for ex-ample *Jacamaralcyon tridactyla*, *Baillonius bailloni*, *Batara cinerea*, and *Ilicura militaris*. Müller (1973) proposed to subdivide the Serra do Mar centre, based mainly on the distribution of several forest birds and reptiles, into the Pernambuco, Bahia, and Paulista subcentres.

Despite the distinct character of the southeast Brazilian forest avifauna its rather close relationship with the Amazonian fauna is shown by the numerous species that are shared by both forest regions (30 non-passeriform species and 67 passeriform species; Müller 1973). With only few exceptions all these species have widely disjunct and often subspecifically distinct populations in Amazonia and in the forests of southeastern Brazil.

Trans-Andean forest region

This area comprises the humid lowlands west and north of the Andes, i.e. the Pacific rain forests of Colombia and Ecuador, the humid portions of Caribbean Colombia, the humid middle Magdalena Valley and the forested lowlands of Middle America. A total of 195 endemic species characterize the trans-Andean forest avifauna and between 112 and 60 of these species (45 to 35 per cent of the local avifaunas) occur in any given area. The numerous Amazonian species found west of the Andes (55 to 65 per cent of the local avifaunas) indicate a close relationship of the trans-Andean and cis-Andean bird faunas. These western species populations are either undifferentiated or at most subspecifically distinct. Even among the 195 endemic trans-Andean species many are closely related to their Amazonian rep-

resentatives and are just at, or barely above, the level of species differentiation.

The most distinct, and perhaps oldest, elements of the trans-Andean forest avifauna are 30 species which belong to 22 or 23 endemic trans-Andean genera, 15 or 16 of which are monotypic. Examples are *Rhynchortyx*, *Androdon*, *Hylomanes*, *Clytoct-antes*, *Xenornis*, *Gymnocichla*, *Phaenostictus*, *Sap-ayoa*, *Zarynchus*, *Erythrothlypis* (Haffer 1975, p. 40).

Most of the trans-Andean species (about 60 per cent) are widely distributed suggesting ecological adaptability of these species and/or ecological uni-formity of the forests in the lowlands west of the Andes (Fig. 5.5(a)); 35 of the endemic species occupy most or all of the humid lowlands of Middle America, 54 species inhabit western Colombia and portions of southern Middle America and 27 species range throughout the trans-Andean forest region. Local concentrations of endemic species occur in the following five areas which may be designated as centres of species endemism (Fig. 5.2): Caribbean northern Middle America (7 species), Caribbean southern Middle America (14 species), Pacific south-ern Middle America (12 species), Cauca-Magdalena region of northern Colombia (14 species), and Pacific Colombia into northwestern Ecuador (32 species). This last area, the Chocó centre is of particular importance because of the highly endemic character of its bird fauna. Haffer (1975, pp. 43–55) published lists of the species assigned to each of the trans-Andean centres of endemism. Chapman (1917), Slud (1960), and Howell (1969) also discussed regional ornithogeographical aspects of the trans-Andean avifauna.

The total number of forest bird species decreases gradually along the Middle American isthmus from 280 species in eastern Panama to 156 species in southeastern Mexico (Haffer 1975, p. 31). The reasons for this decrease may include more effective isolation from the Amazon basin in smaller blocks of forest, as well as a regional filter effect of the reduced available land area. The rate of reduction in species number is particularly high in Nicaragua where pine savannas reduce the width of the lowland rain forest belt and in Honduras where steep mountains reach close to the coast. In addition, the Middle American

Fig. 5.10. Distribution of four superspecies of neotropical forest birds. (After Haffer 1974.) I would now prefer to group the five species of (b) into two parapatric superspecies upper Amazonian *Galbula cyanescens/tombacea/pastazae* and *Galbula galbula/ruficauda* of remaining tropical America. I similarly prefer to group the four species of (d) into two parapatric super-species trans-Andean *Ramphastos sulfuratus/brevis* and mainly cis-Andean *R. vitellinus/dicolorus*.

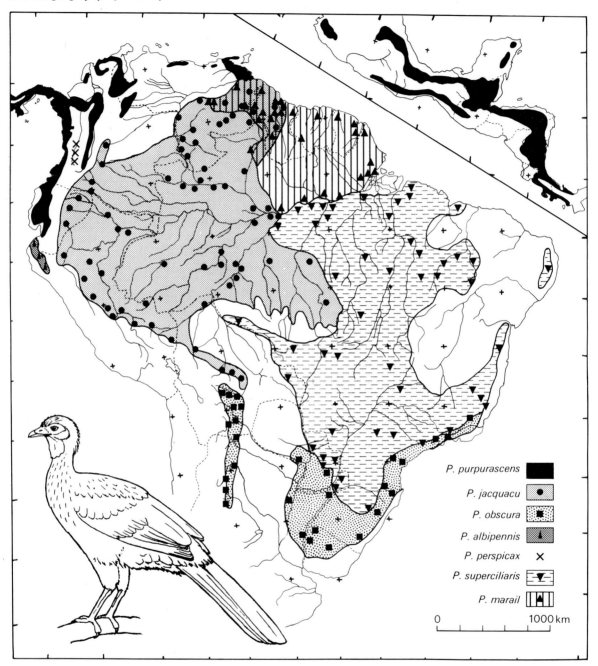

Fig. 5.11. Distribution of the lowland forest guans *Penelope*. (Modified and adapted from Vaurie (1966*a*, *b*).) These dark brown arboreal birds live in pairs or family groups and often feed on the ground. The two eastern species *marail/superciliaris* (hatched) are smaller than the five western and southern species *obscura/jacquaçu/perspicax/purpurascens/albipennis*; each of these two groups forms a superspecies. According to Eley's (1982) analysis the trans-Andean species *purpurascens/albipennis* are the most highly derived ones among the allospecies of the *P. obscura* superspecies. *P. albipennis* was believed extinct until a small population of probably less than 100 birds was rediscovered recently. It is closely related to *P. purpurascens* (Eley 1982, photographs in Dejonghe and Mallet 1978; Williams 1980). Another species of Pacific Colombia and Ecuador is *P.*

forests probably grow progressively less complex structurally and in plant species composition going northward because of climatic reasons. Through the reduction in the number of available niches this situation may lead to the successive northward reduction in the number of forest bird species. However, field-studies are needed to substantiate this assumption. The northward reduction in species numbers of different bird families varies considerably. Species-rich families like the antbirds (Formicariidae) suffer a greater reduction on a percentage basis than families represented by only a few species like the toucans (Ramphastidae) and jacamars (Galbulidae).

Contact zones

Numerous subspecies and allospecies of neotropical forest birds exclude each other geographically, with or without hybridization, along sharp contact zones; some representatives narrowly overlap their ranges. These contact zones are clustered in several areas of the lowlands forming faunal 'suture zones' (a term proposed by Remington 1968) many of which are located between centres of species endemism (Fig. 5.22; see also Haffer 1985, Fig. 4). This is particularly conspicuous in northern Amazonia where no such contact zones occur in the centres of species endemism of the Napo and Guiana regions. A comparatively large number of upper and lower Amazonian species came into contact in central Amazonia. North of the Amazon River, many contact zones are located in the upper Rio Negro and Rio Branco regions as well as in southern Venezuela, i.e. between the Napo and Guiana centres to the west and east, respectively. The broad lower portions of the Negro and Branco Rivers separate on their opposite banks in several cases conspicuously different eastern and western populations. In this way gene flow in hybridizing forms is reduced or ecological competition between allospecies is avoided. The influence of the rivers on the location of contact zones is less conspicuous farther north where these fan out from southeastern Colombia to southern Venezuela. The table mountains of the latter region hindered in various degrees the contact between the

eastern and western populations. South of the middle Amazon River, the contact zones are less concentrated than to the north. The Rio Madeira separates the members of a number of species pairs; again ecological competition is avoided.

Hybrid zones between conspicuously different subspecies of toucans, jacamars, manakins, oropendolas, and other forest birds occur, e.g. in northwestern Colombia (*Pteroglossus torquatus, Galbula ruficauda*) and in central Amazonia (*Ramphastos vitellinus, R. tucanus, Pipra coronata, Psarocolius bifasciatus*) as analysed in detail by Haffer (1967a, 1974).

Other conspicuous faunal suture zones are found in upper Amazonia south of the Río Marañón, i.e. between the Napo and Inambari centres, and in the trans-Andean forest region in northern Honduras, eastern Costa Rica, and northwestern Colombia (Fig. 5.22)

In those species which exclude each other geographically without hybridization it seems very probable that many or all of them would occupy at least parts of the ranges of their respective allies but for the presence of the latter across the zone of contact. It appears inconceivable that the range boundaries of representative species in contact coincide by chance alone. Therefore, I assume that the placement of the mutually exclusive ranges is the result of interaction between the representative species, for example by ecological competition along the contact zone, especially if the species originated in allopatry and are in secondary contact today (see also discussion by Anderson 1977). In areas where climatic gradients are recognized across faunal suture zones or across individual contact zones the gradients are not steep enough to account for the abrupt geographical replacement of members of species pairs. Probably a balanced situation has been reached today with each form at a competitive advantage over its ally in the area occupied. Under the observed ecological conditions, and assuming mutually independent placement of range boundaries, I would expect range overlap, gaps between the ranges as well

ortoni, not shown, which Eley (1982) believed to be related to the *P. marail* superspecies. Notice that all these species are allopatric or parapatric except *P. jacquacu* and *P. marail* (and *pace* the simplified distribution map of *P. superciliaris* published by Delacour and Amadon 1973). Vaurie (1966a, Fig. 1) illustrated a record of *P. jacquacu* just *east* of the mouth of the Rio Madeira (where *P. superciliaris* alone occurs); however, he listed on his page 21 no corresponding locality but only 'Rosarinho, Lago Sampaio' which is located just *west* of the lowermost Rio Madeira suggesting an error on his map. In the Serra do Mar, H. Sick (personal communication) found *P. superciliaris* in the lowlands and *P. obscura* in montane forests; both species occur together at fruiting trees halfway up the slopes. Farther north, in the state of Espírito Santo, *P. obscura* is missing and *P. superciliaris* inhabits the lowlands as well as the cool montane forests. (See note 1, p. 150).

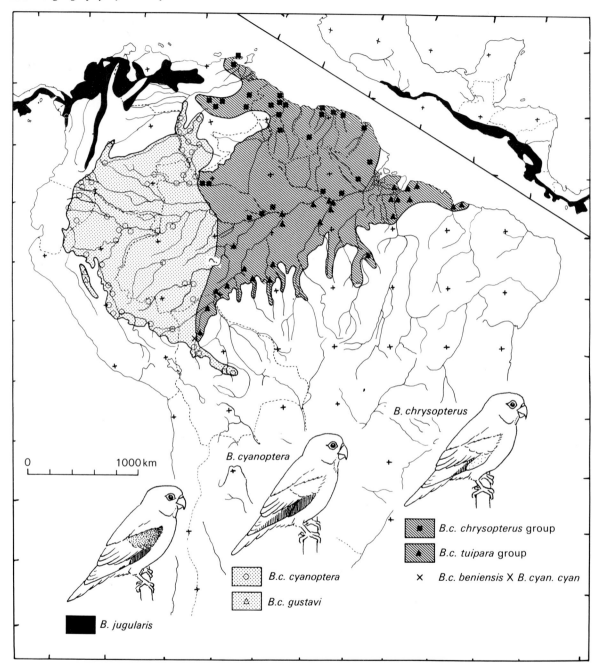

Fig. 5.12. Distribution of the orange-chinned parakeets *Brotogeris*. The two Amazonian species may be combined as the *B. chrysopterus* superspecies; they differ mainly in wing colour which is largely blue in *B. cyanoptera* and green with yellow wing coverts in *B. chrysopterus*. Both species share green body plumage and the orange chin spot with *B. jugularis*, of savanna and open forest in NW South America and Pacific Middle America. *B. c. chrysopterus* group has the primary wing coverts orange, *B. c. tuipara* group has them golden-yellow. The population inhabiting the lower Rio Beni region in NE Bolivia (*beniensis*) is here included with *B. chrysopterus* which hybridizes with *B. cyanoptera* along the middle Beni River (Gyldenstolpe 1945). These species are probably also in contact elsewhere in C. Amazonia.

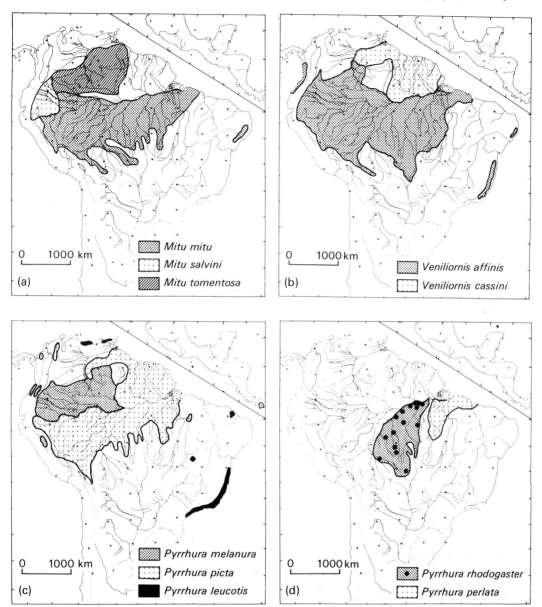

Fig. 5.13. Distribution of several superspecies of neotropical forest birds. (a) *Mitu mitu* superspecies (Cracidae – curassows), adapted from Vaurie (1967); notice that *M. mitu* extends north of the Amazon in SE Colombia (Scheuermann 1977) where it is in contact with the other allospecies. The isolated population of this species in NE Brazil is threatened by extinction (Sick 1980). (b) *Veniliornis affinis* superspecies (Picidae – woodpeckers); the isolated population of the Pacific coast represents *V. affinis* rather than *V. cassini* (see Short 1974); (c) *Pyrrhura picta* superspecies and *P. melanura* (Psittacidae – parrots): isolated populations of *P. picta* exist in NW Venezuela and N Colombia (*subandina, caeruleiceps, pantchenkoi*) and SW Panama (*eisenmanni*), *P. leucotis*, which may actually be conspecific with *P. picta*, inhabits isolated areas in N Venezuela and E Brazil; isolated populations of *P. melanura* exist in S Andean Colombia (*chapmani, pacifica*) and in the Huallaga Valley of E Peru (*berlepschi*); *P. melanura* which lacks red patches on the rump and belly and has red primary wing coverts replaces *P. picta* in NW Amazonia; (d) *Pyrrhura perlata* superspecies (Psittacidae – parrots): *P. perlata* has regionally rather variable plumage coloration.

as exclusion for varying distances along the region of general contact between two representative species.

Unanswered questions in all these cases of contact zones of parapatric species in fairly uniform habitats include 'Which mechanisms assure reproductive isolation of competing species in contact?', 'Are parapatric species prevented from overlapping their ranges through behavioural responses or through resource preemption by the respective partners?', 'Is the location of the contact zones determined mainly by historical or by extant ecological factors?', 'What are the reasons for parapatric species not penetrating each other's range?'. In cases of sympatry the representatives might be expected to maintain interspecific territories or to occupy mutually exclusive patchy areas of varying size. The latter situation has been called 'chequerboard allopatry' for New Guinea forest birds by Diamond (1972, 1973). Detailed field studies of parapatric species pairs along their contact zones would be a valuable contribution towards a better understanding of ecological aspects of the speciation process and of tropical species diversity.

Geographically restricted overlap of the ranges of species within groups of otherwise allopatric or parapatric allies is illustrated by the following examples: the jacamars *Galbula galbula/G. ruficauda* in Guyana and northernmost Brazil (Fig. 5.10(b)), the aracari toucans *Pteroglossus castanotis/P. pluricinctus* in upper Amazonia (Fig. 5.10(c)), the toucans *Ramphastos dicolorus/R. vitellinus* in southeastern Brazil (Fig. 5.10(d)), and the guans *Penelope jacquaçu/P. marail* (Fig. 5.11) in Guyana and southern Venezuela. Haffer (1974, pp. 231, 265, 268, 336) discussed details, as far as known, of the range overlap in these jacamars and toucans. In each of these cases the overlap was probably made possible through differences in size and/or habitat preferences of the members of these species pairs. Comparatively small Marail Guan and large Spix's Guan also may have been able to penetrate part of each other's range because of a fairly conspicuous difference in body size implying different ecological requirements of these species which inhabit the same lowland forest where they are sympatric.

These cases of restricted geographical overlap of species within groups of otherwise parapatric species are of general significance, because they characterize the respective species assemblages as transitional stages between the categories of superspecies and species group. These cases illustrate 'circular overlaps' among closely allied parapatric species. In each

group, the two partially overlapping species are not only reproductively isolated from each other (as the rest of the species are) but, in addition, appear to be ecologically compatible.

The blue-backed manakins (*Chiroxiphia* ssp, Fig. 5.23) inhabit fairly dry woodland, second growth and forest edges with thick undergrowth. They avoid undisturbed primary forests entering wet forests only very locally in Costa Rica (*C. linearis*, *C. lanceolata*) and in southeastern Brazil (*C. caudata*). This explains their fairly spotty distribution, especially in Amazonia, and also the fact that the northern forms are found along the generally dry Caribbean lowlands of Venezuela–Colombia and the Pacific slope of Middle America. The species are sufficiently similar ecologically to be incompatible and to exclude one another geographically, and form a superspecies (Snow 1975). Adult males of three species are black with a blue back whereas in the fourth species (*C. caudata*) the body is mainly blue. All species have a red crown patch except the central Amazonian form (*regina*) of *C. pareola* in which it is yellow. In three species the central tail feathers are variously elongated. As far as known, there is no function for the long rectrices except perhaps to enhance the visual effect of the jumps made in courtship display. Females are inconspicuously coloured and olive green. Snow (1977) analysed and compared the organization of the amazing courtship displays of these birds. Two males (in *caudata* three or more) display together under the command of one dominant bird who, once a female is present, dismisses his subordinate(s) by uttering one or more loud sharp notes prior to the last phase of courtship which leads to copulation.

The four species of *Chiroxiphia* have diverged to varying degrees in the secondary sexual characters and in the first phase of courtship, that is, the duetting calls which advertise the display grounds to females. These calls consist of musical whistles in *linearis* and *lanceolata*, of clicking sounds in *pareola* and of a confused gabbling sound in *caudata*. The later phases of courtship (i.e. alternate jumps and floating flight) are very similar in all species. The duetting calls probably serve as isolating mechanisms, thus preventing hybridization in areas where two species come in contact such as in southeastern Brazil. In this region, *C. pareola* inhabits the forested coastal lowlands south to about the mouth of the Rio Jucu (about 20° 25′ S) in Espírito Santo. D. W. Snow (personal communication) agrees with C. E. Hellmayr

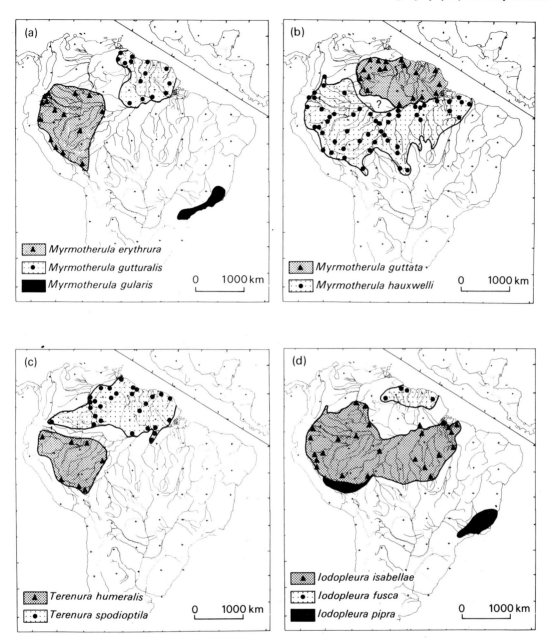

Fig. 5.14. Distribution of four superspecies of neotropical forest birds. (a) Three closely related small antbirds which probably form a superspecies. (b) *Myrmotherula guttata* superspecies. These small antbirds of the forest interior often join mixed species flocks. (c) *Terenura humeralis* and *T. spodioptila* are representative antwrens that also join mixed species flocks in the forest interior. (d) *Iodopleura fusca* superspecies. These small cotingas live in the canopy of open forest and forest edge: the provenances of specimens of *I. pipra* supposedly from Guyana are doubtful and these records are therefore omitted (Haffer 1974, p. 97; Snow 1982, p. 41).

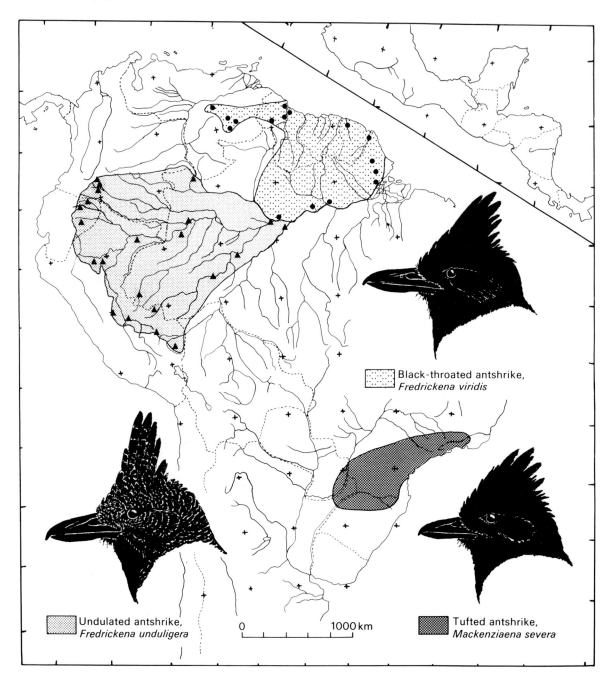

Fig. 5.15. Distribution of three species of antshrikes, genera *Frederickena* and *Mackenziaena* (Formicariidae) which inhabit the understory of the forest interior. The Amazonian genus *Frederickena* probably represents ecologically and taxonomically the two long-tailed species of the SE Brazilian genus *Mackenziaena*.

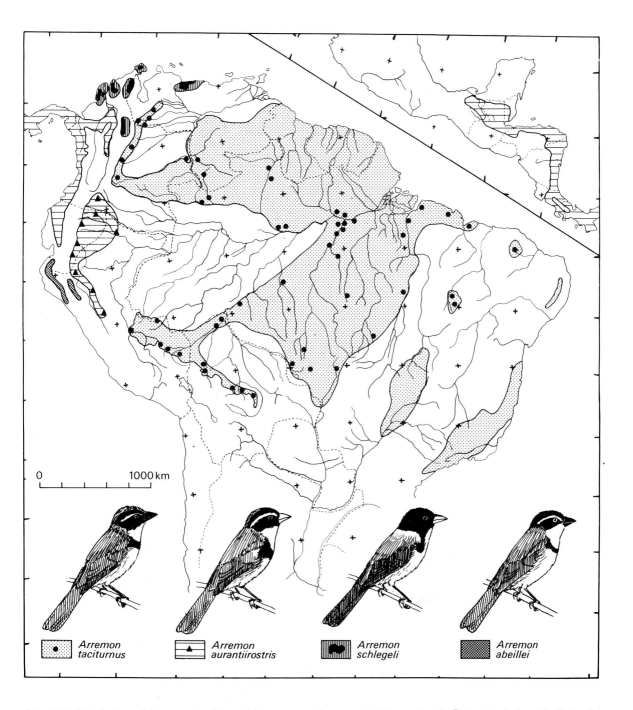

Fig. 5.16. Distribution of four species of terrestrial sparrows, *Arremon*. A fifth species, *A. flavirostris*, is found in E Brazil, Paraguay, Bolivia, and N Argentina not inhabited by *A. taciturnus*. For *A. aurantiirostris* individual records are shown only E of the Andes. (These are from Ecuador: Rio Suno abajo and above Avila, Zamora, San José abajo, Quilano (1150 m), Macas (1050 m) and from NE Peru: boca Rio Curaray, Pomará, Rio Seco (1000 m, 30 miles west of Moyobamba), Guayabamba, and 35 km NE Tingo Maria: J. P. O'Neill, personal communication. For *A. taciturnus*, only selected locality records are shown. (In SE Peru the bird is known from Santa Rosa (upper Rio Ucayali), Balta (O'Neill 1974), La Pampa, Astillero, Rio Tavará (AMNH).) This bird is common in NE and SE Amazonia but is strangely absent from the upper Amazonian lowlands. *A. schlegeli* is a bird of thickets in semihumid regions of the NW.

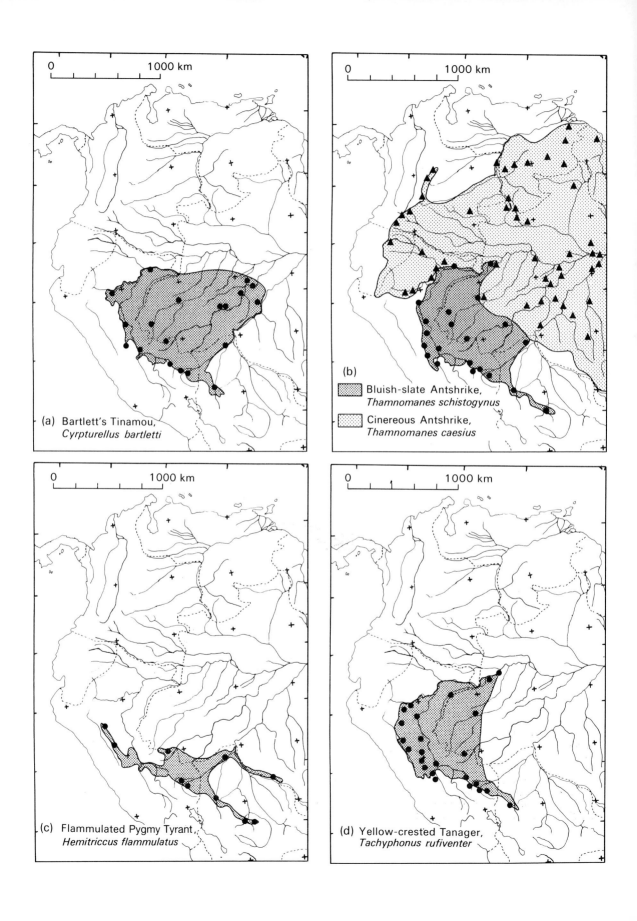

(a) Bartlett's Tinamou,
Cyrpturellus bartletti

(b)

Bluish-slate Antshrike,
Thamnomanes schistogynus

Cinereous Antshrike,
Thamnomanes caesius

(c) Flammulated Pygmy Tyrant,
Hemitriccus flammulatus

(d) Yellow-crested Tanager,
Tachyphonus rufiventer

who doubted that this species ever occurred farther south near 'Nova Friburgo, Rio', a locality record cited by several authorities. *C. caudata* inhabits lowland and hill forests in southeastern Brazil. Where it approaches the range of the lowland species *C. pareola*, *C. caudata* is restricted to the mountains. The specimens of *C. caudata* from the coastal localities Santa Cruz (19°57'S) and Guaraparí (20°40'S), both in Espírito Santo, probably have been collected in the hilly areas inland, although this cannot be proven. H. Sick (personal communication) never found *C. caudata* and *C. pareola* at the same locality in this region of contact between these two species. A similar situation exists in southeastern Peru where *C. p. boliviana* inhabits the lower Andean slope and *C. p. regina* is found in the adjacent lowlands (J. W. Fitzpatrick, personal communication).

The northern species of *Chiroxiphia* probably are not in direct contact. Because of the similarity of their duetting calls, *C. linearis* and *C. lanceolata* might hybridize if they were in contact and may be considered as conspecific. They show a similar divergence from each other as the central and upper Amazonian subspecies *regina*, *napensis*, and *boliviana* do from *pareola*. However, nothing is known about the calls and displays of these Amazonian forms which may, in fact, have reached species status (Snow 1977). *C. caudata*, of eastern Brazil, is less closely related to *C. pareola* and *C. lanceolata/C. linearis* than these species are to one another. Kleinschmidt (1926*a,b*) considered the *Chiroxiphia* forms, except *C. caudata*, as races of one Formenkreis and pointed out that the long-tailed forms *lanceolata* and *linearis* have more rounded wings whereas *C. pareola* with a short tail has more pointed wings.

Non-forest birds

Extensive areas of open non-forest habitats occur in tropical South America to the north of Amazonia in the llanos plains of eastern Colombia to Venezuela and south of Amazonia from northeastern and central Brazil (caatinga and cerrado) to the Argentine pampas. Large portions of the Caribbean and Pacific coastal lowlands of South America, deep mountain

valleys of the northern Andes and the Pacific lowlands of Middle America are also predominantly dry and unforested. Intermediate vegetation types like dense scrub and deciduous woods are often found connecting areas covered with forest and non-forest vegetation whose boundaries are frequently ill-defined. Tongues of open cerrado reach into the Amazon basin. Conversely the forests extend as gallery forests far into the cerrado regions. A great number of isolated savannas and campos cerrados within the Amazon forest region are clustered along the comparatively dry transverse zone of lower Amazonia which extends from central Venezuela to northeastern Brazil (Fig. 2.4, pp. 30–1).

The non-forest fauna of southern central South America inhabits the extensive Brazilian tableland, eastern Bolivia and the dry chaco forest of Paraguay to northern Argentina. Numerous endemic and taxonomically isolated bird species are present in this region (Fig. 5.24), e.g. the South American 'ostrich' *Rhea americana*, the tinamous *Rhynchotus rufescens* and *Nothura minor*, the terrestrial seriema *Cariama cristata*, the ground-dove *Columbina cyanopis*, the woodpecker *Dendrocopus mixtus*, the campo miner *Geobates poecilopterus*, *Pseudoseisura cristata*, the helmeted manakin *Antilophia galeata*, the jay *Uroleuca cristatella*, the tanager *Cypsnagra hirundinacea*, and the finches *Saltator atricollis* and *Charitospiza eucosma* (Sick 1965, 1966).

At least two centres of species endemism may be recognized in this non-forest region (Figs. 5.24 and 5.25) based on the discussions and range maps published by Short (1975) and Fitzpatrick (1980): northeastern Brazil on the one hand and Paraguay to northern Argentina on the other hand. Contact zones, with and without hybridization or with range overlap of the respective geographical representatives, are clustered in the intervening area between these two centres (Figs. 5.22 and 5.25). The northeastern chaco is of particular importance in this respect as a large number of species and well-defined subspecies are here in contact. Short (1975) illustrated and briefly discussed the following additional examples of species and well differentiated subspecies that are in contact in the border region of Bolivia–Paraguay and Brazil. The northern or eastern form is listed

Fig. 5.17. Distribution of bird species restricted to SW Amazonia. The two antshrikes (b) are parapatric and meet near São Paulo de Olivença and in the valley of the upper Rio Juruá. (c) This bird is related to *H. diops* of SE Brazil. (d) A series of 19 specimens from São Paulo de Olivença (CM) and opposite Tonantins (CM) document the occurrence of this species along the Solimões River.

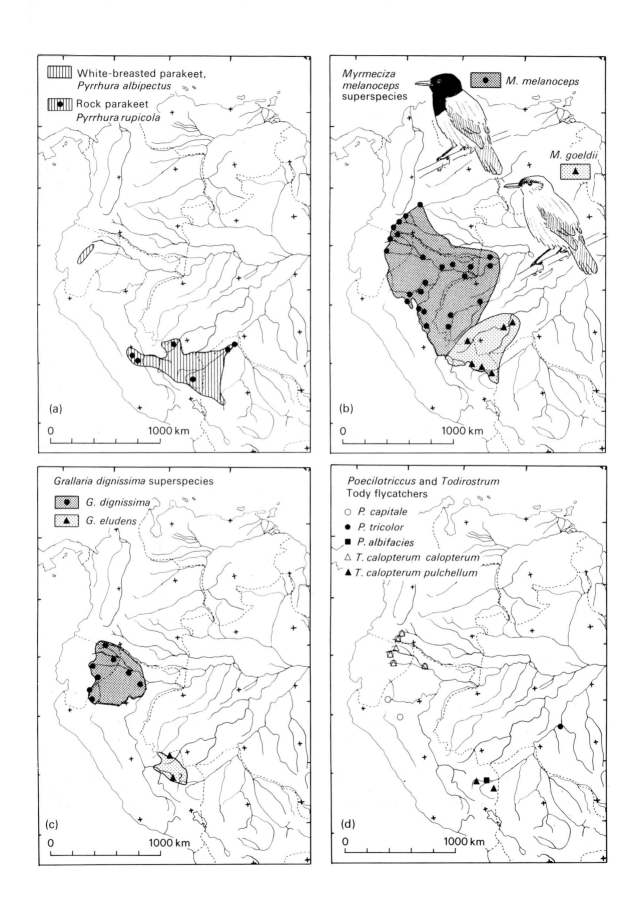

(a)

White-breasted parakeet, *Pyrrhura albipectus*

Rock parakeet *Pyrrhura rupicola*

0   1000 km

(b)

Myrmeciza melanoceps superspecies

M. melanoceps

M. goeldii

0   1000 km

(c)

Grallaria dignissima superspecies

G. dignissima

G. eludens

0   1000 km

(d)

Poecilotriccus and *Todirostrum* Tody flycatchers

○ *P. capitale*

● *P. tricolor*

■ *P. albifacies*

△ *T. calopterum calopterum*

▲ *T. calopterum pulchellum*

0   1000 km

first: the parrots *Ara maracana/A. auricollis*, the parakeets *Pyrrhura devillei/P. frontalis*, the spot-backed puffbird *Nystalus maculatus* (*maculatus/striatipectus* subspecies groups), the checkered woodpecker *Picoides mixtus* (*cancellatus/mixtus* subspecies groups), the campo flicker *Colaptes campestris* (*campestris/campestroides*), the flycatchers *Suiriri affinis/S. suiriri*, the jays *Cyanocorax cyanopogon/ C. chrysops*, the ultramarine grosbeak *Cyanocompsa cyanea** (*sterea/argentina*), the saffron finch *Sicalis flaveola* (*brasiliensis/pelzelni*). Several strongly differentiated species inhabit the region of the campos cerrados, e.g. *Cypsnagra hirundinacea*, *Neothraupis fasciata*, *Porphyrospiza caerulescens*, and *Charitospiza eucosma* indicating the existence of another area of endemism in the central portion of the extensive Brazilian tableland (Sick 1966; Müller 1973). The campos cerrados region of central Brazil represents a centre of endemism as discussed and illustrated by Müller (1973), Cracraft (1985a), and Haffer (1985).

The Colombian–Venezuelan non-forest avifauna is closely related to the Brazilian non-forest avifauna because many forms north of the Amazon forest are differentiated only subspecifically from the respective species populations of central Brazil. Examples are the flycatchers *Machetornis rixosa*, *Idioptilon margaritaceiventer*, *Contopus cinereus*, *Fluvicola pica*, *Xenopsaris albinucha*, the tanager *Thraupis sayaca*, the finches *Sporophila plumbea*, *Sicalis flaveola*, *Coryphospingus pileatus*. The number of endemic species in the Colombian–Venezuelan non-forest avifauna is comparatively small. Moreover, some of these are merely strongly differentiated representatives of Brazilian species forming together superspecies, e.g. *Columba corensis/C. picazuro*. Endemic genera are lacking unless a monotypic genus *Hypnelus* is recognized for the puffbird *Bucco ruficollis*.

In a distributional analysis of 101 bird species of the Colombian–Venezuelan non-forest fauna (excluding water and swamp birds) Haffer (1967b) found 11 species to be restricted to the lowlands east of the Andes, 33 to reach the Caribbean lowlands of northern Colombia and 57 species also to occur in the valleys of the Río Magdalena or the Río Cauca, or both. Thirty-five endemic forms characteristic of these partly dry inter-Andean valleys have reached the subspecies level and some are markedly different morphologically. Forty-six species occur also on the dry Pacific slope of Middle America.

The isolated savannas of the dry transverse zone in lower Amazonia are inhabited by either northern or southern non-forest birds or both. Some of these populations are sufficiently well differentiated to be recognized as subspecies or, in one instance, as a species: *Aratinga pertinax* **chrysophrys*** and *A. p.* **paraensis**, *A. s.* **solstitialis**, *Ara n.* **nobilis**, *Speotyto cunicularia* **minor**, *Chordeiles pusillus* **septentrionalis**, *Poecilurus* **kollari**, *Euscarthmus rufomarginatus* **savannophilus**, *Mimus saturninus* **saturninus**, *Sicalis flaveola* **columbiana**, and *S. f.* **goeldii**, *Coryphaspiza melanotis* **marajoara** (see complete lists given by Haffer 1974, pp. 111–12).

The arid Pacific lowlands of western Ecuador and Peru are inhabited by a very distinctive avifauna that has been analysed in detail by Chapman (1926). Three points are evident. First, the total number of species and genera is small; secondly, the percentage of endemic genera and species is high, and thirdly the relations of the Pacific forms to the Colombian–Venezuelan non-forest bird fauna are obvious. Among the birds with close relatives in non-forested Brazil are the following: the jay *Cyanocorax mystacalis*, the furnariids *Synallaxis tithys* and *Furnarius leucopus*, the tapaculo *Melanopareia elegans*, the antbird

*Valid name now is *C. brissonii* (Lichtenstein 1823); see Eisenmann (1979). *Auk* 96, 766.

*Bold face type indicates the endemic element.

Fig. 5.18. Distribution of upper Amazonian superspecies and species of birds which occur as geographically representative populations in the Napo region of E Ecuador and in the Inambari region of SE Peru. The two *Pyrrhura* species (a) inhabit parts of the Andean foothills; head and wing coloration is similar to those of the lowland parakeet *P. melanura* (Fig. 5.13). (b) The males of these fairly large antbirds are black with (*M. goeldii*) or without (*M. melanoceps*) concealed white mantle feathers; the females (illustrated) are more conspicuously different; these species inhabit the understory of the forest interior and often follow swarms of army ants; in recent years, *M. goeldii* has been found in SE Peru at several localities (Balta, O'Neill 1974; Manu, Collpa, boca Colorado, all FM; Tambopata, Parker 1982). (c) These ant-pittas are secretive birds of the forest floor. *G. eludens* is known from the type locality (Balta) and was questionably recorded at Manu National Park (J. Terborgh *et al.* 1984). (d) Fitzpatrick (1976) had synonymized *Todirostrum albifacies* with *T. tricolor* and considered this combined taxon as conspecific with *T. capitale* of E Ecuador. On the basis of additional specimens and field work he now prefers to list these three representatives of E Ecuador (*capitale*), SE Peru (*albifacies*) and W Brazil (*tricolor*) as allospecies of one superspecies (J. W. Fitzpatrick, personal communication) which he refers generically to *Poecilotriccus*.

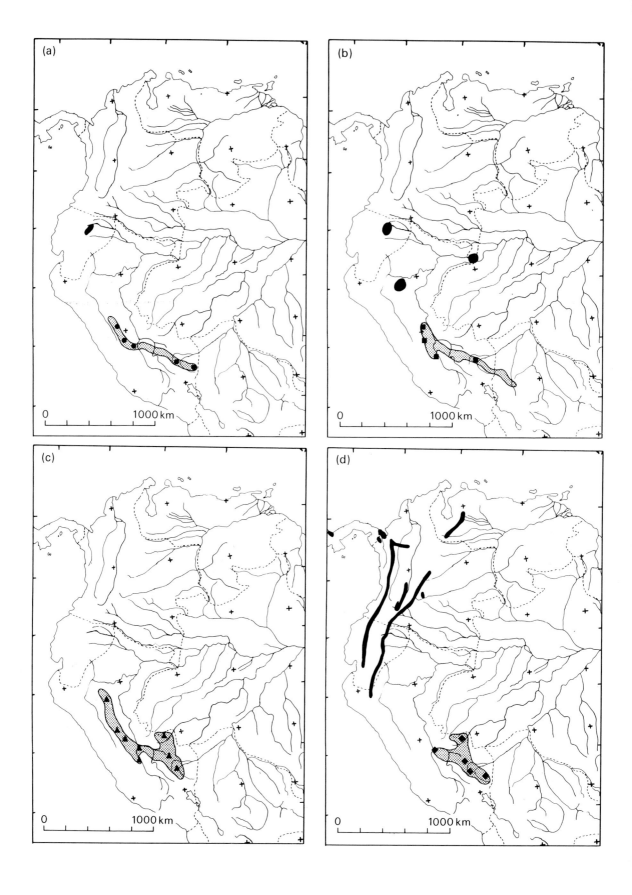

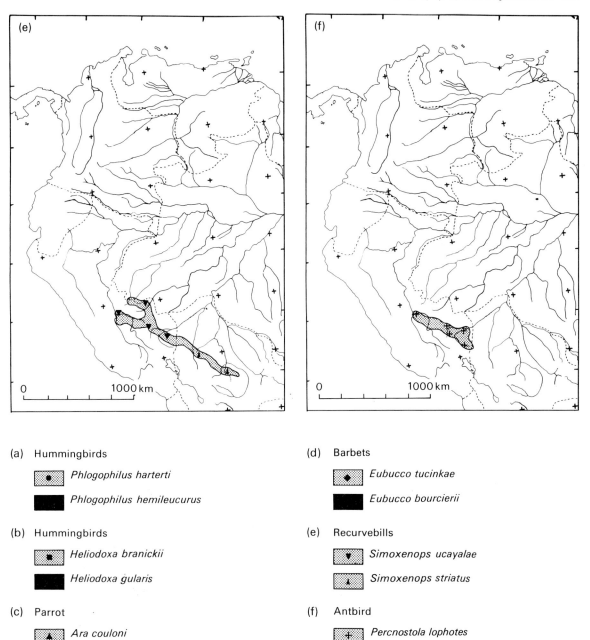

(a) Hummingbirds

    Phlogophilus harterti

    Phlogophilus hemileucurus

(b) Hummingbirds

    Heliodoxa branickii

    Heliodoxa gularis

(c) Parrot

    Ara couloni

(d) Barbets

    Eubucco tucinkae

    Eubucco bourcierii

(e) Recurvebills

    Simoxenops ucayalae

    Simoxenops striatus

(f) Antbird

    Percnostola lophotes

Fig. 5.19. Distribution of six bird species restricted to the Andean foothills in SE Peru. Notice that some species range to the hills along the Brazilian border (see O'Neill 1974). The barbet (d) has a representative, *E. bourcierii*, at lower montane levels of the Andes and of the Middle American highlands. (e) An unlabelled Brazilian specimen of *S. ucayalae* probably came from the border region of the State of Acre (Novaes 1978). (f) Recent localities are Manu, Hacienda Villacarmen, and Collpa (all FM); includes *P. 'macrolopha* Berlioz, 1966' as it is the male of *P. lophotes* (see Parker 1982).

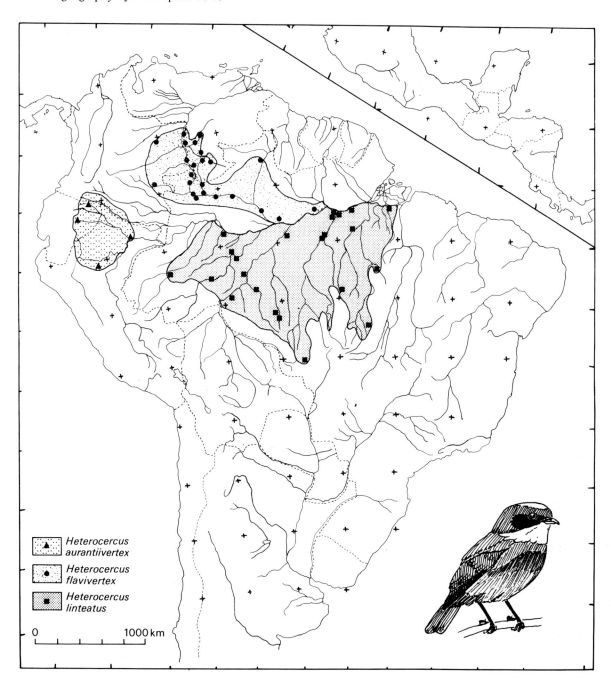

Fig. 5.20. Distribution of the *Heterocercus linteatus* superspecies. These manakins inhabit open forest, campinas, and second growth. The female specimen from Puerto Indiana on the north bank of the Rio Solimões is here referred to *H. aurantiivertex* on geographical grounds, a possibility also discussed by Zimmer (1936). The two northern forms may not be specifically distinct.

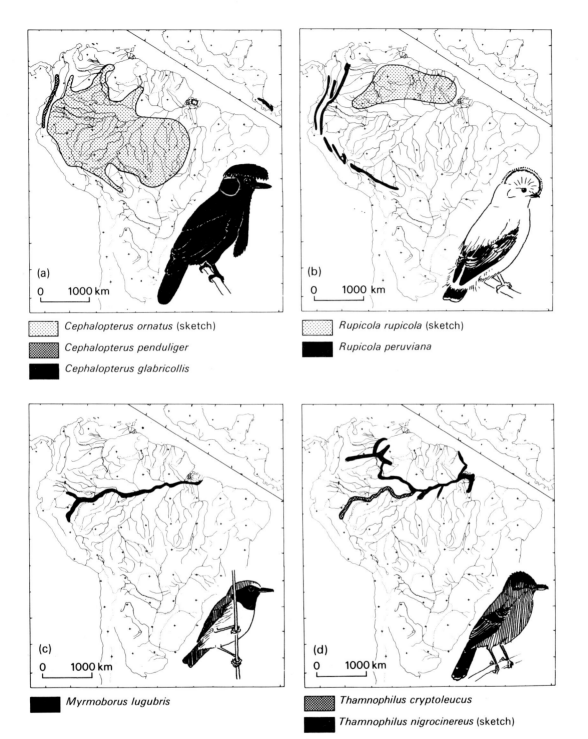

Cephalopterus ornatus (sketch)

Cephalopterus penduliger

Cephalopterus glabricollis

Rupicola rupicola (sketch)

Rupicola peruviana

Myrmoborus lugubris

Thamnophilus cryptoleucus

Thamnophilus nigrocinereus (sketch)

Fig. 5.21. Distribution of selected neotropical forest birds. *Cephalopterus ornatus* superspecies, Umbrella birds (a). These three species differ mainly in the development of the wattle; the two western species inhabit montane forests in contrast to the Amazonian bird. (b) *Rupicola rupicola* superspecies, cocks-of-the-rock: *R. peruviana* inhabits Andean montane forests. Both species require rock exposures and caves to place their nests; this explains why the range of *R. rupicola* is restricted to hilly and mountainous regions of the Guiana Shield. (c) This antbird is found in floodplain and island forests along the Amazon River. (d) These two antshrikes may be conspecific, however, no intergradation is known. They inhabit periodically inundated flood plain forest.

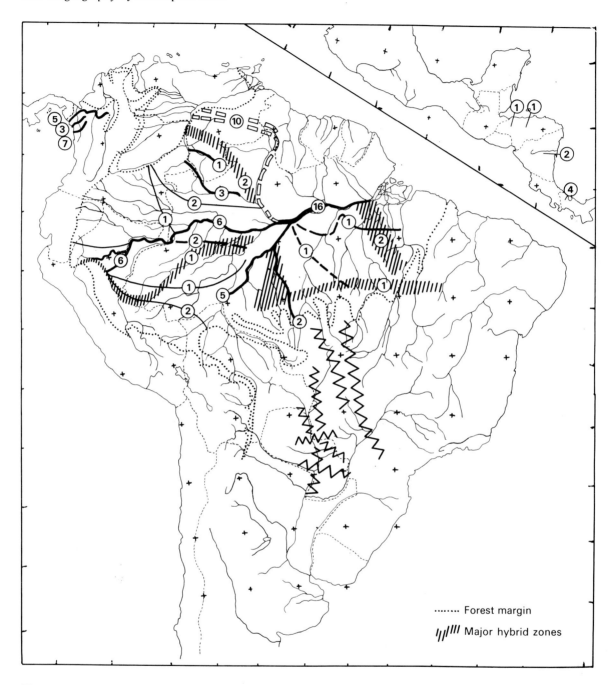

Fig. 5.22. Location of contact zones between parapatric species and hybridizing subspecies of neotropical birds. Circled numerals indicate the number of species or subspecies pairs in contact. Heavy lines and hollow dashes indicate contact zones of parapatric forest species (hybridization is unknown in these cases); zones of hybridization are shown hatched. (After Haffer 1974). Zigzag lines indicate contact zones between non-forest birds in south central South America. (After Short 1975.) Note that contact zones are clustered especially in north central Amazonia and in Paraguay and adjacent Brazil and are absent near the Andes in NW Amazonia and in the Guianan region.

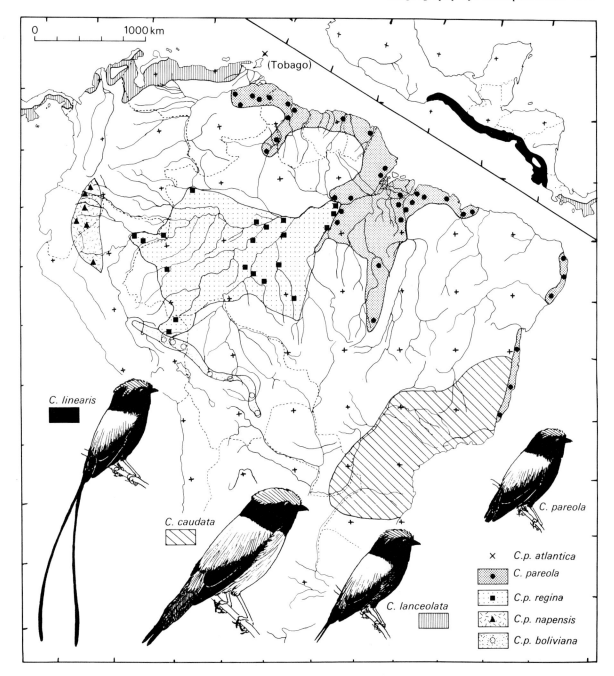

Fig. 5.23. Distribution of the blue-backed manakins, *Chiroxiphia pareola* superspecies. *C. p. regina* has a yellow instead of red crown patch; *C. p. napensis* and *C. p. boliviana* are similar to *C. p. pareola* differing only slightly in coloration and more markedly in measurements. Sketches illustrate adult males. Several unpublished locality records of *C. p. regina* and *C. p. boliviana* have been furnished by J. W. Fitzpatrick (personal communication). The record from Loretoyacu on the Colombian Amazon rests on an immature male (Hellmayr, *Cat. Birds Americas*, pt. 6, p. 57, footnote 2) which, in view of recent records of *C. p. regina* from the Peruvian Amazon further west, is here referred to this form.

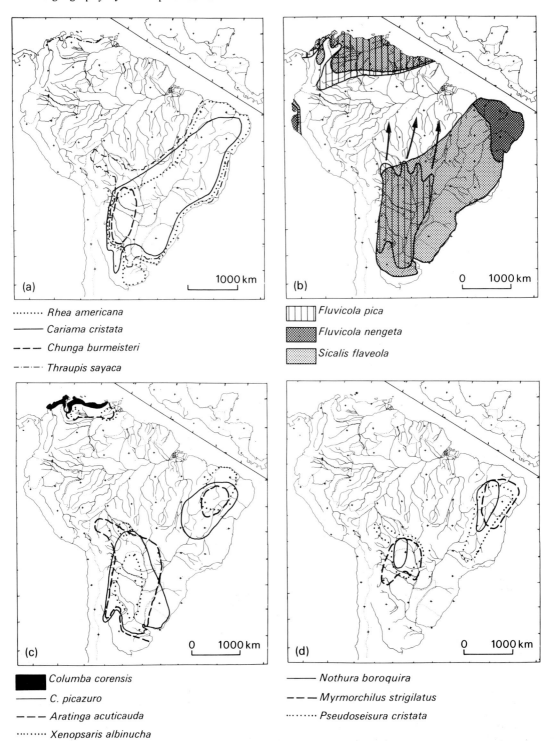

.... *Rhea americana*

——— *Cariama cristata*

— — — *Chunga burmeisteri*

—·—·— *Thraupis sayaca*

||||| *Fluvicola pica*

Fluvicola nengeta

Sicalis flaveola

■ *Columba corensis*

——— *C. picazuro*

— — — *Aratinga acuticauda*

.... *Xenopsaris albinucha*

——— *Nothura boroquira*

— — — *Myrmorchilus strigilatus*

.... *Pseudoseisura cristata*

Fig. 5.24. Distribution of some non-forest birds in South America. Notice wide disjunction of several conspecific populations. (Based on Short (1975).)

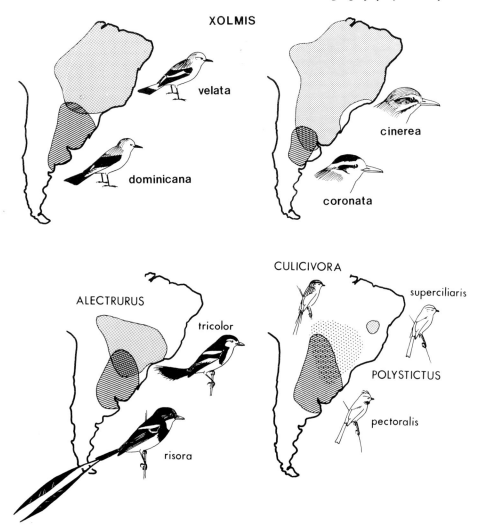

Fig. 5.25. Distribution of four species pairs inhabiting open vegetation in southern South America. Range overlap in three pairs is interpreted as secondary contact. Genera *Culicivora* (one species, *C. caudacuta*) and *Polystictus* (two species) are closely related. *Culicivora*, the result of an earlier speciation, now separates the two latter species and may actually be preventing their contact in south central Brazil. (After Fitzpatrick 1980; see Short 1975, for additional notes on these species.) Isolated populations of *Xolmis cinerea* and *Polystictus pectoralis* occur north of the area shown.

Sakesphorus bernardi, the flycatchers *Euscarthmus melorhyphus* and *Fluvicola nengeta* (Fig. 5.24), the dove *Columbigallina talpacoti buckleyi* and the finch *Sicalis flaveola valida* (Fig. 5.24). A number of widely separated dry pockets in deeply incised valleys in the eastern foothills of the Peruvian Andes form a discontinuous connection between the arid region of the Pacific coast and the dry upper Marañón valley (Dorst 1957) plus the Bolivian–Brazilian non-forested

area. For example, the lower Urubamba valley of eastern Peru, located halfway between the Río Marañón and the Bolivian savanna, is treeless and arid. Its distinctive avifauna ' . . . has evidently been derived through western Brazil . . .' (Chapman 1921, p. 28). Out of a total of 66 species in the Urubamba valley 38 are generally distributed in the tropics; of the remaining 28 species no fewer than 19 are of Brazilian origin (Chapman 1921).

As mentioned above, the taxonomic composition of the neotropical non-forest avifauna is widely different from that of the forest. Certain groups of spinetails and horneros (Furnariidae), tyrant-flycatchers (Tyrannidae), tanagers (Thraupidae), and finches (Fringillidae) are most conspicuous among the birds of open habitats. Only few superspecies or species groups of neotropical birds have been mapped which are composed of both forest and non-forest allospecies or closely related species (e.g. *Phrygilus gayi* superspecies, *Phalcobaenus megalopterus* superspecies: Vuilleumier 1981). This contrasts with the situation in the African bird fauna where representatives of about 9 per cent of the superspecies and species groups are restricted mainly to the Congo forest on the one hand and to the surrounding open vegetation on the other (Hall and Moreau 1970).

Sick (1966) suggested some relations between the forest and non-forest avifaunas of South America by listing several vicariant bird species or ecological representatives of the Brazilian cerrado and of the humid forest as follows (the species of the cerrado habitat is listed first): *Crypturellus parvirostris/C.*
tataupa, *Amazona xanthops/A. aestiva*, *Ramphastos toco/R. vitellinus*, *Lepidocolaptes angustirostris/L. albolineatus*, *Xiphocolaptes major/X. promeropirhynchus*, *Synallaxis albescens/S. scutata*, *Thamnophilus torquatus/T. amazonicus*, *Turdus amaurochalinus/T. fumigatus*, *Saltator atricollis/S. similis*. Of these, the toucans *Ramphastos toco* and *R. vitellinus* are members of the channel-keel-billed species group within the genus *Ramphastos* and in fact are only fairly remotely related (Haffer 1974). The taxonomic relations of the other pairs have not been studied. A possible mechanism for ecogeographical speciation ('the vanishing refuge') leading to habitat shift from forest to non-forest habitat preference or vice versa was suggested by Vanzolini and Williams (1981) who also emphasized recently the need for studies of the differentiation process in geographically isolated populations (Vanzolini 1981).

The weak relations between the neotropical forest and non-forest avifaunas at the species and superspecies level favour the interpretation of a rather early separation of the respective member groups, possibly during the middle or late Tertiary.

ORNITHOGEOGRAPHICAL ASPECTS OF THE NEOTROPICAL MONTANE AVIFAUNAS

Widely distributed New World bird families are most conspicuous in the montane avifaunas of the Andes, for instance hummingbirds, flycatchers, and tanagers. By contrast, the number of species of strictly neotropical families is fairly small. We find few toucans, jacamars, puffbirds, or antbirds, families which furnish many characteristic members of the tropical lowland avifauna. The number of species of North American and Old World families increases rapidly in the mountains of Middle America. The discontinuity of the mountains in northwestern South America and southern Central America may have effectively hindered faunal exchange and thereby prevented the southward immigration of additional northern montane birds. These are poorly represented in the Andes and vice versa.

Andes

Many bird species of the montane rain forests have altitudinally narrow, yet geographically extensive,
ranges because of the ecological uniformity of their habitats along the slopes of the Andes (Chapman 1917, 1926). Species of high Andean vegetation zones, especially above the timber line, have varyingly patchy distributions (Vuilleumier 1970; Vuilleumier and Simberloff 1980). Superspecies are common in both the montane forest and non-forest avifaunas. Habitat discontinuities subdivide or delimit species ranges and in many cases coincide with ecogeographical barriers such as deep valleys or generally lower mountain ranges. In other instances allospecies are in contact without hybridization and no barriers are immediately obvious.

Terborgh (1971) and Terborgh and Weske (1975) adduced strong arguments supporting the significance of direct and diffuse competition in delimiting vertical distributional ranges in bird species along the slopes of the Peruvian Andes. They showed that a large percentage of montane species expand their elevational ranges on isolated mountain massifs (e.g. Cerros del Sira) where many high Andean allies are missing. Among the authors who have studied various biogeographical aspects of the Andean montane avifauna are Chapman (1917, 1926), Vuilleumier (1969, 1970, 1971, 1972, 1980, 1985), Paynter (1972, 1978), Fitzpatrick (1973), Vuilleumier and Simberloff (1980), Graves (1982, 1985), Remsen (1984), and Traylor (1985). The recent studies emphasize the avifaunas of the paramo and puna zones above the timber line. Remsen (1984) and Graves (1985) investigate aspects of the geographic variation of montane birds inhabiting the Andean forest. However, there is no modern biogeographical account of the avifauna of the Andean montane forest zone and its taxonomic and historical relations with the avifauna of the tropical lowland forests. Müller (1973) and Cracraft (1985a) identified and discussed a number of areas of endemism along the Andes mountains which will form the basis of future historical interpretations of the montane avifaunas. Sick (1985) analysed some aspects of the relations between the avifaunas of the SE Brazilian mountains and of the southern Andes.

Northern Venezuelan mountains

The size and ecological variation of the semi-isolated mountains along the Caribbean coast of northern Venezuela decrease progressively to the east. The number of montane bird species decreases correspondingly in going from one cordillera to the next mountain range eastwards, e.g. Mérida Andes 251 species, Cordillera de Caracas 133 species, Cordillera de Caripe 64 species and Cordillera de Paria 33 species (Phelps 1968). About half of the montane birds are differentiated at the subspecies level and 11 per cent of the species are endemic (17 forms) in northern Venezuela. Two species are restricted to each of the Caracas, Caripe, and Paria mountains. Some of these are members of widely distributed superspecies.

Pantepui

The distinctive montane avifauna of the Guiana Highlands in the border region of Venezuela, Brazil, and the Guianas inhabits forests along the steep slopes below the vertical cliffs that delimit the huge mesa mountains (tepuis). This montane bird fauna consists of 96 species (Mayr and Phelps 1967). Thirty per cent are endemic and 57 per cent are subspecifically differentiated; only 13 per cent are undifferentiated. About 30 per cent of the species in the avifauna are related to, and probably derived from, members of the surrounding lowland fauna. About half the montane birds are related to the montane avifauna of the Andes and of the Venezuelan cordilleras (Mayr and Phelps 1967). The flora has no such close relationship (Chapter 3). The number of bird species on individual isolated tepui mountains is independent of the size (area) of the mountain, but is highly positively correlated with its elevation (Cook 1974).

Southern Central America

The mountain massif of Costa Rica and western Panama is strongly isolated to the north and south by the lowlands of Nicaragua and central Panama, respectively. This is reflected by the comparatively high percentage of endemic bird species (35 per cent) and subspecies (56 per cent) (Haffer 1974, p. 117). Of the montane birds, 64 per cent are probably of South American origin. These are mostly species of the lower mountain slopes, as typical faunal elements of the upper montane levels including the treeless páramo zone above 3000 m have not reached Middle America. The South American influence is much reduced in the mountains of Honduras, where almost half the species are typically Central American and North American species (Monroe 1968).

DISCUSSION

In discussing the determinants of the distribution patterns of organisms recent authors (e.g. MacArthur 1972; Vuilleumier 1975; Croizat 1976) have emphasized either the importance of past events (geological factors) or of current ecological factors, such as interspecific competition or the diversity of the environment. On the basis of the Pleistocene climatic-vegetational shifts that have affected the neotropical lowlands (Chapters 1 and 2) and on the basis of the theory of allopatric speciation one would predict some clustering of endemic species and well-defined subspecies of animals and plants in areas of the ecological refugia during adverse climatic periods. Secondary contact zones of differentiated taxa with and without hybridization are expected to occur in areas between former refugia. In these regions populations are assumed to have established contact after spreading from their respective refugia during favourable climatic-vegetational periods.

The available ornithogeographical data of the Neotropical Region summarized in this chapter tend to corroborate the above predictions. Firstly, clusters of endemic species and subspecies do form centres of endemism in superficially uniform habitats, such as the vast tropical lowland forest of Amazonia. Secondly, numerous hybrid zones in that region, if interpreted as secondary contact zones, can be taken to indicate the former existence of ecological barriers which have since disappeared. Maps which show consistent and coterminous distribution patterns in diverse bird families that differ widely in their feeding preferences have been presented in this chapter, namely omnivorous toucans and trumpeters, insectivorous puffbirds, jacamars, woodpeckers, antbirds, flycatchers and icterids, frugivorous guans, parrots, cotingas, manakins, and tanagers. As to the high frequency of parapatric species in Amazonia, we may speculate that fairly rapid morphological–behavioural differentiation of avian refuge populations occurred under conditions of reduced population size and interrupted gene flow. At the same time, the ecological requirements of the refuge populations may have remained rather unchanged, presumably because ecological conditions in the forest refugia remained similar to those existing in the original continuous habitat. The local environmental conditions in the refugia were probably more or less constant through time, thereby possibly slowing the ecological deviation of many avian populations isolated in them. Presumably parapatric species acquired reproductive isolation in these refugia but did not reach ecologic compatibility. Now each partner is competitively superior in the area occupied.

Haffer (1967, 1969, 1970*b*, 1974, 1982, 1985) emphasized the significance of geological-historical factors for the establishment of biogeographical patterns in the neotropics. Similar views are held by Müller (1973) and Cracraft (1985*a*) for the neotropical fauna generally and by Vuilleumier (1969, 1980) and Vuilleumier and Simberloff (1980) for the avifauna of the high Andes, by Short (1975, 1980) for that of the chaco region and for South American woodpeckers, by Dorst (1976) for the whole neotropical avifauna, by Fitzpatrick (1976, 1980) for the avifaunas of Amazonia and southern Latin America. In the case of the tyrant-flycatchers, which he studied in detail, Fitzpatrick (1976) considered that competitively

superior species have spread from refugia into the ranges of competitively inferior allied species, leading to their extinction or the subdivision or reduction of their range, possibly followed by differentiation of some of these isolated populations. Several species he discussed are either very rare, or restricted to low or middle elevations on forested mountains in peripheral portions of Amazonia and, in addition, show some morphological peculiarities, as compared with the remainder of the group. According to Fitzpatrick, these are the characteristics of relict species groups, showing the final stages in an Amazonian speciation-extinction cycle. (See note 2, p. 150.)

Several authors of general zoogeographical works earlier indicated the importance of the repeated appearance and disappearance of vegetational barriers for the speciation and adaptive radiation in many groups of tropical organisms during the Pleistocene (Mayr 1942, 1963; Darlington 1957; Udvardy 1969). The theory of speciation in isolated ecological refugia during the Quaternary was developed during the early part of this century by biogeographers studying the biotas of the north Temperate Zone (review by Haffer 1982) and was later applied to the biotas of Australia (Keast 1961, 1974), Africa (Carcasson 1964; Moreau 1966; Hall and Moreau 1970; Kingdon 1971; Hamilton 1976; Diamond and Hamilton 1980; Crowe and Crowe 1982) and the Neotropical Region.

The general notion of climatic-vegetational fluctuations leading to fragmentation of species ranges and differentiation of isolated populations in ecological refuges has no time connotation. Ecological vicariance probably led to the formation of refuges during various periods of the earth's history. However, the fluctuations of the Quaternary certainly were more extensive than climatic shifts during the preceding Tertiary and Cretaceous periods. The concept of vicariance on the basis of vegetational fluctuations ('refuges') supplements vicariance due to tectonic-palaeogeographical processes and supports the interpretation that biotic differentiation at the species and subspecies levels continued, not only in higher latitudes but also in the tropics, from the Tertiary into the Pleistocene. However, the refuge concept does not propose to explain all of the present biotic diversity of the tropical faunas. It will be the task of future studies to place the biogeographic-evolutionary effects of the Quaternary events in proper perspective by analysing the older patterns of differentiation pertaining to the long time span of the Tertiary period (60 million years) when the ancestors of most modern avian groups evolved and a number of extant species (especially in non-passerine families) originated through palaeogeographical changes in the distribution of land and sea as well as climatic–vegetational fluctuations. Therefore the early history of some or many extant centres of endemism probably dates back to Tertiary, as emphasized by Simpson and Haffer (1978). The latter point was also stressed by Cracraft (1985a) who proposed that the areas of endemism may be considerably older than the Pleistocene and stated that the physiographic evolution of South America was probably as crucial in forming patterns of avian differentiation as were habitat changes caused by cyclical Pleistocene climates. There is no disagreement between Cracraft's view and that expressed above (also Haffer 1985) especially if 'avian differentiation' implies evolution at all taxonomic levels from subspecies and species to genera and families. The concept of Pleistocene refugia is mainly concerned with taxonomically low level and geologically rather recent patterns of differentiation and, therefore, refers primarily to the latest chapter in the history of the neotropical biotas.

An alternative theory contrasting with the above historical interpretation relates the observed biogeographical patterns of the neotropical biotas to the current ecological diversity and corresponding differentiation patterns of the organisms. Thus Endler (1977, 1982) suggested that biotic differentiation across environmental gradients may lead to biogeographical patterns identical to those of coalesced refugia. In particular, he stated that partial barriers or ecological transition zones may produce 'stepped clines' and 'hybrid zones' which are indistinguishable from zones of secondary intergradation. This situation is said to lead frequently to parapatric speciation (which occurs whenever species evolve as contiguous populations in a continuous cline; Bush 1975). It remains to be confirmed that ecological gradients in the Neotropical Region are sufficiently pronounced to produce, through clinal variation, not only in insects but also in birds narrow zones of high variability ('hybrid zones') without prior separation of the populations concerned. Furthermore, different taxonomic groups will respond differentially to environmental gradients and to the ecological diversity of the environment. In assessing the theoretical plausibility and available evidence for non-allopatric speciation, Futuyama and Mayer (1980, pp. 254 and 269) concluded that 'neither the theory nor the evidence are sufficient to justify a major departure from the traditional view: . . . speciation does indeed occur most often by the divergence of population confined to small geographic areas'.

The ecological diversity of the Neotropical Region is more complex than currently acknowledged and may eventually be shown to determine some of the biogeographical patterns observed in birds. However, the basic pattern, which consists of (i) the clustering of endemic species and subspecies in certain restricted regions of, e.g. Amazonia; (ii) the existence of widely disjunct populations of representative species with poor dispersal capabilities; (iii) the occurrence of terrestrial non forest animals on the isolated savanna enclaves in Amazonia; and (iv) the occurrence of conspicuous hybrid zones, appears to be more consistent with an influence of the changing palaeogeographical and climatic–vegetational history of the Neotropical Region than with other causal mechanisms proposed. Ecological aspects of the environment and the community (interspecific competition) probably have been important in determining the details of distribution areas of avian species, especially the placement of distributional limits, and patterns of abundance. Similarly, Vuilleumier and Simberloff (1980) emphasized that, as determinants of distribution patterns in high Andean birds, 'both ecology (short-term ecological factors, working their effects in ecological time, i.e. very short periods of absolute time, very few generations) and history (in this case the consequences of climatic changes during the Pleistocene and their effects on the vegetation, or the long-term evolutionary time) have played roles together at all times'. The Quaternary temperature fluctuations during glacial and interglacial periods led to repeated upward and downward displacements of the montane zones along the Andean slopes. Lateral range extension and faunal interchange was facilitated during glacial periods when upper montane zones and their faunas were located along less dissected lower mountain slopes. Conversely, the upward shift of the climatic zones during interglacial periods led to an increasing dissection of montane forests and other habitat zones along the higher slopes where speciation again occurred in small isolated populations (Vuilleumier 1969; Simpson Vuilleumier 1971; Haffer 1970b, 1974). A different situation may have existed in the large altiplano region of the central Andes (Vuilleumier and Simberloff 1980). In this region, speciation possibly occurred predominantly during the glacial rather than interglacial phases when enlarged glaciers and glacial lakes in the high plains fragmented species ranges. These were reconnected

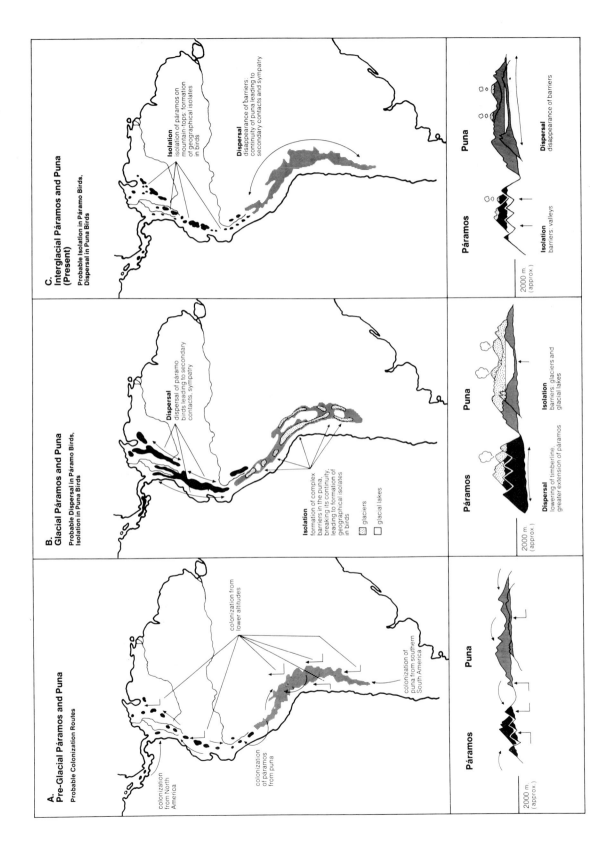

A.
Pre-Glacial Páramos and Puna

Probable Colonization Routes

colonization from North America

colonization from lower altitudes

colonization of páramos from puna

colonization of puna from southern South America

Páramos

Puna

2000 m (approx.)

B.
Glacial Páramos and Puna

Probable Dispersal in Páramo Birds, Isolation in Puna Birds

Dispersal
dispersal of páramo birds leading to secondary contacts, sympatry

Isolation
formation of complex barriers in the puna, breaking its continuity, leading to formation of geographical isolates in birds

glaciers
glacial lakes

Páramos

Puna

Dispersal
lowering of timberline, greater extension of páramos

Isolation
barriers: glaciers and glacial lakes

2000 m (approx.)

C.
Interglacial Páramos and Puna (Present)

Probable Isolation in Páramo Birds, Dispersal in Puna Birds

Isolation
isolation of páramos on mountain-tops: formation of geographical isolates in birds

Dispersal
disappearance of puna leading to secondary contacts and sympatry

Páramos

Puna

Isolation
barriers: valleys

Dispersal
disappearance of barriers

2000 m (approx.)

during interglacial periods when the glaciers and lakes retreated (Fig. 5.26).

Dispersal across the tropical lowlands from the Andes to isolated mountain systems such as the Guiana Highlands (Pantepui) took place either by 'hopping' across the lowland barrier (active dispersal; Mayr and Phelps 1967; Cook 1974; Mayr and Diamond 1976) or through lowland forests during periods of barrier reduction when montane life zones were lowered. Small mountains between the Andes and Pantepui in southeastern Colombia as low as 500 m elevation may have had a montane (subtropical) summit during cool periods and probably served as stepping stones for montane species to reach far distant mountains. Individual colonizing flights may not have been longer than 20 to 150 km in northern South America. The low table mountains of the forested southeastern Colombian lowlands may have facilitated the immigration of many Andean birds into the southern Venezuelan highlands during glacial periods (Haffer 1970b, 1974) but these remain largely unexplored. A few montane birds of the forest interior may have reached widely isolated mountains through the lowland forests. The isolated mountains served as refugia upon the return of warm and adverse conditions in the lowlands. The varying rate of differentiation of isolated bird populations on widely separated mountains may have subsequently concealed the fact that the immigrating portion of the fauna was acquired during several more or less well-defined periods when the tropical climate was somewhat cooler (Haffer 1970b, 1974).

It will be difficult to 'prove' that active dispersal of bird species between presently isolated mountains across lowland barriers has taken place in the past. However, if conspecific populations of upland animals inhabit mountains that are geologically independent in their history (e.g. Andes and Guianan Highlands) then dispersal from the area of origin must be assumed to have occurred because (i) the montane species for climatic–ecological reasons never inhabited the intervening lowlands to form a continuous range between these isolated mountains and (ii) it is very unlikely that a montane species could have been differentiated from one lowland ancestor independently several times in these isolated mountain regions. The ancestors of many widely distributed

Andean species probably originated during the early uplift of the mountains from restricted populations which had colonized the newly available montane habitat and differentiated in geographical isolation. Horizontal range expansion ('passive' or biotic dispersal; not across barriers) of these newly differentiated montane species occurred when environmental conditions reached lateral continuity through continued uplift of the mountains.

Croizat (1958, 1976) mapped the ranges of many neotropical birds and other animals and attempted to explain certain distribution patterns at the species and subspecies level on the basis of tectonic movements in offshore regions of South America and of the erosion of mountain systems. There seems little doubt that Tertiary palaeogeographical changes determined the early evolution of present-day species and groups. But there have subsequently been the major vegetational shifts during the Quaternary described in Chapter 1. The evidence presented and reviewed in this book strongly suggests that it is these changes of the past two million years which have very effectively influenced evolution at the level of species and subspecies thus causing the establishment of current distribution patterns.

Numerous, well-differentiated avian subspecies possibly date back to the late Quaternary only, whereas many species, especially of passerine birds, may have originated during the early or middle Quaternary periods of geographical isolation. This does not mean that a Quaternary age is assumed for all species of birds; members of certain non-passerine groups like storks, vultures, and some raptors probably date back to the late Tertiary.

The origin of several groups of neotropical birds may be traced to geological events during the Tertiary or the Cretaceous. Thus, the separation of South America from Africa and Antarctica during the Cretaceous isolated the western portion of the Gondwana fauna which included many of the supposed ancestral groups of the contemporary South American fauna. Birds which were affected by this separation perhaps included primitive members of the pantropical parrots (Psittacidae), barbets (Capitonidae) and trogons (Trogonidae), the ancestors of the curassows (Cracidae), jacamars (Galbulidae), toucans (Ramphastidae), and several families of

---

Fig. 5.26. A scenario of latitudinal (above) and altitudinal (below) changes in the distribution of high Andean páramo and puna vegetation during the Pleistocene and of effects on isolation and dispersal of birds. See text for further details. (After Vuilleumier and Simberloff 1980.)

suboscine Passeriformes (Cracraft 1973). These groups differentiated extensively during the course of the Tertiary when the Guianan and Brazilian Shields formed vast land areas separated by the shallow waters which then occupied the Amazon basin. Several taxonomically isolated species or groups of species restricted to the regions of the Guiana Shield (e.g. *Rupicola rupicola*, *Perissocephalus tricolor*) and the Brazilian Shield (e.g. *Rhea americana*, *Cariama cristata*) may have originated in these areas during the Tertiary. The Andes were slowly emerging throughout the Tertiary and were fully uplifted only at the end of the Pliocene when the final connection with tropical North and Middle America was established and led to extensive faunal interchange; for recent reviews see Webb (1978), Vuilleumier (1985), and Stehli and Webb (1985).

Continued geoscientific and biogeographical analyses will lead to improved understanding of the significance of Tertiary fragmentation events for the origin of extant species in comparison with the effects on evolution of the dramatic fluctuations in the vegetation during the Quaternary. For this purpose analyses of the branching sequence of various groups of closely related neotropical birds are needed to establish the degree of concordance between taxic cladograms of different groups and thir concordance (if any) with area cladograms of tectonic-geological as well as vegetational units of South America (Cracraft 1982, 1983, 1985*b*; Haffer 1985).

NOTES ADDED IN PROOF

1. The ranges of *P. jacquaçu* and *P. superciliaris* may overlap in Central Brazil, as Roth (1982) reported both species from Aripuana (North Mato Grosso).

2. Bigarella *et al.* (1981) and Prance (1982*c*, 1985) summarized the phytogeographical and zoogeographical evidence supporting the Pleistocene refuge hypothesis regarding the Neotropics, and Mayr and O'Hara (1986) regarding the African rain-forest region. The results of a statistical analysis of the location of the boundaries of bird species distributions in Amazonia did not support the refuge theory (Beven *et al.* 1984). It would be more meaningful in this respect to investigate statistically the location of range boundaries between hybridizing subspecies and parapatric species in secondary contact (these are located between the major centres of endemism, Haffer 1985). A further statistical analysis by Connor and McKenney (1985) focused on the distributional areas of localized Amazonian bird species and confirmed that their ranges are markedly clumped in a number of areas of endemism rather than being independently and randomly distributed in Amazonia. This result is consistent with a major role for vicariant speciation within the Amazon basin. For this purpose analyses of the branching sequence of various groups of closely related neotropical birds are needed to establish the degree of concordance between taxic cladograms of different groups and their concordance (if any) with area cladograms of tectonic–geological as well as vegetational units of South America (Cracraft 1982, 1983, 1985*a*; Haffer 1985).

6

THE EARLY HISTORY OF MAN IN AMAZONIA

INTRODUCTION AND SUMMARY

It seems clear that environmental oscillations were an important factor in creating the neotropical forest biome as we know it. Episodes of aridity in Amazonia during and since the Pleistocene are indicated by geology, geomorphology, and palynology, and implied by the distributional patterns of plants and animals. Archaeological, ethnographical, and linguistic data from tropical lowland South America have patterns of distribution similar to those observed among plants and animals. The coincidence between centres of dispersal and enclaves of diversity among languages and several centres of diversification and regions of secondary contact among plant and animal taxa is particularly interesting because different time scales are presumably involved and because the distributions of cultural traits have traditionally been explained by processes independent of climatic or environmental change.

Recognizing and interpreting patterns in cultural phenomena comparable to those exhibited by plants and animals suffers from several kinds of handicap. Some, such as the sparsity, unevenness, and differential reliability of the data, are shared by many disciplines. Others originate from the relatively recent incorporation of *Homo sapiens* into the South American biota, the distinctive relationship between culture and environment, and the absence of a general theory of culture. The shortcomings of the data require no special comment beyond the warning that filling areas now blank may alter the patterns described herein.

A major difference between the evidence for forest refugia provided by plants and other animals and by man is the timetable. Floral and faunal ranges can be interpreted to reflect events stretching backward through the Quaternary and perhaps even further. Man appears to have arrived in South America only between about 20 000 and 14 000 years ago, and direct evidence for his presence in Amazonia is restricted to the past 5000 years. Although landscapes were rearranged on a small scale during the Holocene, these oscillations are considered by biologists to have played a minor role in determining contemporary distributional patterns.

Also, in contrast to biologists, anthropologists have not developed a general body of theory that can be used for generating hypotheses and evaluating observations about cultural behaviour. The absence of such a framework makes it easy to dismiss resemblances in patterning between biological and cultural phenomena as either coincidences (similar effects of different causes) or fortuitous products of unrepresentative or incomplete data. A third possibility is that cultural behaviour is in fact a sensitive medium for natural selection, making *Homo sapiens* more vulnerable to short-term local environmental fluctuations than has been recognized.

The significance of the refuge model for anthropologists is the guidance it provides for organizing existing cultural data and for identifying the regions and kinds of information most relevant for establishing the extent to which human adaptation to the tropical lowlands can be explained by general principles of natural selection. In this chapter, I shall summarize the principal differences between cultural and biological forms of adaptation and then describe several archaeological, linguistic, and ethnographical patterns that may reflect environmental stress on human populations in the past. For additional examples, reference should be made to Meggers (1974, 1975, 1977, 1979, 1982) and Meggers and Evans (1973).

THE ENVIRONMENT IN HUMAN PERSPECTIVE

As with other organisms, the viability and density of human populations depend on the availability of subsistence resources. In this respect, *Homo sapiens* has two major advantages: (i) the biological capacity to consume a wide variety of plants and animals and to change the composition of the diet temporarily or permanently, literally from one meal to the next, and (ii) the cultural capacity to increase the productivity of certain foods by domestication. As a consequence, the more obvious features by which landscapes are classified – temperature, rainfall, elevation, topography – do not correlate with general types of human societies. Indeed, the existence of hunter–gatherers in desert, arctic, forest, and savanna habitats has been cited as evidence that the environment is not a primary determinant of culture.

When the special character of human subsistence is used as a criterion for classification, however, four general kinds of environments can be distinguished, each providing different potentialities and limitations for food production and thus for increase in cultural complexity. These are: (i) environments unsuitable for agriculture; (ii) environments with limited agricultural potential; (iii) environments with potential that can be increased by various technological and managerial inputs; and (iv) environments with unlimited natural potential (Meggers 1954 gives details).

The first three of these occur in tropical lowland South America. The manner in which they correlate with the types of vegetation distinguished in Chapter 2 must be specified to provide a common ground for comparing human and non-human biogeography.

Environments unsuitable for agriculture

Several types of vegetation denote environments unsuitable for agriculture, the most important being caatinga, cerrado, Amazonian savanna, igapó, and permanent white-water forest (Table 2.3, p. 29). The utility of these biomes for other forms of human exploitation varies. Hunter–gatherers have evolved subsistence cycles that integrate seasonal plant foods with those available throughout the year. Game, fish, insects (grubs, larvae), turtles, honey, and eggs are among other wild resources often consumed. Many of the latter are obtained from adjacent forests, which are richer in large animals than open vegetation (Bourlière 1973, Tables 4 and 5).

Dependence on wild foods places a ceiling on the size and sedentariness of human communities, which in turn limit elaboration of other aspects of the cultural configuration. In Amazonia, where individuals of the same species among both plants and animals tend to be widely dispersed, population density of human foragers is generally less than 0.1 person per square kilometre. Under special circumstances, bands number up to 150 individuals, but groups of less than 50 are more typical. Two general patterns of wandering prevail, both usually practised within a band territory: (i) almost continuous mobility, with pauses of a few days or weeks where the local food supply permits, and (ii) expeditions issuing from a main camp, to which the group returns seasonally or periodically over a number of years.

The interaction between members of a band is regulated by ties of kinship. Division of labour by sex is confined to a few basic tasks, such as hunting or care of infants; most other activities can be performed by men and women. Differences in individual capabilities are appreciated, but seldom rewarded. The accumulation of material goods is inhibited by the requirement of mobility and, except for weapons and a few other essentials, objects are made when needed and discarded after use. Simple shelters of poles and thatch are erected for protection from rain. Concepts of the supernatural are usually vaguely defined; illness, death, bad luck in hunting, and other distressing events are generally attributed to spirits offended by broken taboos.

This general way of life was universal to our species until the past few millennia. It survives mainly in environments too cold, too wet, too dry, or having some other quality incompatible with the adoption of agriculture. It also persists in some regions where shifting cultivation could be practised and most hunter–gatherers in such circumstances undertake rudimentary forms of gardening. Hunter-gatherers cannot be recognized archaeologically in the humid tropical lowlands of South America because their high mobility and use of perishable raw materials leaves nothing to be found. Their presence in the surrounding dryer habitats dates from at least 14 000 BP.

Environments with limited agricultural potential

Terra firme rain forest is the principal type of vegetation associated with slash-and-burn agriculturalists. Gallery forest is also used, as well as enclaves of more fertile soil in regions dominated by the various kinds of white sand forest. Trees are usually cut at the beginning of the dry season and burned just before the rains start, when planting is done. Depletion of soil nutrients generally lowers productivity sufficiently that fields are abandoned after about three years.

The principal crops are manioc and sweet potato;

yams, maize, and peanuts are secondary in importance. Additions subsequent to European contact include bananas, plantains, and sugar cane. Wild fruits, seeds, berries, roots, and nuts are harvested, but their contribution to the diet varies greatly. Game and fish are the principal sources of protein.

This subsistence pattern is correlated with semi-permanent settlement. Typically, villages are moved at intervals of about five years, when the fertility of the land in the vicinity is exhausted and the game depleted. Whether a settlement consists of a single communal house or several communal houses arranged around a plaza, it is an independent political and economic unit. The population of such villages is usually less than 200, but occasionally reaches 500 or more. Density averages about 0.3 persons per km^2.

Social relations are defined by ties of kinship, but division of labour by sex is more extensive than among hunter–gatherers. Sedentary life permits the elaboration of material goods, including the use of pottery. The only specialized occupation is shamanism; shamans perform cures, conduct rituals, and identify sorcerers with the aid of supernatural helpers. These activities usually do not relieve them of duties and obligations allocated to members of their sex, however. Puberty rites, victory celebrations, harvest festivals, and other social events provide opportunities for arranging marriages, as well as for feasting and dancing. Raids are often stimulated by the need to avenge the death of an adult member of the community, which is attributed to sorcery. Prisoners (particularly females and children) and head trophies are sometimes taken. Acquisition of material goods or territory is not a motive for aggression, but inter-village trading is characteristic.

This 'basic tropical forest culture', as it was labelled by Steward (1948), combines the maximum permanency of residence with the minimum irreversible damage to the environment. Populations are prevented from becoming too large, too concentrated or too sedentary by a variety of cultural practices. Among the most common are infanticide, abortion, taboos on sexual relations, blood revenge, and warfare (Meggers 1971, pp. 103–13).

Archaeological remains identifiable with this subsistence and settlement pattern are small and shallow accumulations of refuse consisting mainly of fragments of pottery, little of which is decorated. In Amazonia, they appear about 3300 BP, two millennia later than in the Andean region.

Environments suitable for intensive cultivation

The only type of vegetation in the humid Amazonian lowlands associated with soils used aboriginally for perennial cultivation is várzea forest (Table 2.3). Nutrients removed by leaching and harvesting are replaced in silt deposited during the annual inundation, but effective use of this environment requires careful scheduling of planting. Maize, which reaches maturity in three months, can be grown on lower land than manioc, which requires at least six months. Such allocations are reported among groups that were exploiting the várzea at the time of European contact (Meggers 1971, pp. 125–6, 134; Roosevelt 1980, pp. 156–7). The crop most often mentioned is maize, for which two to six annual harvests were obtained. Wild rice (*Oryza perennis*), which grows in profusion, was gathered. Other wild foods, both plant and animal, were available in abundance.

This resource base supported larger settlements and a higher level of sociopolitical complexity than could be maintained in the terra firme rain forest. At the time of European contact, towns along the margins of the Amazon often contained over a thousand inhabitants and population density reached one person per square kilometre. Although kinship remained an important cohesive factor, social relations were transformed by the emergence of social stratification and full-time occupational specialization. Several towns were often united politically and ruled by a high chief. Two chiefdoms sometimes formed temporary alliances for defence or aggression. Priests and craftsmen were professionals, who exchanged their skills or products for food and other necessities. Concepts of the supernatural were more formalized than among shifting cultivators; idols, kept in temples or shrines, were attended by priests and entreated with offerings and prayers. The dead were often buried in cemeteries and provided with articles befitting their status. Warfare was waged to obtain captives, plunder or territory. Captives were relegated to servile status or sacrificed, in contrast to the practice of adoption prevailing among shifting cultivators.

Archaeological evidence for this level of cultural development consists principally of large habitation sites, deep refuse accumulations, differential treatment of the dead, and pottery exhibiting standardization and elaboration in form and decoration. The earliest known Amazonian occurrences date about the beginning of the Christian era.

CULTURAL PATTERNS INDICATING ENVIRONMENTAL INSTABILITY

The substitution of cultural behaviour for biological responses as the primary mechanism of adaptation has both advantages and disadvantages for an effort to establish whether human populations in the tropical lowlands experienced subsistence stress as a consequence of environmental fluctuations since they arrived in the region. As with other animals, some characteristics are subject to stronger selective pressures than others. Those most critical for survival, such as subsistence techniques, social organization, and religion, will be altered during adaptation to new kinds of resources. In order to recognize whether groups with similar behaviours have heterogeneous origins, attention must be directed to categories of traits irrelevant to adaptive success. Languages and ceramic traditions fall into this category and have the additional advantage that their differentiation can often be placed in a chronological framework, which can be compared with the estimated dates for the inceptions and terminations of episodes of climatic change.

The existence of a general relationship between cultural elaboration and the subsistence potential of the environment provides another means of recognizing disturbances. Although increasing complexity is not inevitable, the history of cultural development everywhere has been characterized by the replacement of hunter–gatherers by agriculturalists if the environment permits and the technology is available. Other things being equal, therefore, we should expect the three kinds of environments represented in lowland South America to have been associated with 'climax' types of subsistence adaptations and cultural configurations at the time of first reporting (European contact). Situations where the potential was not realized deserve attention, as do those where a culture possesses greater complexity than appears compatible with the environment in which it occurs. A hiatus in an archaeological sequence, which implies a severe decline in population density or abandonment of the region, is another significant kind of anomaly.

In the following pages, I will review some archaeological, ethnographic, and linguistic patterns suggesting that human populations in the tropical lowlands experienced several severe disruptions during recent millennia. Comparing these episodes with one another, with the kinds of distributional patterns observed by biogeographers, and with the dates and durations of climatic oscillations inferred by geomorphologists and palynologists provides a basis for assessing the probability of a cause and effect relationship.

ARCHAEOLOGICAL EVIDENCE

Throughout aboriginal human occupation of the tropical lowlands, most of what people made and used was perishable. Prior to the adoption of pottery, the likelihood of encountering cultural remains is therefore slight. Fortunately, ceramics are well suited to serve as tracers of prehistoric movements and developments because of their durability, abundance, and relative freedom from adaptive constraints. As long as a vessel can be used for cooking, eating, storage, or some other intended function, it can vary tremendously in almost every detail of composition, surface finish, rim and base form, body shape, and decoration. Even the most elaborated ceramics incorporate only a small proportion of the potential variation, making it possible to recognize complexes and traditions. Differences in the relative frequencies of characteristic elements appear to differentiate communities sharing a ceramic tradition (Meggers and Evans 1981). Since complexes and styles have specific geographical locations, their dates of inception and termination in each location can be estimated from carbon-14 determinations. Comparing local chronologies makes it possible to infer centres of origin and routes of dispersal. Aspects of technology, use, and disposal permit reconstructing general levels of cultural complexity.

Distributional patterns

The oldest and most common approach to interpreting the archaeological remains from the Amazon Basin has been to trace the distributions of ceramic traits. In 1885 Netto suggested a unitary origin for the mound-builders of Marajó and the Mississippi Valley based on similarities in the pottery (1885, p. 419). A more detailed comparison was made by Palmatary (1939, 1950), who concluded that both the Santarem and Marajó styles shared more features with the eastern United States than with any intervening region. Other investigators, among them Verneau (1920) and Uhle (1920), pointed out affiliations with the Andean area and considered the Amazonian expressions derived from the highlands. Howard (1947) divided the lowlands into four geographical regions, analysed their ceramic styles, selected the features most useful for making comparisons, and noted the presence or absence of the diagnostics of each style in every other style. This procedure showed the closest resemblances between Amazonia and the Orinoco–West Indian area. In contrast to other authors, who either refrained from postulating a direction of movement or considered Amazonia the recipient, Howard concluded that 'the evidence would indicate that Amazonia has been a centre of influence for the total area' (1947, p. 86; cf. Lathrap 1970, pp. 74–5; Rouse and Allaire 1978, p. 476).

A more precise way of analysing these distributions is by plotting the locations of sites representing the same ceramic tradition on a map. Three are sufficiently well known and widely distributed to warrant examination: the zoned hachure tradition, the polychrome tradition, and the incised-and-punctate tradition. Each is defined by a core of characteristics, but the total assemblage varies in time and space. Mixed complexes result from adoption of new ceramic elements without relinquishing older ones. In Amazonia, complicated mosaics appear to exist, in which relatively pure assemblages are interspersed with hybrids of varying composition and with independent local styles. In discussing distributions, only the diagnostic features of the three major traditions will be considered.

Zoned hachure tradition

Decoration consisting of zones defined by broad incision and filled with fine cross or parallel hachure is characteristic of several early ceramic complexes (Fig. 6.1). Two variants can be recognized. Variant A, in which the zones tend to be irregular and filled with cross hachure, has been reported on the coast of Ecuador (Valdivia D), the lower Amazon (Jauari), and the island of Marajó (Ananatuba). Variant B, in

which the zones are often narrow bands and the hachure is typically vertical or diagonal, occurs in eastern Ecuador (Pastaza, Yasuń) and eastern Peru (Waira-jirca, Tutishcainyo). On the Caribbean coast of Colombia, broad incisions bordering zones containing punctations, gashes or parallel straight or curved lines are characteristic of a succession of complexes (Puerto Hormiga, Canapote, Barlovento). The eastern Andean occurrences originate prior to 4000 BP; the Marajó complex begins about 3300 BP. The Colombian sequence extends between about 5000 and 3000 BP. The likelihood that the absence of sites from the central lowlands reflects limited knowledge of the archaeology is enhanced by the recent report of a ceramic complex dating from c.4000 BP at La Gruta on the middle Orinoco, which may constitute a link (Vargas 1979, pp. 226–7; Roosevelt 1980, Fig. 56). In any case, these distributions imply long-range dispersals about the beginning of the fourth millennium BP.

Polychrome tradition

Red and/or black painting on a white-slipped surface is a widespread decorative technique in the lowlands. Two general methods of execution occur: (i) positive, in which the painted lines and zones create the motifs, and (ii) pseudonegative (Fig. 6.2), in which they fill the background and the motifs are produced by the unpainted portions of the surface. Although this distinction may have chronological and geographical significance, existing data are insufficient to assess this possibility. A variety of other decorative techniques is often associated, including excision, incision, and grooving combined with red and white slips. In several local complexes, manufacture of pottery by specialists and use for non-domestic purposes are implied by the standardization and elaboration of form and decoration. Anthropomorphic vessels were often used as burial urns.

The archaeological evidence for a level of cultural complexity more advanced than that characteristic of slash-and-burn agriculturalists is consistent with the distribution of the sites along the várzea, where intensive cultivation can be practised. The tradition was established at the mouth of the Negro by 1800 BP and spread rapidly up the Madeira and down the Amazon (Fig. 6.2). Outside Amazonia, the earliest reliable dates for similarly painted pottery are from sites representing the Tocuyanoid series near Lake Maracaibo in northwestern Venezuela and range

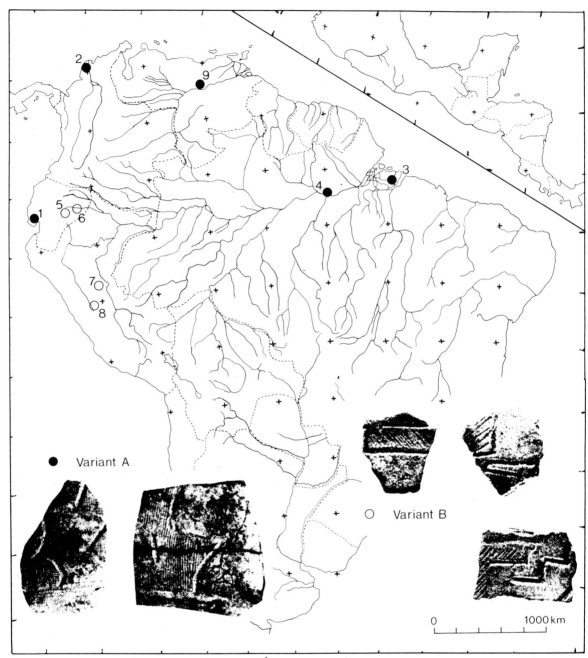

Fig. 6.1. Distribution of sites representing the zoned hachure ceramic tradition imply a long-range dispersal between about 4000 and 3000 BP, probably from a source in northwestern South America. The probability that the gap between the Andean occurrences and the lower Amazon reflects failure to encounter sites rather than a true disjunction is suggested by the presence of a complex possessing several features of the tradition at La Gruta on the middle Orinoco (Roosevelt 1980). Solid circles: Variant A, open circles: Variant B. 1. Valdivia D (Meggers, Evans, and Estrada 1965, Pls. 113–14). 2. Puerto Hormiga, Canapote, Barlovento (Willey 1971, pp. 268–71). 3. Ananatuba (Meggers and Evans 1957, Pls. 38–40). 4. Jauari (Hilbert 1968, Pls. 7 and 9). 5. Pastaza (Porras 1975). 6. Yasuni (Evans and Meggers 1968). 7. Tutishcainyo (Lathrap 1970, pp. 84–9). 8. Wairajirca (Izumi and Sono 1963). 9. La Gruta.

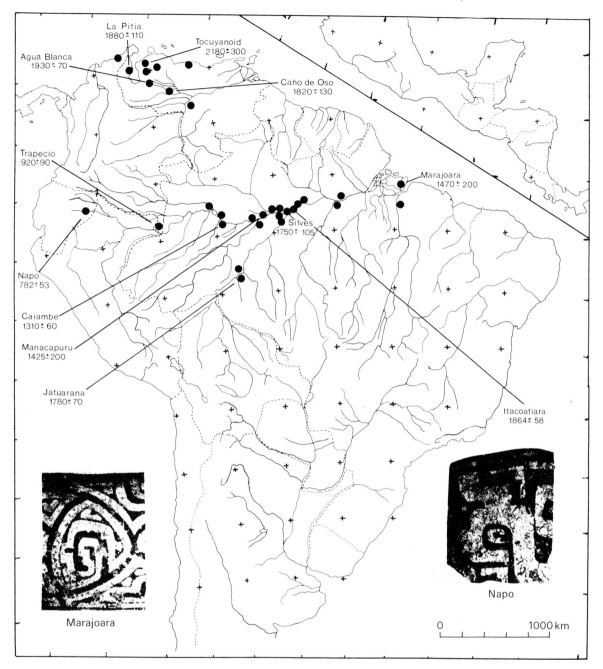

Fig. 6.2. Distribution of sites of the polychrome ceramic tradition prior to c. 1100 BP. Existing carbon-14 dates suggest intrusion from the northern Andes and dispersal up the Madeira and down the Amazon from the mouth of the Negro. Examples show the pseudonegative technique of painting and scroll motif. Tocuyanoid series sites (Cruxent and Rouse 1958–9; Rouse and Cruxent 1963; Gallagher 1976; Rouse and Allaire 1978). Middle and lower Amazon (Meggers and Evans 1957; Evans and Meggers 1968; Hilbert 1968). Rio Negro (M. F. Simões, personal communication). Rio Madeira (E. Th. Miller, personal communication).

between about 2100 and 1800 BP. The tradition has not been encountered in eastern Venezuela, the Guianas, or the upper courses of the southern tributaries of the lower Amazon. The extant pattern thus suggests an intrusion from northwestern Venezuela or adjacent Colombia into central Amazonia around the beginning of the Christian era.

Incised-and-punctate tradition

The hallmark of the incised-and-punctate tradition is a narrow decorative band divided by one or more parallel, diagonal incisions of alternating slant into triangular areas containing punctations (Fig. 6.3). Variations include parallel slanting lines filling triangles, an undulating line incompletely separating trianguloid zones, and crossed lines creating four triangular fields. Incisions are often perfectly straight, evenly spaced, and uniform in depth. Anthropomorphic, zoomorphic, and geometric adornos are frequently associated (Lathrap 1970, Figs. 41 and 42). Ceramic complexes representing this tradition occur on the Greater Antilles, the central coast of Venezuela, the middle and lower Orinoco, the coasts of Guyana and Surinam, the lower Amazon, the Xingu, Tapajós, and lower Madeira, and in lowland Bolivia (Fig. 6.3).

The Arauquinoid series, which represents this tradition on the middle Orinoco, is considered to have evolved locally during several centuries (Rouse and Allaire 1978, p. 442) and the time slope of the existing carbon-14 dates favours this region as the immediate centre of dispersal over the lowlands. The diagnostic features were present on the middle Orinoco by c. 1400 BP; by c. 1265 BP, they had spread to Surinam and by c. 1150 BP, to the middle Amazon. The chronological position in lowland Bolivia also seems to be late. Throughout most of its distribution, the tradition persisted until European contact. It displaced or influenced representatives of the polychrome tradition in the corridor across the lower Amazon, where Am and Aw climates prevail (Fig. 1.3). As a consequence the polychrome tradition has a disjunct distribution after about 1200 BP (Fig. 6.3).

Cultural–environmental disconformities

Early accounts of the mounds on Marajó Island attribute them to intruders who possessed a higher level of cultural complexity than could be sustained by local subsistence resources (Penna 1879, p. 53; Netto 1885, p. 262). More recent investigations support the inference that the social organization of this representative of the polychrome tradition was characterized by social stratification and occupational division of labour (Meggers and Evans 1957, p. 593). The centre of dispersal and the histories of local manifestations of the tradition are consequently of particular interest. An origin outside Amazonia in northwestern South America, implied by existing carbon-14 dating, is in keeping with evidence for the widespread occurrence of a similar level of sociopolitical development in the Andean and Caribbean areas. Dispersal seems to have been rapid along the lower Amazon and up the Madeira, where fertile várzea soils and rich aquatic fauna provided a reliable food supply. On Marajó, the Marajoara Phase declines from an initially stratified society to an unstratified one, as would be predicted if it were displaced from a region suitable for intensive cultivation to one with limited agricultural potential (Organização dos Estados Americanos 1974, p. 10). The short duration characteristic of large settlements representing the polychrome tradition along the Napo in eastern Ecuador implies that, here too, the level of cultural complexity of the invaders exceeded what could be sustained by local subsistence resources. In both these instances, cultural simplification seems attributable to inherent inadequacies in the local environment rather than temporary deterioration. The questions that must be answered, and which may involve climatic alteration, are why the immigrants left their homelands and why they settled in regions where the level of cultural complexity they had achieved could not be maintained.

Pottery of the incised-and-punctate tradition, like that of the polychrome tradition, is associated with a chiefdom level of sociopolitical development. Populous towns were reported around the mouth of the Tapajós by the first European explorers (Meggers 1971, p. 133). Figurines and other ritual paraphernalia are characteristic of archaeological sites here, as well as in Venezuela and the Greater Antilles. Whether a similar level of general social complexity was associated with pottery of this tradition in the Guianas is uncertain; on the upper Xingu, however, it is clear that only a few elements of ceramic decoration were adopted by slash-and-burn agriculturalists. Along the lower Amazon, bearers of this tradition displaced or absorbed bearers of the earlier polychrome tradition (Fig. 6.3). The strong correlation

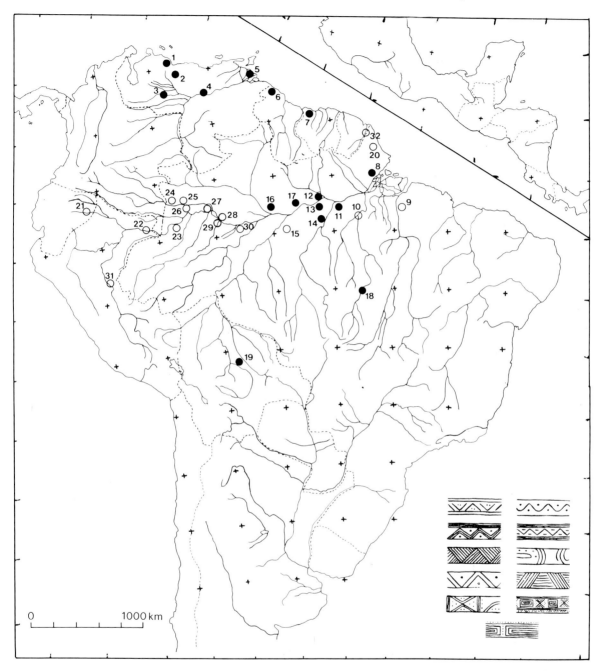

Fig. 6.3. Distribution of sites of the incised-and-punctate ceramic tradition (closed circles and insert) which dispersed around 1200 BP. In Amazonia, it is closely correlated with the weakly seasonal Am and Aw humid tropical climates and avoids the perhumid Af areas (cf. Fig. 1.3). Intrusion, apparently from the north, created a disjunction in the distribution of the Poly-chrome tradition (open circles) which persisted at the mouth of the Amazon and spread westward to the base of the Andes in Ecuador and Peru. Sites and carbon-14 dates in years BP.

1. Tucacas
2. Valencia 1000 ± 70
3. Caño Caroní 1095 ± 55
4. Camoruco 1200 ± 85
5. Guarguapo 300 ± 50
6. Mabaruma
7. Hertenritz 1265 ± 60
8. Mazagão

of sites within the driest portion of the Amazonian lowlands (Figs. 1.2, 1.3, and 2.3) raises the question whether a somewhat different mode of subsistence adaptation may have provided a competitive advantage in this environment.

Continuity and discontinuity

Another aspect of the archaeological record relevant to assessing environmental instability is the presence or absence of continuity in local sequences. Lathrap (1970, pp. 131 and 139) has distinguished at least six distinct successive ceramic complexes on the central Ucayali and suggested they reflect colonists of different ethnic affiliations (Fig. 6.4). Four different traditions follow one another on the Napo (Evans and Meggers 1968) and five on the island of Marajó (Meggers and Evans 1957). Multiple replacements have also been reported on the middle Orinoco (Roosevelt 1980), in lowland Bolivia (Dougherty, personal communication), and southern Rondônia (E. Th. Miller, personal communication). Two sequent occupations have been recognized in northwest Guyana, in northern Amapá, and on the lower Xingu. By contrast, surveys along the Japurá (Hilbert 1968), the upper Orinoco and Ventuari (Evans, Meggers, and Cruxent 1960), in southern Guyana (Evans and Meggers 1960), on the upper Xingu (Simões 1967), and on the headwaters of the Juruá and Purus (O. Dias and Carvalho, personal communication) have led to identification on only a single ceramic complex. Some of the intruders are affiliated with one of the major ceramic traditions; others represent localized styles associated with slash-and-burn cultivators.

The situation along the middle Amazon is difficult to interpret because although considerable information exists, none of the local sequences appears to be complete. Sites represent not only the zoned hachure, polychrome, and incised-and-punctate traditions, and various combinations of the latter two, but also complexes not affiliated with a major tradition. In many of these cases, neither the distribution of the complex nor its chronological position is known. Discontinuity and heterogeneity are clearly appropriate labels for the prehistoric cultural remains from this region.

Another form of discontinuity has been observed in the refuse accumulations of rock shelters on the margins of the rain forest. On the Sabana de Bogotá in Colombia and in southwestern Goiás, human occupation began before 10 000 BP and continued without substantial interruption until about 5000 BP when the shelters appear to have been abandoned. About this time shell middens began to proliferate both on the coast of Brazil and along northern South America, implying more intensive exploitation of marine resources. Around 2500 BP in Colombia and 1000 BP in Goiás, the shelters were reoccupied by groups who made pottery and presumably derived part of their subsistence from domesticated plants (Correal and Van der Hammen 1977, p. 188; Schmitz 1980, Fig. 5).

More refined dating of the durations of the ceramic complexes on Marajó (J. Danon, personal communication) has brought to light a possible hiatus in the sequence there, as well as correlations between intrusions and episodes of climatic change (Fig. 6.5). The arrival of the Ananatuba Phase coincides with the end of an arid episode. The terminal dates for the Mangueiras Phase are separated by nearly a millennium from the initial dates for the Formiga Phase, implying that the region was abandoned by agriculturalists. Most of this interval was characterized by aridity. The Formiga intrusion occurred when humid conditions returned. The arrival of the Marajoara and their replacement by the Aruã seem to correlate with briefer episodes of aridity.

In summary, the archaeological evidence indicates that human occupation in the tropical lowlands during the past three millennia was characterized by: (i) long-range dispersals from the northwestern part of the continent by groups representing at least

9. Nazaré dos Patos	10. Independência 780±60	11. Santarém	12. Konduri
13. Parintins	14. Castanha 1010±30	15. Borba	16. Paredão
17. Sanabani 1010 ± 85	18. Diauarum 830 ± 90	19. Masicito	20. Aristé
21. Napo 782 ± 53	22. Yanayacu	23. Pirapitinga	24. Mangueiras
25. Paraiso	26. São Joaquim	27. Mapari	28. Lago Amaná
29. Caiambé 1310 ± 60	30. Coarí 1170 ± 65	31. Caimito 630 ± 60	32. Aristé

Sources: 1-5, Rouse and Allaire (1978); 6, Evans and Meggers (1960); 7, Bruijning *et al.* (1977); 8, 20 Meggers and Evans (1957); 9, F. A. Costa (personal communication); 10, C. Perota (personal communication); 12, Hilbert (1955); 11, 13-16, 23-30, Hilbert (1968); 17, M. F. Simões (personal communication), 18, Simões (1967); 19, Nordenskiold (1913, p. 241); 21, 27, 28, Evans and Meggers (1968); 22, Myers (1970); 31, Lathrap (1970); 32, J. Petit-jean-Rojet (personal communication).

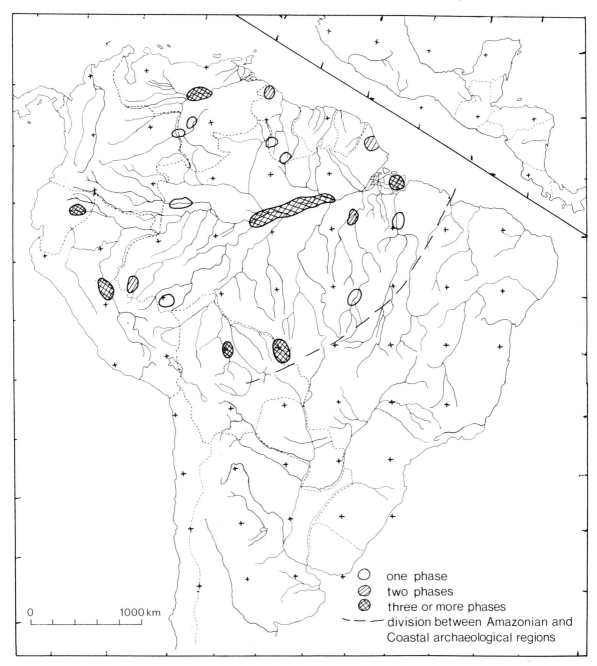

○ one phase
◍ two phases
⊕ three or more phases
– – – division between Amazonian and
Coastal archaeological regions

Fig. 6.4. Continuity and discontinuity in archaeological sequences from various parts of Amazonia. Only in these regions has sufficiently intensive survey and excavation been conducted to provide a reliable ·indication of the prehistoric situation. The areas where a single ceramic complex has been identified correlate generally with areas of high linguistic diversity (Fig. 6.9) and with biological zones of secondary contact (Fig. 5.22). The areas where pottery making has a long history are negatively correlated with centres of differentiation of the major linguistic stocks (Fig. 6.8) and represent successive intrusions by groups with different cultural antecedents. (Evans and Meggers 1968, p. Fig. 8; O. Dias, B. Dougherty, C. Perota, F. Costa, personal communication.)

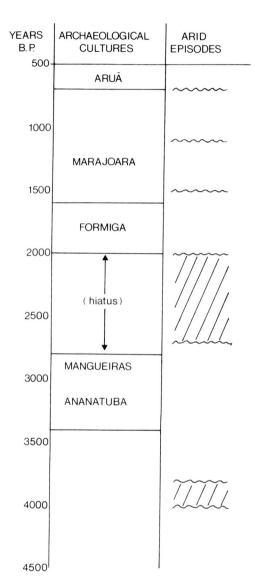

Fig. 6.5. Correlation between a hiatus in the archaeological sequence of Marajó and an increase in aridity in Amazonia between about 2700 and 2000 BP suggesting abandonment of the island by slash-and-burn agriculturalists. Briefer episodes of aridity about 1500 and 700 BP, tend to correlate with the introduction of populations associated with distinct ceramic traditions. Additional local chronologies of this kind are needed to evaluate the impact of environmental stress on human groups. Archaeological (thermoluminescence) dating, J. Danon (personal communication), archaeological sequence, Meggers and Evans (1957), climatic reconstruction, Absy (1979, and personal communications).

three different ceramic traditions; (ii) decline in the complexity of intruding groups with more highly developed cultures than could be sustained by the subsistence resources of terra firme rain forest; and (iii) repeated disruption of local continuity in widely separated portions of Amazonia and on its northwestern and southeastern margins.

ETHNOGRAPHIC EVIDENCE

Nordenskiold, who devoted considerable effort to establishing and interpreting the distributions of a variety of ethnographic traits, called attention to some of the pitfalls of this approach to historical reconstruction (1924, pp. 10–14). Aside from lapses in reporting and inadequate descriptions, the most important are: (i) changes in the locations of groups since European contact; (ii) changes in expression as a consequence of European influence; and (iii) differential losses through acculturation. The first and second can distort indigenous patterns of distribution; the last can create gaps that might be mistaken for disjunctions. Independent invention is another variable that aggravates the confusion, since a trait dispersed from two or more centres may present several disjunctions or achieve a continuous distribution that obscures its multiple origin.

Distribution patterns

One object whose distribution may have historical significance is the babracot, a wooden frame used for roasting meat. This may be either a tripod, three poles joined at their upper ends, or a tetrapod, constructed by inserting four cleft sticks into the ground (Fig. 6.6). The tripod is reported from the central Guianas and four widely separated locations between the Madeira and the Tocantins, whereas the tetrapod is used throughout the lowlands, on the Brazilian coast, and in the Antilles. Nordenskiold (1924, p. 127) interpreted this pattern as evidence for greater antiquity for the tetrapod. Both types appear to be equally efficient, making it unlikely that the tripod would replace the tetrapod without dispersal of its original users.

The distributions of three kinds of hunting devices illustrate the difficulty of interpreting ethnographic patterns. Pole snares and simple nooses are primitive instruments and are so widespread that they have been considered survivals of one of the earliest waves of immigration over the continent (Rydén 1950, p. 315). Traps with a trigger mechanism set off by the prey have a more restricted distribution, implying less antiquity. Their greater efficiency would make them prone to adoption, however, and perhaps to multiple invention. Without additional evidence it is impossible to decide whether their distributional pattern is the consequence of coalescence between centres of independent invention, diffusion from a single centre,

or population displacement (Meggers 1982, Fig. 26.2).

Steward, whose editorship of the *Handbook of South American Indians* gave him a comprehensive grasp of the ethnographic data, prefaced his discussion of the features used to distinguish subareas within the tropical forest area with the observation that 'to a very large extent, their occurrences seem random and capricious, inexplicable in environmental, historical, or functional terms' (1948, p. 883). He was impressed by the combination of uniformity and heterogeneity characteristic of lowland groups:

From a technological and ecological point of view, the basic Tropical Forest culture is strikingly uniform so far as present data reveal . . . The more conspicuous and the most often mentioned differences between the Tropical Forest peoples are such readily observable items as dress, ornaments, body painting, tattoo, and featherwork. These external features, however, distinguish tribes and individuals even more than major areas; the cultural elements involved have highly diversified distributions. The same is probably true of ornamentation, form, and other secondary features of bows, basketry, ceramics, and the like . . . (1948, pp. 885–6).

Cultural–environmental disconformities

Several culture area classifications have been proposed for South America, the best known being a fourfold scheme popularized by Steward. His tropical forest area generally corresponds to the region dominated by terra firme rain forest (Fig. 6.7). Various attempts have been made to subdivide the tropical forest culture area using different sets of criteria. Steward (1948, Map 8) distinguished 11 regions, one of which has three partially disjunct geographical variants. Murdock (1951, Fig. 1) also recognized 11 subareas, but few of his boundaries agree with Steward's. Galvão (1960) pointed out that some of the disagreement arises from combining data obtained over the four centuries since European contact. To achieve greater chronological uniformity, he restricted his sources to the period between 1900 and 1959. After eliminating the margins of the Amazon, the Madeira, and the lower portions of other major tributaries, where aboriginal groups were completely acculturated or extinct, he distinguished 11 regions within Brazil as a whole. Because he included changes resulting from acculturation, his areas are less similar to those proposed by Steward and Murdock than the latter are to one another.

An interesting aspect of Steward's map is the

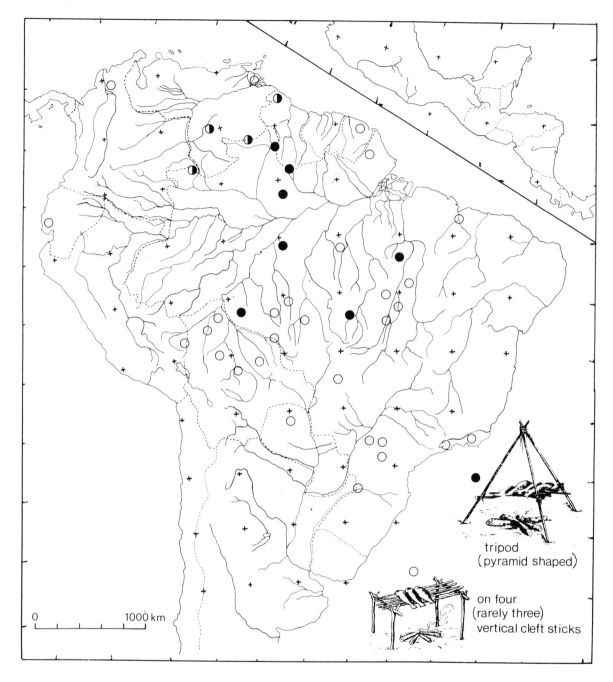

Fig. 6.6. Distributions of two types of babracots. The tetrapod is widespread, both within Amazonia and beyond its margins. The tripod is concentrated in the Guianas and also occurs on several southern tributaries of the lower Amazon. Both types are used on the northern frontier of the rain forest. Since the two varieties do not appear to differ in efficiency, the spread of the tripod probably implies population expansion. Few of the recorded distributions of ethnographic traits have patterns that can be interpreted in this fashion. (Nordenskiold 1924, map 15.)

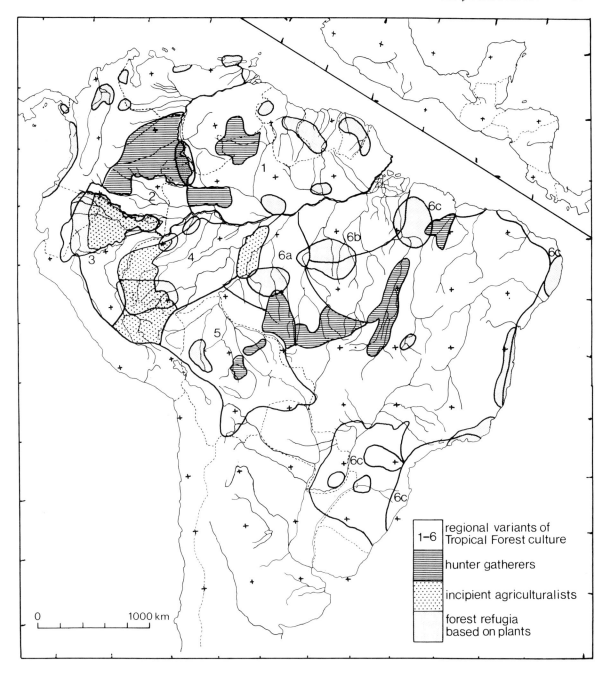

Fig. 6.7. Subdivisions of the tropical forest culture area recognized by Steward, compared with forest refugia inferred from plant distributions (Fig. 3.6). The numbered regions are variations of tropical forest culture; Region 6 has a disjunct distribution on the lower Amazon and along the Brazilian coast. See text. (Steward 1948; map 3; Steward and Faron 1959, p. 275 give a slightly different version.)

existence of several large regions dominated by hunter-gatherers or incipient agriculturalists (Fig. 6.7). On the assumption that domesticated plants and associated traits would have been adopted when they became known, Steward attributed the absence of more than rudiments of the tropical forest pattern of culture among these groups to their remoteness and isolation (1948, pp. 883–4). This led to the further inference that 'what is thought of as a typical Tropical Forest or silvan culture . . . flowed along the coast and up the main waterways, stopping where streams were less navigable and leaving the hinterland tribes on a more primitive level' (1948, p. 883). In keeping with this hypothesis, he ruled out any significant diffusion from the Orinoco Basin or the Andes, and suggested that tropical forest traits reached the western margin of the lowlands by spreading around rather than across regions occupied by hunter-gatherers (1948, p. 885). Métraux, noting that many of the hunter-gatherers of the Rio Negro Basin spoke unaffiliated languages, considered they might be 'the last representatives of an ancient people who occupied vast areas of the Amazon Basin before they were exterminated or assimilated by the

Carib, Arawak, and Tucano, the carriers of a more advanced culture based on farming' (Métraux 1948, pp. 865 and 861).

The validity of Steward's interpretation can be assessed from two perspectives. One is environmental: are these habitats unsuitable for more than rudimentary practice of shifting cultivation? The other is historical: does this subsistence pattern represent deculturation of former slash-and-burn farmers, who have been pushed into inferior habitats? Archaeological evidence for the earlier presence of pottery making eliminates the first alternative in several regions. The second may apply to some groups, but others exhibit a degree of linguistic divergence consistent with the hypothesis that they are relicts of populations that inhabited Amazonia prior to the appearance of agriculture.

In summary, the ethnographic evidence is characterized by heterogeneous distributions of a variety of adaptively neutral traits, inexplicable in terms of existing environmental conditions, and by enclaves of hunter-gatherers and incipient cultivators, some of which occupy habitats where shifting cultivation can be practised.

LINGUISTIC EVIDENCE

Potentially, linguistic data have several advantages over other categories of cultural evidence for detecting movements of people in the past. The processes of linguistic and genetic change are similar (Sankoff 1973) and linguists have adopted a uniform approach to analysis and a hierarchical type of classification that facilitate comparisons. Similarities in vocabulary, regular shifts in sounds, and resemblances in structure are the principal criteria employed to establish relationships, which can be expressed as family trees. Original centres of dispersal can be postulated, but the locations of speakers at times of subsequent divergence are more speculative (Diebold 1960). Degrees of relationship (lexicostatistics) can be converted into absolute time scales (glottochronology), using a constant based on the average of rates of change observed among languages that have been written for several millennia.

Linguistic data are particularly important for reconstructing prehistory in Amazonia because they provide greater time depth than archaeological re-

mains. The realization of their potential contribution is hampered, however, by deficiencies in information. Numerous languages have perished unrecorded. Word lists are often impaired by distortions in meaning, ambiguities in transcription, and briefness of vocabulary. The major language families (Arawak, Tupi, Carib) have undergone revision, but authorities do not agree on some of the affiliations. Two continent-wide classifications, by Mason (1950) and Loukotka (1967), are valiant efforts to bring order out of chaotic data, but their reliability is uneven. Many 'isolated' languages might be included in a family if more information were available; others have been assigned an affiliation although not a single word has been recorded. Greenberg (1969) and Swadesh (1959) have boldly grouped into a small number of phyla what Loukotka separated into 107 families and 27 unclassified languages, but their alignments also differ considerably (Rodrigues 1974, pp. 55–6). In spite of these handicaps, several kinds of general patterns can be observed.

Long-range dispersals

A striking characteristic of the four major language stocks of lowland South America is their wide dispersal (Fig. 6.8). Arawak languages are spoken in the southern Peruvian highlands, large parts of the western lowlands, Venezuela, the coast of the Guianas, and the Antilles. Carib is concentrated in the northeast, but has outliers in western Venezuela, eastern Colombia, and southern Amazonia. Tupi dominates the region east of the Madeira and south of the Amazon, as well as the Brazilian coast. Macro-Pano-Tacanan is mainly in the southwest with a northern outlier. Degrees of similarity have been established for many pairs of languages in each stock (Noble 1965; Rodrigues 1955, 1958; Durbin 1977; Migliazza 1982). These data have been reviewed and amplified by Migliazza (1982) and interpreted in the context of the refugia model.

Rodrigues (1958) classified Tupi languages into seven families and one isolate, Purubora (Fig. 6.8(c)). Six are concentrated south of the Madeira and north of the Guaporé, suggesting this was the locus of differentiation. One family, Tupi-Guarani, spread not only over large parts of southeastern Amazonia and upriver to the Ucayali, but also to what is now Paraguay and along the Brazilian coast. Migliazza (1982) has distinguished two general periods of diversification. The first begins about 5000 BP, when Proto-Tupi separated from related languages, and ends about 2000 BP, when the separation into families was completed. The second, between 2000 BP and European contact, is characterized by differentiation within families and by long-range dispersals of speakers of Tupi-Guaranian (Fig. 6.8(c)).

Noble (1965) constructed a family tree to show the sequence of diversification of the languages he assigned to Arawak (Fig. 6.8(a)). He distinguished seven families and estimated their initial separation about 3300 BP. As occurred with Tupi, one family achieved a very wide distribution. Speakers of Maipuran reached the Venezuelan coast on the north, the mouth of the Amazon on the east, the headwaters of the Rio Paraguay on the south, and the Andean highlands on the southwest. Noble suggested that the centre of dispersal was in the southeastern lowlands of Peru, where Proto-Arawak differentiated between 5000 and 3500 BP, and that the Maipuran expansion occurred around 2000 BP.

Migliazza (1982) has proposed a somewhat different phylogeny, placing Guajiro (spoken on the Guajiro peninsula of Colombia) and Taino (spoken in the Greater Antilles) in one branch and the remaining families in the other. This implies two widely separated loci of dispersal after c. 4500 BP, one centring on Lake Maracaibo and the other in southwestern Amazonia (Fig. 6.9). Such a disjunction might reflect loss of earlier continuity over the intervening region or a two-way dispersal from an intermediate homeland. The subsequent chronology of diversification is similar to that of Tupi, with most of the families originating by 4000 BP and subdividing after 2000 BP.

Carib languages have a more compact distribution, extending from central Venezuela across the Guianas to the north bank of the lower Amazon (Fig. 6.8(b)). There are disjunct occurrences in northern and eastern Colombia, immediately south of the lower Amazon, and at the headwaters of the Tapajós and Xingu. Durbin (1977, p. 34) identified some fifty languages and made a preliminary classification. He recognized four general subdivisions in the Guianas; of these, Southern Guiana Carib is most closely related to the languages in southeastern Colombia and the upper Xingu. Initial internal differentiation is estimated to have occurred c. 4500 BP in the region between Venezuelan and French Guiana. Some 56 per cent of the separations took place between 3400 and 2400 BP, and only 14 per cent between 2300 and 1000 BP (1977, pp. 35-6). Migliazza (1982, Table II) provides a somewhat different chronology, placing crystallization of the major groups between 4000 and 2000 BP, and their internal diversification after 2000 BP. In contrast to Arawak and Tupi, the long-range dispersals of Carib appear to date prior to 2000 BP.

A fourth stock analysed by Migliazza, Macro-Pano-Tacanan (Fig. 6.8(d)), is of particular interest because it appears to belong to a phylum that includes Chibchan (spoken in Colombia) and Fuegian languages (spoken at the southern extreme of the continent; Migliazza 1982, p. 510). Proto-Pano-Tacanan speakers settled in southwestern Amazonia around 5000 BP. Tacanan, Panoan, and Yanomama had differentiated by about 2500 BP. The first two language families underwent further diversification after c. 1500 BP, but did not spread far from the homeland. Speakers of Yanomama, by contrast, moved to north-central Amazonia between the Rio Branco and the headwaters of the Orinoco. This stock thus shares with Arawak an ancient long-range dispersal of the proto-language along the western margin of the continent. Unlike Arawak, however, later expansion over the lowlands was minimal.

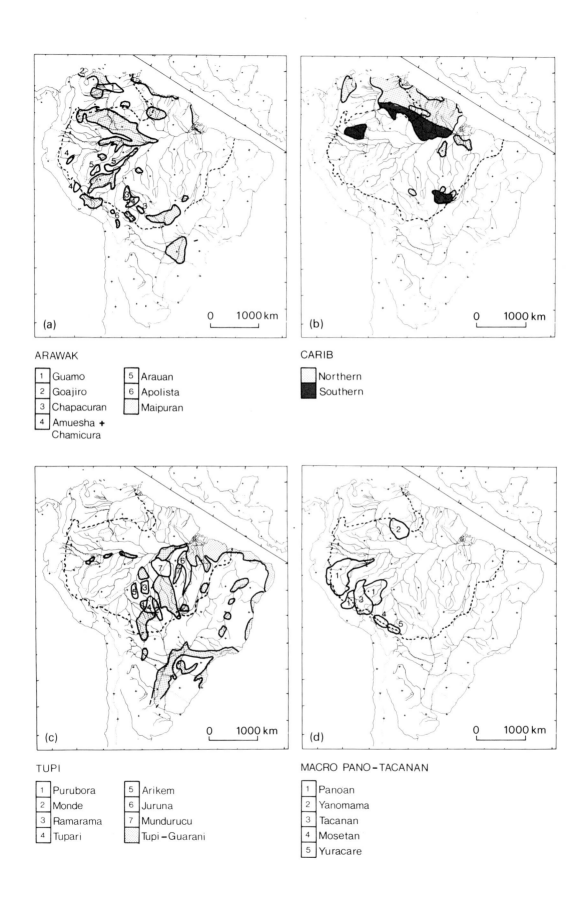

ARAWAK

1	Guamo	5	Arauan
2	Goajiro	6	Apolista
3	Chapacuran		Maipuran
4	Amuesha + Chamicura		

CARIB

☐ Northern
■ Southern

TUPI

1	Purubora	5	Arikem
2	Monde	6	Juruna
3	Ramarama	7	Mundurucu
4	Tupari		Tupi – Guarani

MACRO PANO – TACANAN

1	Panoan
2	Yanomama
3	Tacanan
4	Mosetan
5	Yuracare

Disjunct distributions

All the major and several of the minor linguistic stocks of the lowlands contain families with disjunct distributions (Fig. 6.8). In Arawak, only two of the seven continental families occupy a contiguous region. In Carib, languages belonging to the two major divisions are widely separated geographically. In Tupi, five families have small and compact territories, one is divided by a narrow gap, and one is spoken in disjunct locations over much of the southeastern lowlands and on the upper Amazon. Macro-Pano-Tacanan contains five families, one of which occurs in three disjunct localities. Minor families with two or more disjunct occurrences include Tucanoan and Puinavean-Macú, spoken in northwestern Amazonia (Mason 1950, Map). The existence of disjunctions within stocks, within families, and within languages implies several episodes of disruption of the communities of speakers.

Isolated languages

All classifications contain a residual category of 'isolated' languages. Nimuendajú, whose identifications were based on considerable first-hand knowledge, recognized 42 stocks and 34 isolated languages in the tropical lowlands, leaving a residue of hundreds of unclassified languages (Mason 1950, pp. 166-7). Some of the latter could probably be assigned to families if sufficient information were preserved; others may be relics of ancient stocks or products of early separations in the phyla to which the more widespread stocks belong. Many isolated languages are associated with very small populations but some, for example Jívaro, are spoken by several thousand people inhabiting large regions.

Centres of diversity

Several parts of the lowlands are characterized by remarkably high local diversity, shown on Fig. 6.9. In the Uaupés drainage, more than 25 languages are spoken by groups that share a common culture (Sorensen 1967). Three stocks are represented, Eastern Takano with 13 languages, Arawak with two

languages, and Tupi with one language. Five or more additional languages are of uncertain affiliation. Most of the inhabitants speak two or more indigenous languages as well as Spanish or Portuguese. They are careful to preserve linguistic fidelity and there is no tendency toward either diminution of diversity or loss of 'purity'.

Another complex mosaic occurs in the western Guiana highlands (E. Migliazza, personal communication). Around AD 1800, two major language families dominated the region: Arawak with 13 languages or dialects and Carib with 20. Interspersed were enclaves of speakers of Saliban (four languages), Yanomama (four), and six independent languages.

A third region exhibiting high diversity is the upper Xingu. In 1948, a total population of less than a thousand was divided among 14 tribes, of which six spoke Carib languages, four Arawak, two Tupi, one Ge, and one an isolated language (Oberg 1953, p. 4). As on the Uaupes, linguistic diversity contrasted with cultural uniformity and socioeconomic integration (Basso 1973, pp. 3-4).

A similar situation prevails in eastern Bolivia, which has been characterized as 'one of the most diversified cultural and linguistic areas of any part of South America' (Steward and Faron 1959, p. 349). In addition to a number of localized languages, three major stocks are represented: Pano with five languages, Tacanan with two, and Arawak with at least three.

Phyletic reconstructions

Aside from the limitations of the data base, which prevent using a single standard word list and the same structural criteria for all comparisons, reconstructing phylogeny is hampered by the increasing ambiguity that accompanies increasing time depth. When the degree of similarity between two languages is reduced to 5 per cent shared cognates, the possibility that these result from chance rather than genetic relationship is difficult to evaluate. Depending on the rate of retention employed, the antiquity attributable varies from c. 2900 (65 per cent per millennium) to c. 9200 BP (85 per cent per millennium). At the rate generally adopted as a constant, 80.5 per cent, two languages

Fig. 6.8. Distributions of the four major linguistic stocks of lowland South America. Arawak (a) and Tupi (c) each contain one family that dispersed widely beginning c. 2500-2000 BP. Groups of Carib speakers (b) moved to distant locations in the south and west a few centuries earlier. Yanomama (d) separated from other members of the Macro–Pano–Tacanan stock c. 2500 BP and moved to the north central lowlands. (After Migliazza 1982, Figs. 27.1, 27.3, 27.6 and 27.7.)

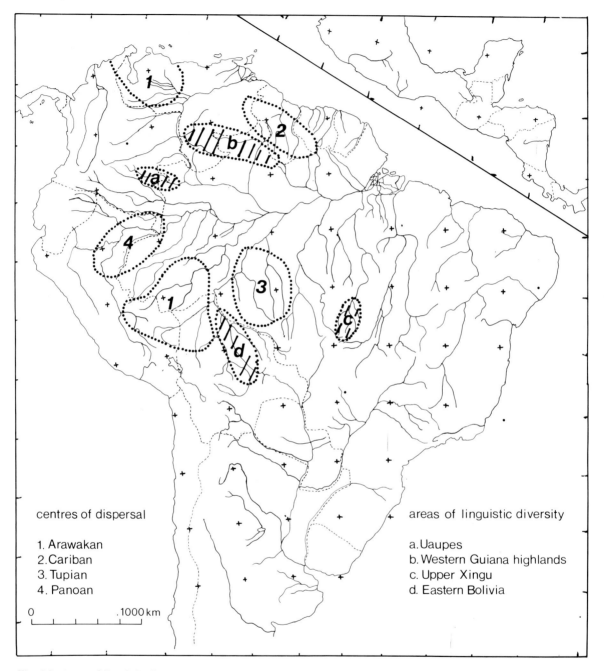

centres of dispersal

1. Arawakan
2. Cariban
3. Tupian
4. Panoan

0 _____ .1000 km

areas of linguistic diversity

a. Uaupes
b. Western Guiana highlands
c. Upper Xingu
d. Eastern Bolivia

Fig. 6.9. Areas of linguistic diversity and centres of dispersal of major stocks; the latter are inferred from the phylogeny of the stocks and the recent locations of speakers (Fig. 6.8). Arawak differentiation preceeded from two widely separated regions, established by c. 4500 BP in northwestern Venezuela and southwestern Amazonia. The centres of dispersal for Tupi and Macro–Pano-Tacanan lie to the east and northwest of the southern Arawak centre. The Carib centre is on the opposite margin of the basin, in the Guiana highlands. Centres of dispersal after Migliazza (1982, Fig. 27.9). Centres of diversity: Uaupés (Sorensen 1967), West Guiana highlands (Migliazza, personal communication), Upper Xingu (Oberg 1953), Eastern Bolivia (Steward and Faron 1959).

sharing 5 per cent' cognates have an antiquity of separation of about 6900 years. This is less than half the duration of human occupation of South America, as indicated by archaeological evidence.

Few efforts have been made to establish relationships among South American language stocks and the results are inconsistent. The best known is by Greenberg, who recognized three phyla: (i) Macro-Chibchan with two major divisions; (ii) Andean-Equatorial with four; and (iii) Ge-Pano-Carib with six (Greenberg 1960; also Steward and Faron 1959, pp. 22–3). His Equatorial division of Andean-Equatorial includes Arawak and Tupi, as well as numerous smaller families and isolated languages. Macro-Ge, Macro-Carib, and Macro-Panoan comprise three divisions in Ge-Pano-Carib. Greenberg has not published details of his analysis and some alignments have been disputed on the basis that the similarities are typological rather than genetic (Sorensen 1973, pp. 336–7). Rodrigues (1974, pp. 55–6) considered Tupi related to Carib rather than Arawak, and Davis (1968, p. 47) suggested Ge may be related to Tupi. Migliazza (1982) compared Tupi and Carib, applied lexicostatistics, and arrived at a tentative time of separation c. 4000 BP. He also compared Arawak with other American proto languages and found a closer correspondence with Mayan than with any family in South America.

In summary, the languages of lowland South America display the following kinds of features: (i) long-range dispersals; (ii) disjunct distributions within stocks and families; (iii) relict occurrences associated with small populations; and (iv) pockets of high diversity.

SIMILARITIES BETWEEN CULTURAL AND BIOLOGICAL PATTERNS

Archaeological, ethnographic, and linguistic evidence all display features similar to those interpreted by biogeographers as implying discontinuity in the environment of the tropical lowlands during the Quaternary. Like some of the biological data, the patterns exhibited by different kinds of cultural phenomena do not seem to correlate with one another. The times, centres, and routes of dispersal of the archaeological ceramic traditions are completely different from those inferred for the major linguistic stocks. Since the distributions of both kinds of cultural phenomena seem likely to involve movements of people, these discrepancies are puzzling. Subsequent linguistic dispersals might eliminate languages associated with earlier ceramic traditions, but the dates for the spread of the incised-and-punctate tradition are sufficiently recent that a linguistic counterpart ought to be detectable if one existed. Extinction of most of the inhabitants along the lower Amazon before their languages were recorded and the post-European intrusion of speakers of different languages from the Atlantic coast may prevent its recognition. Differences between the centres of dispersal of this ceramic tradition (Fig. 6.3) and those of Arawak and Carib languages (Fig. 6.8) seem to rule out affiliation with either of these stocks. Sites of the polychrome tradition occur where Tupi, Arawak, and Carib, and various other languages were spoken after AD 1600 (Figs. 6.3 and 6.8).

When each category of cultural data is compared separately with patterns derived from biogeography, a few similarities can be observed. Müller's reconstruction of arboreal and non-forest dispersal centres leaves a corridor stretching eastwards from western Venezuela to the lower Amazon, bounded on the south by the Rio Negro (Fig. 7.2). This corridor would have been an appropriate route of introduction for the zoned hachure tradition from northern Colombia to the mouth of the Amazon (Fig. 6.1). It also incorporates early sites of the polychrome tradition, which has affiliations with older pottery from northwestern Venezuela (Fig. 6.2).

The three principal rain forest centres of terrestrial vertebrates proposed by Müller correspond to the regions dominated by the three principal linguistic stocks of the lowlands, as can be seen by comparing Fig. 7.2 with Fig. 6.8. His Centre 22 (Guianas) equates with Carib, Centre 24 (east of the Madeira and south of the Amazon) with Tupi, and Centre 25 (western Amazonia between the Negro and the Madeira) with Arawak. These regions were also important for differentiation among forest birds (Fig. 5.2 J, I, and F + G). Müller (1973, p. 201) interprets his vertebrate dispersal centres as centres of origin and as forest refugia, which also agrees with the linguistic reconstructions suggested by Migliazza (compare Figs. 6.9

and 7.2). The Carib dispersal centre equates with Müller's Centre 22, the Tupi centre is in the southern part of Centre 24. Migliazza's Panoan and southern Arawakan dispersal centres are now the western part of Müller's Centre 25. There is in addition a general similarity between the directions of linguistic expansion proposed by Migliazza (1982, Fig. 27.9) and those suggested by Haffer (1974, Fig. 9.13) for forest faunas.

Superposition of the forest refugia as deduced from the distribution of a variety of plants (Fig. 3.6) on to Steward's map of subdivisions within the tropical forest culture area (Fig. 6.7) shows regions dominated by hunter–gatherers to be between or peripheral to those considered permanently forested. This negative correlation might signify that these groups are remnants of an early population adapted to the more open environment prevailing at the end of the Pleistocene, who retained this way of life during subsequent climatic oscillations. The fact that many speak unaffiliated languages favours ancient divergence rather than recent deculturation. A positive correlation exists between the two largest refugia and the two regions in western Amazonia where wild foods remain more important than domesticated plants (incipient agriculturalists). In this case, one might speculate that long-term occupation of a relatively stable rain forest habitat allowed the evolution of a mixed subsistence strategy that was equally or more reliable and productive than slash-and-burn agriculture.

Zones of linguistic diversity tend to correlate with zones of contact among birds. The region between the Branco, Uaupes, and middle Orinoco, where a large number of upper Amazonian and Guianan bird species have established secondary contact (Fig. 5.22) is characterized by linguistic heterogeneity (Fig. 6.9). Another zone of secondary contact shown by birds runs across southern Amazonia close to the upper Xingu, which is noteworthy for its linguistic complexity. Only in eastern Bolivia does a faunal counterpart for linguistic diversity seem lacking, but major zones of secondary contact occur somewhat to the north and west. Since zones of secondary contact result from interaction between forest populations formerly separated, they should correspond to the last frontiers in coalescence of the forest. Archaeologically, this should be reflected in relatively slight time depth for shifting cultivators and hence for ceramic sites. Indeed, there is good correspondence between regions where archaeological remains are

recent (Fig. 6.4) and linguistic diversity is high (Fig. 6.9).

The existence of methods for assigning time scales to linguistic diversification and archaeological sequences provides an opportunity for examining whether periods of environmental change correlate with significant cultural events. The climate is believed to have been more arid than at present between c. 5000 or 4000 and c. 2500 or 2000 BP, allowing expansion of open landscapes (as was described in Chapter 1). There is a general correspondence between these dates and the estimates for linguistic differentiation (Fig. 6.10). The Arawak, Tupi, and Macro-Pano-Tacanan stocks emerged at the beginning of the episode and had separated into families by its end. Long-range dispersals over the lowlands followed resumption of humid conditions.

The attempt to discern more specific correlations is frustrated by the scarcity of detailed local chronologies. It is generally accepted that climatic oscillations during the past five millennia were more restricted geographically and shorter than during the Pleistocene. Regional variation is evident when the sequences reconstructed for the lower Magdalena Valley of Colombia and the central Amazon, discussed in Chapter 1, are compared with those observed on the coast and interior of southern Brazil (Fairbridge 1976; Bigarella 1971). Where refined local chronologies are available, gaps in the archaeological records appear to coincide with periods of aridity. On Marajó, for example, a succession of five prehistoric cultures is broken between about 2700 and 2000 BP, suggesting the island was abandoned by pottery making groups during the corresponding arid episode (Fig. 6.5). A hiatus has also been reported in the occupational debris of several rock shelters in southern Goiás between c. 4500 and 1500 BP, when the supply of wild foods exploited by hunter–gatherers was reduced as a consequence of climatic change (Schmitz 1980, p. 222).

There are other important factors to consider in interpreting cultural data, which biologists are seldom obliged to confront because of less precise methods of dating events. The duration between the initiation of divergence and the emergence of a distinct ceramic tradition or language is unknown and likely to be variable. How should this variation be incorporated into correlations with climatic oscillations? The differential sensitivity of hunter–gatherers and agriculturalists to short-term aridity is another consideration. Existing dating indicates the major

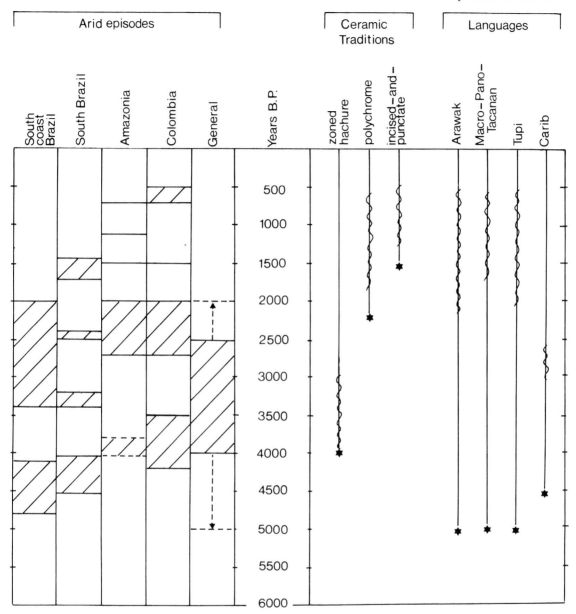

Fig. 6.10. Relationship between arid episodes during the past 5000 years and the dispersals of languages and ceramic traditions. In spite of differences in the estimated dating of climatic and linguistic events, the emergence of the major languages tends to coincide with the inception of the last arid episode and their dispersal to follow restoration of humid conditions. Among ceramic traditions, the zoned hachure dispersal correlates with drier conditions and the polychrome and incised-and-punctate dispersals with humid conditions. See text for further details.
*differentiation from proto-language or tradition. ———— differentiation into families. ᐱᐱᐱ dispersal through lowlands.

ceramic traditions were introduced into Amazonia from the northwest, perhaps reflecting population dispersals provoked by local drought and famine. In Amazonia, the agricultural subsistence base of the invaders may have given them a competitive advantage over groups depending on wild foods or lowland cultigens in certain types of habitats, especially the várzea.

Similar patterns are exhibited by cultural and biological phenomena in spite of the different time scales involved and the many variables affecting their genesis, preservation, and reporting. The correspondences between regions of high linguistic diversity and zones of secondary contact among forest fauna (Fig. 6.9) are particularly noteworthy. In the upper Xingu, linguistic diversity has been attributed to the expulsion of indigenous groups from the surrounding region by modern colonists. This may be a factor elsewhere. Why, then, is there a correlation with much older biological 'frontiers'? If humans are not subject to the adaptive constraints acting on other organisms, the cultural and biological records would be expected to exhibit distinct patterns. The fact that correspondences can be observed in spite of gross deficiencies in the data implies the refuge model is relevant for explaining aspects of human history. The operation of many other kinds of factors certainly also contributed to the results, including considerations reviewed in Chapter 7.

7

CONCLUSIONS, SYNTHESIS, AND ALTERNATIVE HYPOTHESES

SUMMARY

Neotropical biogeography is still in its infancy. Data are woefully scanty on distribution, biosystematics, natural history, ecological interactions and evolutionary rates and trends of both plants and animals. By contrast, geoscientists can claim to have made real progress in the broad understanding of the restructuring of neotropical landscapes during the late Quaternary. Knowledge of geology, topography, geomorphology, hydrology, present and past climates, and physicochemical limnology has broadly advanced in the last twenty years together with some understanding of palynology, palaeontology, and the historical dynamics of vegetation types. These data, which indicate great variations in the extent of humid climates and forests during the Quaternary, help focus on the important questions to ask about neotropical biogeography.

Biogeographical patterns which have been described for various groups of neotropical plants and animals are briefly reviewed (Figs. 7.1–7.3). It is noted that, where the data are scanty, simplistic conclusions have sometimes been presented about evolution and species diversity. It is common, for example, for collections to come mainly from near to rivers and major settlements. Data are missing or scarce for large areas, giving biased biogeographical patterns.

Species are thought to have evolved at different rates within and between groups. For example, most modern insect species are thought to date from the Tertiary whereas many modern bird species are thought to have evolved in the Pleistocene. Some modern frog species diverged from their closest relatives in the early Tertiary while others are Holocene in age. Changes in forest extent during the Pleistocene will have affected different species differently, depending upon their ecological amplitudes and genetic plasticities. Therefore no one species or group can reflect in its biogeographical patterns more than a fraction of the possible influence of past landscape restructuring. The dispersal capacity of a stock will also influence the number of its subunits recognized today, thus confusing matters further. This again emphasizes the necessity to base biogeographical conclusions upon broad sampling with good population and ecological data to build up a full picture.

Hence there are many reasons why different groups should show radically different patterns of differentiation but they do not. Despite the inadequacies of the data there is a good correlation between the patterns of endemism in plants, butterflies, birds (Fig. 7.4) and other organisms (Figs. 7.1–7.3). Such close similarity begs explanation. The superimposition in Fig. 7.5 of the biogeographical patterns (Fig. 7.4) on the geoscientific model of palaeoecological forest refugia (Fig. 2.8) shows a close fit. Varying ecological preferences and capacities for genetic response to the environment notwithstanding, there appear to exist real and definable centres of endemism for a variety of organisms which collectively show a strong correlation with an independent palaeoecological model of discrete refugia to which the humid tropical forest was confined at the end of the last, Wisconsin–Würm, glaciation.

'Refuge biogeography' has sometimes been uncritically espoused for groups on which information is scanty. This has laid it open to criticisms. Two main alternative interpretations exist. 'Vicariance biogeographers', epitomized by Leon Croizat-Chaley, deny that Pleistocene climatic fluctuations have had important effects on biogeographical patterns in the neotropics, which they consider arose from wide diffusions followed by isolations resulting from geotectonic events over a much longer time span. In fact, refuge biogeography, when carefully applied, is but a recent chapter of vicariance biogeography. The other view comes from population biologists. They have pointed out that a combination of parapatric, sympatric, and stasipatric differentiation in a complex, heterogeneous tropical landscape could lead to the observed biogeographical patterns with no need to invoke past landscape restructuring. Niche or chromosomal speciation is believed possible even under negligible environmental pressure. In Fig. 7.6 zones of environmental conformity and rapid

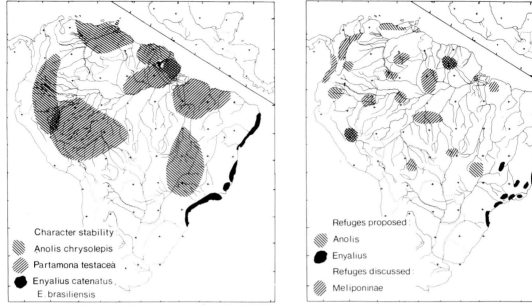

Fig. 7.1. Biogeographical patterns revealed by quantitative morphometric analysis of variation in neotropical forest animals. Left: areas of relative stability of characters examined in lizard (*Anolis, Enyalius*) and bee (*Partamona*) species. Right: proposed refuges corresponding to these evolutionary patterns and others discussed for Meliponine bees and favourable geoscientific evidence. Transects were taken for 177 localities in *Anolis* and 61 in *Partamona*; samples from 113 localities were analysed in *Enyalius*. (Vanzolini and Williams 1970; Vanzolini 1970; Jackson 1978; Camargo 1978, 1980.)

transition are superimposed on the observed endemic centres and hybrid zones for butterfly subspecies (Fig. 4.15) and contact zones for birds (Fig. 5.22). The fit is about as good as between endemic centres and forest refugia in Fig. 7.5. The conclusion drawn is that both interpretations are probably valid but to different extents and in different ways, with vicariance biogeography too on a longer time scale. Refuge biogeography must take its place with other models to help to explain the complex evolutionary history of any taxon observed today.

It is shown that evidence which has been published to contradict the geoscientific model can be reconciled. Even at the height of dry periods rivers were fringed by gallery forest (Chapter 2), so palynological samples from river valleys are very likely to show forest continuity. Different simulations for the 18 000 BP Atlantic pressure cells considerably alter wind direction and rainfall over South America. But the changes in the geoscientific refuge model, Fig. 2.8, are very slight as rainfall is only one of the five data sets used in its construction. The geoscientific model and biogeographical patterns are still crude. Further effort is needed to substantiate and refine them. In particular, there is still very little palynological data from lowland Amazonia, and little is known about relations between local vegetation and pollen rain. The impact of past landscape restructuring on man remains enigmatic. We need a much better picture of environmental variation for the period since he entered Amazonia. At present biogeographers and anthropologists are largely talking about different time scales of change.

The most important practical result of the ideas and data presented in this book has been and will continue to be in planning conservation and land-use in the neotropical forests. It has been discovered, for example, that heterogeneous landscape mosaics may contain the highest genetic diversity. Conservation areas need to encompass both the nuclei and the periphery of centres of endemism. Some tropical populations occur at low density and suffer unpredictable local extinction but are good at recolonization. Big enough continuous areas are needed to allow for these phenomena. Reserves are being created on the basis of the data reviewed in this book; since 1975 conservation areas totalling 100 000 km² have been established (Fig. 7.7). This is over half the total area of the neotropics at present under conservation. These new areas encompass 28 of the 44 endemic centres of butterflies (Fig. 4.15) and 26 of the 62 forest refugia (Fig. 2.8). Further proposed areas are still needed, followed by consolidation and then the preparation of management plans.

THE IMPORTANCE OF GEOSCIENTIFIC DATA IN BIOGEOGRAPHY

The different strands of geological and biological evidence for palaeoecological phenomena, which are united and interpreted in this book, show that neotropical biogeography is still in its infancy. Indeed, it is abundantly evident that very basic data – biosystematics, natural history, broad geographical samples, information on fundamental ecological interactions, evolutionary rates and trends – are still woefully scanty for almost all neotropical organisms. Each of the preceding four chapters mentions the problem of inadequate information on the organisms covered, even though these are among the best known in the region. Lamentably few investigators have the will or the way to expand the inadequate biogeographical data-sets, which by their small size, inexactitude, bias, and deceptive simplicity continue to nurture popular and attractive myths about neotropical ecology. As recently as March 1979, an eminent ecological correspondent in *Nature*, commenting on Simpson and Haffer's review (1978), expressed surprise that equatorial South America has been subjected to major climatic change, pointed out the importance of this to ecological theory as if it were a major new discovery, claimed that 'no start has yet been made on pollen analysis in the region' and concluded that ' . . . the important part of the stability of tropical systems is still that areas must always have persisted in which seasons were minimal, because without them the biota of the wet tropics as we know them cannot have survived' (Colinvaux 1979).

In contrast, the geosciences can claim to have made real progress in the broad understanding of the restructuring of neotropical landscapes during the late Quaternary. An examination of textbooks on soil, climate, and vegetation written from twenty to thirty years ago (e.g. Bunting 1965; Money 1965; Mohr and Van Baren 1954) suggests that geoscientists are many years ahead of biologists in understanding the dynamics of tropical ecological processes. Knowledge of geology, topography, hydrology, geomorphology, present and past climates, soil units, palynology, physicochemical limnology, and palaeontology, as well as vegetation types and structures related to these physical factors (as presented in Chapters 1 and 2), offers a reliable basis for interpretation of biogeographical patterns which,

for the most part, are still cryptic. Thus, it is not unexpected that the first major statements of the dynamic viewpoint of Quaternary neotropical biogeography, as discussed in this book, were made by a geologist (Haffer 1967*a*, 1969) and a geomorphologist (Ab'Sáber in Vanzolini and Ab'Sáber 1968; Vanzolini 1970), each resident in South America and deeply interested in the biological reflections of unfolding realities in the earth sciences.

Besides suggesting the most appropriate questions to ask about neotropical biogeography, geoscientific data can give answers to numerous inappropriate questions which somehow continue to distract neotropical biogeographers. Studies of palaeoclimate show that high-rainfall areas have moved about through glacial and interglacial periods, as changing winds interacted with stable orographic features, helping to explain why endemism patterns in humidity-requiring and humidity-shunning organisms do not always correlate with modern isohyet maps. Palaeontology of insects gives no support for massive species formation or extinction during the late Quaternary, but does substantiate some long-distance movements and dispersals of species and faunas, as they tracked down and concentrated in the most favourable environments (Coope 1970, 1978, 1979); this casts doubt on many phylogenetic assumptions based on morphological succession in palaeohorizons. Biotas which have moved about in this way may have created the contradictions between local diversity and endemism values (butterflies, Chapter 4). Soil texture and chemistry can be used accurately to predict fine biogeographical parameters in animals, including some groups of nectarivorous butterflies (Chapter 4), over a wide range of present climate, vegetation types, and latitudes. The strange structure of some aquatic communities in central Amazonian rivers can be related to cyclical chemical and geological changes in their headwaters, thousands of kilometres away. Landforms, revealed by examination of a side-looking radar image may suffice to conclude that a forested region suffered a rapid transition from dry to wet climate while under sparse vegetation, sometime between 10 000 and 15 000 years ago (Tricart 1974; Journaux 1975). Even greater palaeoecological detail can be seen in any of the numerous cores of Quaternary sediments

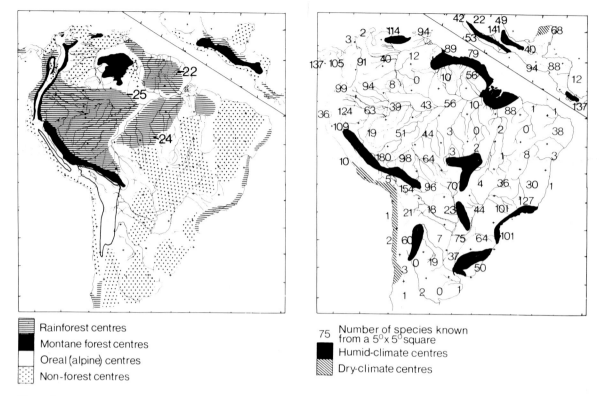

Rainforest centres
Montane forest centres
Oreal (alpine) centres
Non-forest centres

75 Number of species known
    from a 5°x 5° square
    Humid-climate centres
    Dry-climate centres

Fig. 7.2. Biogeographical patterns revealed by 'projection analysis' of restricted ranges in neotropical animals of various habitats. Left: 4817 species and subspecies of vertebrates (Müller 1972, 1973 using his numbering). Right: 446 species and subspecies of hawk moths (Sphingidae) (Schreiber 1978; island species and centres eliminated; Sphingidae are strong flyers, 'depend on montane biomes', and 'differentiate best on islands'). In this method, poorly and well-collected regions (see species numbers known from each 5° square on right map) tend to be respectively underestimated and overestimated, in the determination of the centres of dispersal and evolution. These thus tend to reflect the diversity or distribution of collectors as much as of organisms. The method also shows both the advantages (easy collection of large data-sets) and the disadvantages (spotty area coverage, untrustworthy records) of relatively uncritical literature and museum surveys, unaccompanied by rigorous taxonomic revision and new field work.

in the neotropics taken for palynological study (see Graham 1978 and 1979 for lists of nearly fifteen hundred references on neotropical palaeobotany).

Biologists today are often tempted to incorporate limited biogeographical information, often strongly skewed by fragmentary collecting, into the picture of dynamic Quaternary changes in the neotropical landscape, and yet not refer to the abundant geoscientific data which indicate recent reorganization.

'Such proposals should not be confused with paleoecological refuges based on geoscientific data. We suggest that biologists use terms such as "center of endemism" and "center of species diversity" to describe their data, since even "center of evolution" implies processes which elude direct scientific investigation' (Brown and Ab'Sáber 1979, p. 4); ' . . . the former existence and changing size of the refugia ultimately can be traced only through detailed palynological, pedological, and geomorphological studies rather than through zoogeographical analyses. After all, the refugia are geological–palaeoclimatological, not biological, phenomena' (Haffer 1979, p. 121); 'The location of refuges . . . is the province of geomorphology and paleopalynology rather than systematics' (Vanzolini 1981).

In spite of these viewpoints, 'palaeoecological forest refuges' are sometimes proposed from biological data concentrated along major rivers, or in areas with white-sand podzols, or in lowlands which were under water during much of the present interglacial. It is earnestly hoped that the day will come when abundant palaeoecological data will be included in any discussion of low-level biological differentiation in the neotropics, just as plate tectonic data already are *de rigeur* in discussions of higher-level divisions (genera to families), or modern ecological factors are used in the understanding of adaptations in local populations. The research summarized in this book suggests that, in ten years' residence and field work in the neotropics, a scientist may greatly extend almost any biogeographical data-set, and contribute also to taxonomy, ecology, and practical aspects of conservation and land management. It is urged that such practical experience be regarded as a required component of all in-depth systematic studies which purport to describe the historical biogeography of the region and the evolution and distribution of its biotas.

ADDITIONAL BIOGEOGRAPHICAL DATA FROM OTHER STUDIES

Numerous biogeographical studies on birds, butterflies, and plants have already been discussed in Chapters 3, 4, and 5, in many cases on admittedly inadequate information. Examples in Chapter 3 were specifically noted, in which supposedly 'point endemic' plant species were later collected thousands of kilometres away. Only nine of the 28 recently revised tropical butterfly groups mentioned in Table 4.1 were regarded as showing an acceptable geographical sample. On the other hand, the very substantial coverage already achieved in some bird and butterfly groups, and for plants in Mexico and Venezuela, has begun to reveal a multitude of subcentres of endemism, each with only a small fraction of its local flora or fauna recognizably differentiated. Such studies approach a population-level or ecological focus on local adaptation, in which broader historical factors, which should provide the baseline for modern patterns, are partially obscured by the 'noise' in the local systems. Several recent publications have, however, attempted a broad view of neotropical biogeography, including extensive sampling and analysis of differentiation at the infrageneric level (which is likely to reflect Quaternary evolutionary history). These include detailed research on fruit fly differentiation, beetle species, moth species and subspecies, bee and lizard morphometrics, frog subspecies, monkey semispecies, and geographically restricted vertebrates in general, as well as several important plant groups (Prance 1983).

Vanzolini and Williams (1970; Vanzolini 1970), in a pioneer work on geographical differentiation in reptiles, identified six 'core areas for evolution' (Fig. 7.1) for quantitative morphological characters in the *Anolis chrysolepis* group. In these regions, the variability in the characters was relatively low, while between the regions and especially across central Amazonia it was much higher. This work was extended to the lizard genus *Enyalius* in the eastern part of Brazil in a more recent paper (Jackson 1978) (Fig. 7.1). A very similar 'transect' and morphometric method was also applied more recently to an entirely different group of organisms: large Meliponine bees of the genus *Partamona* (Camargo 1978, 1980; Fig. 7.1). These three studies indicated areas of relative character stability, and then attempted to relate them to geoscientific evidence for recent restructuring of neotropical landscapes. Little consideration was given to the selective significance of the characters chosen for measurement. Vanzolini identified three areas not recognized in Haffer's simultaneous work (1969) and helped to call the attention of later workers to these, as well as to infraspecific variation as a biogeographical tool. Jackson helped to define the complex endemic patterns in the Atlantic forests from Pernambuco to Santa Catarina, reflected in many organisms, and Camargo filled in data for the Amazon basin.

Geographical variation in the *Drosophila willistoni* group of fruit flies was studied by Spassky, Richmond, Pérez-Salas, Pavlovsky, Mourão, Hunter, Hoenigsberg, Dobzhansky, and Ayala (1971) and Winge (1973). They identified several well-defined geographical 'core areas' (in the sense of Vanzolini, with appreciable overlapping) which they considered might be related to the refuges proposed by Haffer. Five such areas were mentioned by Spassky *et al.* for *D. paulistorum* semispecies: Centroamerican, Orinocan, Amazonian, Interior, and Andean–Brazilian (coast). Their work approximated to the broader patterns often seen in species of neotropical organisms, but was more clearly tied to relatively recent evolutionary events visible at a lower taxonomic order. Winge's analysis focussed on finer patterns of

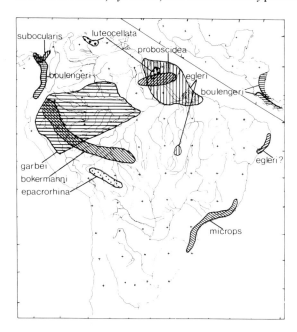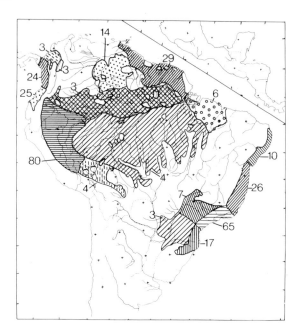

Fig. 7.3. Biogeographical patterns revealed by analysis of limited ranges in members of the neotropical herpetofauna. *Left*: ranges of some tree frogs in the genus *Hyla* (Duellman 1972; Duellman and Crump 1974; modified with data from I. Sazima and J. Lescure). *Right*: endemic regions identified for forest Amphibia and the numbers of species in each (Lynch 1979).

subspecies in *D. equinoxialis*; and chromosome inversions in *D. willistoni*; her detailed data and conclusions have yet to be published.

In 1972, Müller published the results of a broad analysis of geographically restricted ranges for neotropical terrestrial vertebrate species and subspecies (reptiles, birds, mammals, and amphibians), derived principally from a literature survey. Known minimum ranges were superimposed in a 'projection analysis' (Lattin 1957), which led to the recognition of a large number of 'centres for dispersal and evolution', presumably related to a recent regressive phase in the case of forest biotas. Centres were also specified for non-forest and temperate biotas in the neotropical realm (Fig. 7.2). Again, many areas were identified which corresponded to those perceived by other authors, and a few important new ones were recognized and amply justified (Müller 1972, 1973). A similar method was later applied to hawk moths (Sphingidae), identifying 18 centres of distribution and dispersal (Fig. 7.2) based on over 74 000 museum specimens from 1006 localities (Schreiber 1978). While these studies have the advantage of very ample organism and area coverage, they are compromised by erroneous systematic, ecological, and locality data transferred uncritically from collections and the literature.

Since 1972, many papers have appeared relating biogeographical patterns in the neotropics to one or another of the 'forest refuge' models developed by Haffer (1967*a*, 1969, 1974), Vanzolini (1970), Müller (1972, 1973), Lamas (1973, 1976), Prance (1973, 1982*a*), or Brown *et al.* (1974; Brown 1976*a*, 1979). A few of these papers have suggested important new areas for endemism or evolution, or discussed more deeply some of the areas already suspected. Special mention is due to Whitehead (1976) on curculionid beetles and Middle American zoogeography; Duellman (1972; Duellman and Crump 1974; Fig. 7.3), Heyer (1973, 1975), Silverstone (1975, 1976), and Lynch (1979) on amphibians (Fig. 7.3); Hoogmoed (1973, 1979) and Dixon

(1979) on reptiles, and Kinzey and Gentry (1979; Kinzey 1982) on *Callicebus* and other monkeys. Ávila-Pires (1974) and Hershkowitz (1978) gave thorough analyses of monkey differentiation with occasional reference to, but in the latter case rejection of, refuges. Plowman (1979) discussed possible refuge-related differentiation in species of the plant genus *Brunfelsia* (Solanaceae) that possess the added interest of being specific food-plants of *Methona* butterflies (Fig. 4.3; Lamas 1973). He remarked (p. 489) that 'comparisons of the natural ranges of *Brunfelsia* and *Methona* reveal striking similarities, particularly in reference to the isolation of species and subspecies in forest refugia', thus giving a significant co-evolutionary dimension to the differentiation processes. Other interesting discussions of plant distributions and their relationships to palaeoecology are seen in Simpson–Vuilleumier (1971; Simpson 1975) on *Polylepis* (Rosaceae); Simpson (1972) on Rubiaceae; Langenheim, Lee and Martin (1973) on *Hymenaea* (Leguminosae, Caesalpinoideae); Moore (1973) on palms; Soderstrom and Calderón (1974) on bamboos; Morley (1975) on Memecyleae (Melastomataceae); Toledo (1976, 1982) on Central American plants; Sastre (1977) and Gran-

ville (1978, 1981) on French Guyana plants; Lleras (1978) on Trigoniaceae, see Fig. 3.8; Gentry (1979, 1982) on Bignoniaceae; Steyermark (1979, 1982) on Venezuelan plants, and Andersson (1979) on effects of different glacial periods giving 'multi-layered refugia' in the Marantaceae (see Chapter 3 and Prance 1983). Four volumes from international symposia have appeared recently presenting numerous additional papers in neotropical biogeography (Descimon 1977b; Duellman 1979a; Eisenberg 1979; Prance 1982b); these also include analyses of open-vegetation, high-mountain, and temperate biotas. Cerqheira (1982) applied refuge theory to South American mammals.

Mention of geoscientific data-sets can only be found in a minority of recent papers; only a few include broad visions of the entire neotropical realm, and these show characteristic gaps, with minimal sampling in Amazonian interfluvials and other less accessible regions. Concentration of data, resulting from uneven collecting, are not infrequently confused with evolutionary processes (distribution of differentiated organisms), which in turn are frequently and erroneously equated with species diversity as was discussed in Chapter 4.

SYNTHESIS FOR PLANTS, BUTTERFLIES, AND BIRDS

In view of the general deficiency of geographical, systematic, and ecological data for neotropical organisms, the disorganized sampling, the varying choice of taxonomic characters, and the inherent diversity in evolutionary rates, adaptive resources, dispersal capacities, and the environmental predilections of the better-analysed groups, it could only be judged as extraordinary to find any consistent correlation in the respective biogeographical patterns. Such a correlation is nevertheless easily discernible between plants, butterflies, and birds (Fig. 7.4, cf. also Figs. 7.1-7.3) and begs explanation. Several approaches towards this explanation will be discussed here, after a brief survey of the variability in fundamental factors which introduces much 'noise' into the fundamentally correlated system.

Variability in evolutionary strategies and velocities, species concepts, and character analysis

The reproductive strategies of forest trees, land birds,

and flying insects are dramatically different, but no less so than their ecology, their dispersal capacities and rates, the selective pressures at work, and evolutionary trends. The groups selected for the more complete analyses reported in Chapters 3, 4, and 5 were not initially chosen for a particularly close fit to each other or to an 'ideal group' for determining biogeographical patterns, but only because they were satisfactorily ordered systematically, well enough known by resident biologists, and sufficiently well sampled, to suggest that they would show some sort of pattern, other than simply an artefact of inadequate collecting. Almost all were well-defined as humid-forest organisms; occasional relatives with other habitat preferences were eliminated from the picture (a step missing in some other analyses). All were examined at a low but recognizable level of regional taxonomic differentiation, as well as at higher levels to provide contrast in the overall picture which extended to the limits of the neotropical forests.

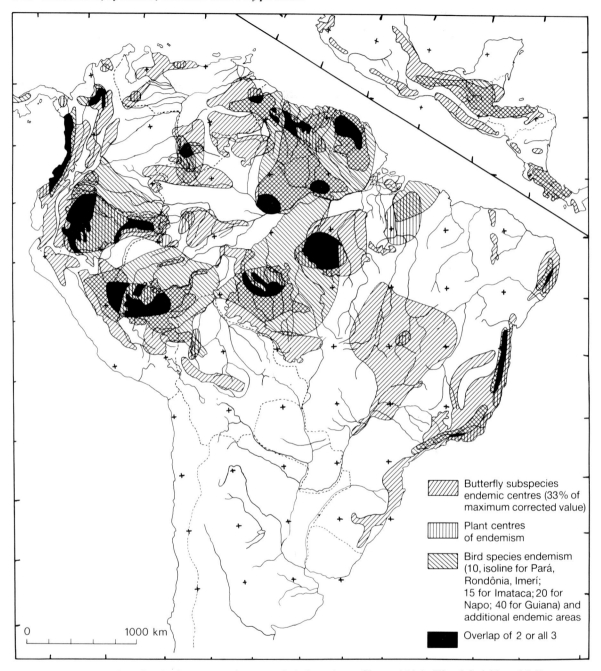

Fig. 7.4. Superimposition of endemism patterns in neotropical plants, butterflies, and birds (Figs. 3.8, 4.20, and 5.2).

It seems clear, from the fossil record, that most modern species of insects date back to before the Pleistocene climatic fluctuations (Coope 1979). On the other hand, there seems to be a common view-point, reflected in the bird analysis (Chapter 5), that many modern vertebrate species are more recent in origin. The cultural differentiation discussed for man (Chapter 6) is infrasubspecific and almost surely post-Pleistocene. Some modern frog and lizard species were probably formed under 5000 years ago; others, by albumin immunological analysis, undoubtedly date from the middle Tertiary (Vanzolini and Ab'Sáber 1968; Heyer and Maxson 1982; Maxson and Heyer 1982; P. E. Vanzolini and W. R. Heyer, personal communication).

It seems neither possible nor desirable to harmonize the current and conflicting views on rates and mechanisms of evolutionary divergence and speciation (achievement of reproductive isolation) in plants, insects, and birds, nor these with cultural evolution in man. A butterfly fossil from the Oligocene could belong to a modern species in the genus *Hypanartia*. Subspecies of *Heliconius erato* which probably have experienced no effective interchange since the mid-Tertiary still regard each other as potential mates. On the other hand, semispecies of warblers, manakins, and finches which may have diverged during the Pleistocene use habitat choice and behavioural displacements to make introgression very rare or impossible. Many flowering plants depend upon animals, cued by specific chemical signals, to help maintain their reproductive isolation, which could thereby be established or destroyed in ecological time. Yet, forest species in all these groups show a similar habitat-related differentiation pattern of geographical varieties, regional subspecies, or semi (allo-)species (Fig. 7.4) at the lower taxonomic levels. This seems to emphasize the importance of some extrinsic factor (such as reorganization of the landscape) in the formation of these patterns.

It *is* desirable to encourage evaluation of geographical differentiation based on carefully chosen characters whose ecological functions and selective regimes (within their integrated genotypes) are known, or at least presumed and clearly stated. If the characters are subject to selection regimes tending to promote rapid local divergence (for example, food plant preference in butterflies, flowering season and hour in plants, nesting material in birds) or which are more stable characters, typical of species-groups (such as wing venation, types of secondary compounds, and global feeding habits, in the three groups respectively), the taxonomic structure may not correspond to a useful biogeographical pattern. Ideal characters for biogeographical analysis at a level corresponding to Quaternary climatic cycles are those which are regionally selected by broad, generalized factors of the physical or biotic environment, such as mimetic colour-pattern or chromosome

complement in butterflies, total flower structure or growth habit in plants, and song structure and plumage in birds (Chapters 3, 4, and 5).

Variation in response to ecological change

Individual species and populations within the same group of organisms also vary greatly in their response to environmental pressure. By definition, we are observing today those populations which have been able to deal with the constantly changing Pleistocene environments, through adaptation, migration, chance residence in more stable habitats, or in some cases fragmentation into new and competing species. Which one or combination of these or many other conceivable mechanisms that a particular population, geographical subunit or entire species has found most successful for survival, is a delicate question of optimization of costs and benefits, which in the heterogeneous tropical environment will obviously differ from place to place, time to time, and species to species. With specific reference to populations confined in enclaves, the complexity of mechanisms for response to environmental changes, and the many possible results of this response, have been discussed by Turner (1977, 1982), Vanzolini (1981), and Haffer (1982).

Recent comparative work on birds, butterflies, and trees in dense virgin rain forest north of Manaus, Brazil (part of the Minimum Critical Size of Ecosystems Project, World Wildlife Fund/National Amazon Research Institute) has shown convincingly the different responses of these three groups to ecological transition in forest environments. Passing from poor yellow latosols on dissected tablelands to richer red-yellow podzolic soils in hilly terrain, within the same endemic centre for all groups, led to complete restructuring of the forest system to give an open-canopy, palmy, well-lit undergrowth vegetation with few major plant species in common with the tall dense forest on the former substrate. Butterfly species lists more than doubled and whole new groups appeared, not seen in forests on yellow latosols. Bird communities remained almost unchanged in composition, diversity, and abundance (see also Lovejoy 1975). These detailed data support the suspicion that any correlated regional biogeographical patterns in these three groups are likely to exist in spite of, rather than because of, their interactions with the environment, and thus be more readily explained by historical factors. Even the regionally correlated endemic centres of different groups of forest butterflies have differing shapes, related to diverse ecological preferences (Fig. 4.16; Brown 1982*b*).

The variation in biogeographical patterns between species and groups, which results initially from different responses to a wide diversity of environmental selection, should lead to a picture in which no one species or small group of closely related species can reflect more than a fraction of the complex picture of landscape restructuring in the past. The more the characters used in taxonomy correspond to selective pressures differing from subunit to subunit, and the higher the survival rate of these subunits, so many more taxonomically recognized stocks would be perceived in the resulting polytypic species or superspecies-complex. Thus the importance of a substantial data-base is emphasized.

The number of subunits detected (i.e. the degree of geographical differentiation seen today) will also be sensitive to the dispersal capacity of the stock, which can vary geographically. Species that are subject to colonization–extinction cycles in the present, or inhabit the sun-baked forest canopy and do not distinguish it from other bright and dry habitats, or which can disperse freely across wide rivers or through gallery or seasonal transition forests, will include widely separated patches of habitat in their ranges. As a consequence they may be expected to differentiate rather poorly in isolated patches of acceptable habitat (Turner 1977). In these species, present-day gene-flow may also help to eliminate any boundaries which may have developed by differentiation in the past. Species or superspecies which show 'relict' distributions today, with many sharply isolated subspecies or semispecies still visible after intensive geographical sampling in and near potential hybridization zones, could have been fragmented by a variety of historical or modern barriers which prevented random gene-flow. This complication adds to the variability in adaptive responses and makes the overall picture still less clear (or deceptively simple, for some), especially when only a few species are examined.

The emerging consensus

Thus, there are many reasons why the groups of neotropical forest organisms which have been examined in depth should show radically different geographical patterns of evolution. However, they do not, and as more and more species are added to each group examined, representing ever finer, more complex, and more restricted patterns of regional geographical differentiation, the pictures have approached each other, and also approached the appropriate geoscientific model for recent restructuring of the landscape (Fig. 7.5; see also Fig. 4.20). The same could be predicted to occur with better sampling and ecological understanding of other groups, including not only forest-restricted organisms but also those typical of other habitats such as páramo, montane rain forest, thorn scrub, field, or any other widespread and at some time discontinuous major vegetation type (with its appropriate palaeoecological model analogous with Fig. 2.8 for tropical forest at the end of the last ice age).

It seems safe to state that there exist real and definable centres of endemism for a variety of organisms in the humid neotropical forests, which collectively show a strong correlation with an independent geoscientific–historical model (Fig. 2.8).

The better known biogeographical patterns (Fig. 7.4) show a basic agreement with many less well-known patterns in other organisms, but cannot be said to validate them. It is easy to demonstrate that a sufficiently limited data-set can be found to correlate satisfactorily with almost any model, since the numerically few (if often statistically important) exceptions can be written off as due to insufficient sampling, leaving the eventual 'true' patterns to suggestion, reinforced by and based upon more ample data sets.

REFUGIA, SPECIATION, AND DIVERSITY

The growing consensus over the importance of Quaternary climatic and vegetational fluctuations in conditioning modern patterns of geographical distribution of neotropical forest organisms has led many specialists to apply refuge biogeography to two areas where there is still very little consensus: speciation and tropical diversity. The divergence of viewpoints on these phenomena among the authors of this book, who have studied different groups of organisms, is evident in comparing the discussions of Chapters

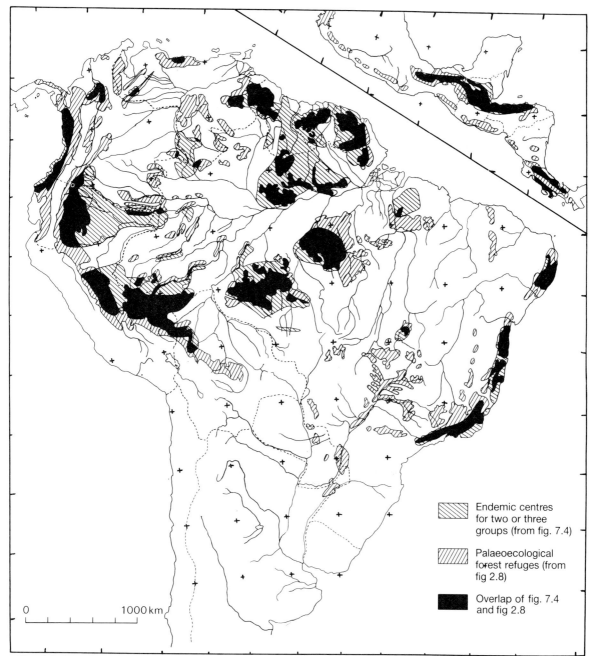

Endemic centres
for two or three
groups (from fig. 7.4)

Palaeoecological
forest refuges (from
fig 2.8)

Overlap of fig. 7.4
and fig 2.8

Fig. 7.5. Superimposition of combined endemism in neotropical plants, butterflies and birds (Fig. 7.4) with the palaeoecological refuge model derived from considerations of palaeoclimate, topography, geomorphology, soils, and vegetation structure
(Fig. 2.8). See also Fig. 4.20 (superimposition for only butterfly endemism). Both this Fig. and Fig. 4.20 show appreciably
larger endemic areas around many of the palaeoecological forest refuges, which is expected; both also show a number of refuge
areas not reflected by endemism (more here), which is an index of poor correlation with the model, except for unsampled or
poorly isolated areas (see discussion in Chapter 4).

3-6. Chapter 3 finds refuge-related differentiation
both in morphology (like that studied by Vanzolini,
Jackson, and Camargo; see Fig. 7.1) giving widespread

plant ochlospecies, and also at species and even
generic levels; like most other plant work, it gives
much importance to species diversity (see Gentry

1982) as well as to endemism (Steyermark 1982) in direct proposal of 'refuges' based on biogeographical data. Chapter 4 concentrates on 'subspecies' defined by regional differentiation in mimetic wing-colour pattern; it finds essentially no influence of refuges on species distributions, which are found to correspond with major physiographical barriers in the neotropics, resulting from Tertiary orogenies; the data presented also suggest a negative correlation of species diversity with refuges (see also Brown 1982*b*). Chapter 5 concentrates on the biogeography of superspecies ('zoogeographic species'), eliminating from consideration the geographical differentiation in polytypic species, and presents a longer view of repeated refuge formation and speciation cycles through all of Quaternary time; it strongly implies that tropical bird species diversity may partly be related to these phenomena. Chapter 6 focuses on cultural differentiation within a single subspecies (race) of man, and carefully weighs problems of cultural diversity against extensive migrations, mixtures, and habitat specialization. One of the pioneers in neotropical refuge biogeography, Paulo Vanzolini (1970), takes a broad view of speciation from the Tertiary to the present, regards Quaternary refuges as very important in this process, and offers an attractive proposal (1973) of diversity enhancement due to repeated isolation of congenerics in small enclaves in which divergence and speciation would be aided by exalted competition – a theme developed further, with added emphasis on local extinctions, by another pioneer in the field, geneticist John Turner (1977, 1982).

There is no question that isolation of populations can and often does lead to divergence. There are still some who believe that this is the only method for animal speciation (see Futuyma and Mayer 1980). Most modern biologists accept it as a 'mainstream model', admitting occasional participation of other mechanisms. On the other hand, White (1978) places heavy emphasis on chromosomal models of speciation, which are practically independent of environment and isolation. The controversy over modes of animal speciation is only partly of interest to botanists, who deal with species one-third of which may have a hybrid origin (Stebbins 1971) and which often acquire adaptive traits through interspecific introgression.

The 1981 volume of the *Annual Review of Ecology and Systematics* is an excellent source for a wide variety of opinions about speciation (half the chapters discuss it). One article (Templeton 1981) is devoted exclusively to the question, and represents by far the

most serious treatment of speciation mechanisms since Bush (1975). Templeton examines speciation from a population-genetics viewpoint with the help of 250 references, and gives equal time to allopatric (classical), clinal (parapatric), habitat, chromosomal, and hybridization modes. He concludes that the first and the last modes are dominant in animals and plants, respectively, and that superficial examination (not predictive but only explanatory) and preconceived notions about speciation mechanisms are highly inadvisable and often misleading. Especially helpful is his clear distinction between the speciation process and its results (the species differences, which are often overemphasized in discussions). Templeton's paper is recommended reading for. those interested in this fundamental and still very controversial area of biological thought.

With relation to neotropical species diversity (also discussed in half the chapters of the 1981 *Annual Review of Ecology and Systematics*), the question must be approached through at least three distinct components. The long-term evolutionary component, seen on a continental scale, has no doubt been exalted by repeated cycles of isolation of populations into smaller areas. It is not established, however, whether the effects of founder speciation, competitive displacement, and skewed selection in these 'refuges' has led more to species multiplication or to species extinction in the frequently supersaturated communities. Whether the inevitable divergence and differentiation (anagenesis) has necessarily been accompanied by extensive speciation (cladogenesis) surely depends upon a myriad of highly variable external factors (including time and total space available, and all sorts of biological selective pressure), as well as intrinsic population parameters (genetic variation, responses to selective pressures, generation time, dispersal capacity, and chromosome architecture), and the size and niche dimensions of each organism. The vast majority of available data on neotropical diversity deal with this very crude level, and have almost no meaning; they only really speak to the broadest potentiality of species available for colonization of given regions or habitats.

The regional multihabitat (beta) component must have also been profoundly influenced by landscape restructuring. The 'relative richness of endemic centres' data presented in Chapter 4 indicates that the preponderant result of this restructuring has been the extinction of species and not their multiplication, at least in butterflies of the core areas of 'refuges' (see Turner 1977, 1982). The local community (habitat or alpha) component of diversity is clearly determined by ecology and not evolutionary factors. It seems to be strongly dependent on the fine structuring of tropical forest systems in different food webs, and on mild disturbance which opens up new niches (Connell 1978; Brown 1978, 1982*b*; Fox 1979; Gilbert 1977, 1980). Thus, very good and voluminous ecological data, in a large number of carefully prescribed points, are now necessary before any convincing association can be derived between species diversity in neotropical systems and past restructuring of the landscape.

Dimensions and limits of the neotropical refuge model

The emerging consensus, which relates low-level endemism in neotropical forest organisms to a geoscientific model for repeated forest reductions, best documented in the late Wisconsin–Würm ice age (Figs. 7.5 and 4.20), has at least two legitimate partners: the fragmentation of species at higher altitudes as their habitats expanded, coalesced and once again fragmented during the glacial–interglacial cycles (Simpson-Vuilleumier 1971; Simpson 1979; Adams 1977; Duellman 1979*b*; Vuilleumier and Simberloff 1980; Haffer 1981) and the local modern evolution of species adapted to open habitats within and around forests, which are restricted today in contrast to their extent in the last ice age (Ducke and Black 1953; Eden 1974; Sarmiento 1975; Brown and Benson 1977; Prance 1978).

Vanzolini (1974, 1976, 1980, 1981; Williams and Vanzolini 1980; Vanzolini and Williams 1981) has dedicated special attention recently to the dynamics and derivation of open-vegetation species, proposing important concepts such as the 'vanishing refuge' which may cause an ecotone-adapted forest population to move over into open vegetation and subsequently speciate. This concept bridges the geographical and ecological speciation models in a single plausible and undoubtedly frequently occurring mechanism. At any rate, all observed biotic patterns may be tied to the same overall model for landscape restructuring, as are related patterns already amply observed in subtropical Central America, the West Indies, and parts of the Old World tropics (Hamilton 1976; Livingstone and van der Hammen 1978; Whitmore 1981). Fuller data from subtropical Patagonia are still awaited to complete the American picture (see Formas 1979; Cei 1979; and Gallardo 1979).

These patterns have also acquired a number of illegitimate partners, in part conceived through partial and erroneous viewpoints about ecological processes in the tropics. The most persistent of these is the attempt to relate a legendary richness of species in tropical systems to such recent restructuring of the landscape, minimizing the ecological components of species diversity. Thus, Gentry (1982, p. 115) affirms that 'one of the prime indicators of refuge areas is the number of ·taxa occurring in a region'; see also Erwin and Adis (1982) and various other chapters in the same volume. Another dangerous partner is a blithe assignment of higher-level taxa to late Pleistocene evolutionary phenomena (Hoogmoed 1979, p. 264); see discussion of this problem by Heyer and Maxson (1982) and Brown (1982*b*). Even the meagre fossil record clearly shows this is uncommon. Nevertheless, the combined volume of premature, and often almost data-free, publications which espouse such views is such that many scientists have become thoroughly disenchanted with the late-Pleistocene refuge model of biotic evolution in the neotropics. Some of them have proposed and discussed other important models. These in fact mostly serve, correctly, to discredit the illegitimate but common abuse of the 'palaeoecological forest refuge model', rather than invalidate its careful application as a geoscientific aid to understanding and predicting some limited aspects of neotropical biogeography. These other models must be carefully examined here.

TWO ALTERNATIVE VIEWS

Vicariance biogeography

A mode of biogeographical theory and interpretation at present popular relates cladistic principles to known or surmised events in the geotectonic history of the Earth, especially on a continental scale. Often classed as 'vicariance biogeography' (Nelson and Rosen 1981) this school has grown largely from the monumental works of Leon Croizat-Chaley in Venezuela, the phylogenetic systematics of the late Willi Hennig in Germany, the scientific philosophy of Karl Popper, and the synthesis of these by Nelson, Platnick, and others in the American Museum of Natural History in New York. Present-day tropical American distribution patterns are accounted for by past geotectonic events separating once continuous, widely diffused populations. Vicariant species are formed, as allospecies.

Some writings of this school (e.g. Croizat 1976) have made vigorous attacks on Haffer's suggestion that Pleistocene climatic fluctuations have had important effects on biogeographical patterns in the neotropics. Other more moderate analyses (e.g. Nelson 1979) have justifiably complained that 'refugiasts' often forget about the far-reaching effects of pre-Pleistocene vicariance phenomena in their analyses of species distributions. A very few papers have admitted that refuge biogeography, when correctly applied, is but a recent chapter of vicariance biogeography on a sub-continental scale, with a more ecological emphasis, observing low-level differentiation, and safely based on abundant and reliably recent geoscientific data-sets.

The authors of this book all agree with this last position, and hope that 'vicariance biogeographers' will join with them in collecting larger and somewhat more ecologically oriented data-sets in the still little-known neotropics. Only in this way may we perceive more of the true results of the many vicariance events on different scales throughout all the Mesozoic and Cainozoic, tempered by strong ecological pressures in recent times, which have constantly shaped populations and determined large and small geographical replacements. The modern biogeographical patterns represent a combination of major past evolutionary events and continual ecological reorganizing forces. Periods of evolution which are less subject to ecological restructuring (such as the tearing apart of continents) can be perceived even though they lie very far in the past. But, in general, it is recent major events which determined the complex patterns which have survived the forces of ecological randomization.

Population ecology

Vicariance biogeography and its 'recent palaeo-ecological refuge' chapter presume that major differentiation takes place in geographically isolated populations or larger subunits of formerly continuous gene-pools. They also tend to employ traditional and often rather arbitrary definitions of morphological characters in classification. But standard and widely accepted taxonomic characters in many groups have been repeatedly shown to have no important or discernible correlation with, for example, mating realities or ecological adaptations of the organisms in the field.

Disharmonic character clines (non-coincident gradients along ecological transects) are nearly universal, rendering biogeographical patterns progressively less clear as more and more populations and characters are adequately sampled. Even 'well-behaved' polytypic species often show stepped clines, as more thorough collecting and character analyses become available, and the steps (regions of more rapid change) are in different places for different character-sets.

Most genera include some species with 'relict' subspecies distributions along with others for which there are no discernible subspecies boundaries. Some species demonstrate one type of division in part of their range, the other elsewhere.

Furthermore, it is increasingly recognized that, in some organisms, differentiation and even speciation can take place along any ecological gradient (Endler 1977, 1982), leading even to clear 'step clines' near regions of rapid environmental change, such as are abundantly present in the neotropics (see Chapters 1 and 2). In the presence of microenvironmental heterogeneity – a universal and fundamental reality in neotropical ecosystems – sympatric fragmentation of populations is also possible which, in the absence of extensive gene-flow, can lead rapidly to speciation (Bush 1975; see Chapter 3). In some groups, with relatively unstable chromosome complements, differentiation or species fragmentation can even occur locally under negligible environmental pressure (White 1978). Plant speciation can occur through hybridization, nearly independent of environmental selection (Templeton 1981; Grant 1981).

Thus, population biologists have pointed out that a combination of these parapatric, sympatric, and stasipatric differentiation models, with the emerging picture of environmental heterogeneity in the tropics, makes past landscape restructuring completely unnecessary to explain the complex biogeographical patterns observed – especially when these are often based upon arbitrarily chosen characters and deficient geographical and populational data (Endler 1977, 1982; Lynch 1979; Benson 1982). They have proposed to relate the verifiable patterns to a process of differentiation along clines and across environmental gradients, transitions, and barriers, including broad rivers, and have accumulated impressive data to support their claim that the correlation of biogeography with modern ecology is at least as good, and is far easier to verify, than that with presumed (and conveniently non-testable) palaeoecological hypotheses.

To the extent that the population biology models are testable by careful studies of modern populations and their environments, they are very important for biogeographical science, especially in the neotropics, where, happily, the vast majority of these populations and environments are still essentially unaltered. A fairly simple (if not definitive) test can be performed by superimposing important major ecological transitions, barriers, and unfavourable habitats in the interior of the forest region on the 'contact zones' or zones of hybridization observed for birds (Fig. 5.22) and butterflies (Fig. 4.15) (Benson 1982). Using Benson's criteria for areas of rapid environmental change (1982, personal communication), I have drawn a more complete picture of such areas from the data in Chapters 1 and 2. This is combined with hybrid zones in Fig. 7.6 which also indicates the butterfly endemic centres (isoline for one third of maximum) for comparison with regions of 'habitat conformity' as defined by Endler (1982), that is regions not crossed by any obvious environmental transitions (Brown 1982*b*). I am sure that readers will want to come to their own conclusions as to whether Fig. 7.6 shows a more or less convincing correlation between biogeographical patterns (represented as endemic centres and hybrid zones) and the population–ecology model, than is shown by Fig. 7.5 (or the more complete Fig. 4.20) between centres of endemism and regions of probable palaeo-ecological forest continuity. The 'fits', in terms of level of overlap of observations with the model, are indeed similar. If the relative amount of conflict between biological data and the model is measured in each case, there appears a somewhat greater amount of transitional area within endemic centres in Fig. 7.6 than of regions of forest continuity not known to show endemism in Fig. 7.5 (and even less in Fig. 4.20; Brown 1982*b*). However, the differences are well within the inaccuracies of the construction of the respective models and representation of the biogeographical data.

Testing the models

On a more detailed level it is possible to make some simple tests of historical and allopatric *versus* modern and parapatric models for evolutionary divergence using the clear distinction seen in aposematic butterflies between endemism (in characters with a well-studied ecological function) and species diversity (even after elimination of very different subspecies of a single species which might be regarded as separate species in a classical analysis) (Brown 1982*b*). A number of objective predictions have been presented (Brown 1979, 1982*a*) which point out areas favoured in the palaeoecological reconstruction (Fig. 2.7) but judged to be less favourable to forest organisms today, and vice versa. These areas should have characteristics in their butterfly endemism which indicate which model has been more important in determining regional evolution. Contrasting predictions by Benson (1982) deal with the population characteristics to be expected in certain species and areas, which should differ if one model or the other has been more important in giving rise to these characters. Both these sets of predictions have stacked the cards in their favour, since each narrowly defined the phenomena to be observed and evaluated (population characters, corrected endemism values) around those already analysed in accord with the respective model. Nevertheless, each approach helps greatly in defining the scope of application of the respective models, and in the collection of new data in the field, which is at present a necessary prerequisite for any new proposals or tests of different hypotheses.

It is thus evident that the palaeoecological refuge model is strongly compressed on the one hand by 'classical' vicariance biogeography, relating higher taxonomic categories to older geotectonic events; and on the other by population ecology, emphasizing the importance of relating differentiation patterns to modern environments, gradients and barriers. The model could, in theory, be squeezed enough from both sides so as to lose its relevance altogether. It seems more likely, however, that it will be found to be important for a particular and reasonably well verified period, the late Wisconsin–Würm ice age, with some dimmer reflections (varying from group to group) of earlier ones, like parts of the Illinois–Riss glaciation, and to particular and well-studied phenomena (especially selection-mediated and low-level regional differentiation in forest organisms), thus taking its place among available and accepted models for some widely recognized biogeographical features in the neotropics. Vanzolini (1981) expresses

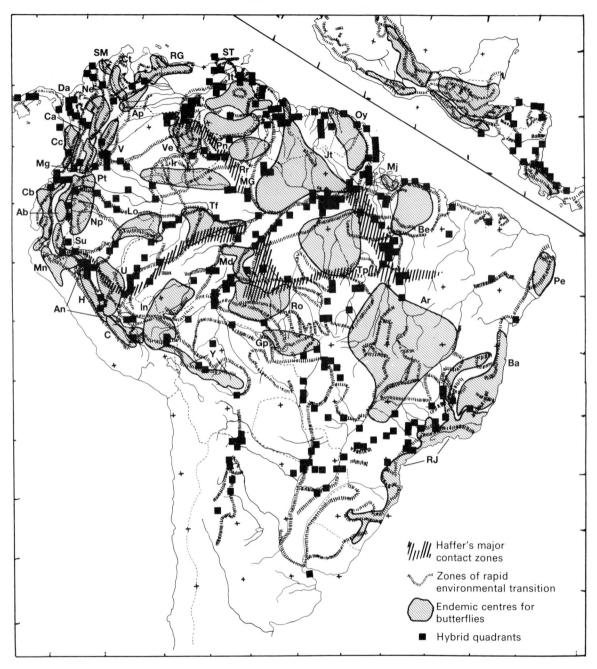

Fig. 7.6. Superimposition of subspecies endemic centres and hybrid zones in neotropical forest butterflies (Fig. 4.15) and contact zones in forest birds (Fig. 5.22) with zones of environmental conformity and rapid transition (Endler 1977, 1982; Benson 1982; Brown 1982*b*). Environmental data from Benson (1982 and personal communication), Brown (1979, 1982*a,b*), and Chapters 1 and 2; transitions are drawn in areas of rapid change for at least two of the major factors (climate, soils, and vegetation). Of the 300 quadrants for butterfly hybridization, 206 (69 per cent) are on or adjacent to ecological transitions, which however occur in many areas within the centres also. The avian hybridization zones show less overlap with the regions of rapid environmental change. (See Fig. 4.15, p. 98 for explanation of abbreviations).

this best when he remarks that the refuge model 'is not of universal application; its use should be restricted to species of rigid ecological fidelity that can be shown to have undergone geographic differentiation in the near past. Within this scope it is a maximum parsimony model.' Once again, it is necessary to emphasize that small data-sets can probably be made to conform to any number of models, and also that it is likely that many of the available models have in fact acted to different extents and in different ways through the complex evolutionary history of any taxon observed today. The broad geographical and ecological study of well-defined but reasonably diversified taxonomic groups and the thorough study of many taxa in a limited geographical region should help to decide the issue.

Intensive regional studies

Several recent studies of the latter type have confirmed their great value. Steyermark's analysis of plant endemism in Venezuela (1976, 1979, 1982) and Toledo's lucid discussion of tropical forest refuges in southern Mexico (1976, 1982) were mentioned in Chapter 3. Further important regional studies of endemism include those of Gentry (1982) and Terborgh and Winter (1982) on Chocó plants and birds, de Granville (1978, 1982) on the plants of French Guyana, Hoogmoed (1973, 1979) on Guianan reptiles, and Lamas (1976, 1982) on Peruvian butterflies. An important recent work by the late David R. Gifford (in revision at the time of his death) analyses ecological preferences of the butterfly subspecies assigned by Brown (1976a, 1979) to the Araguaia 'refuge', and concludes that a large part could have differentiated parapatrically in recent times from close relatives to the north and south, while the remainder did not need a postulated forest refuge for their differentiation since they survive happily in relict and gallery forests today. Nevertheless, Gifford also presents much climate, soil, and vegetation data in support of the existence of an Araguaia-centred palaeoecological forest refuge much as that shown in Fig. 2.8, which may be useful in the prediction and explanation of other biogeographical patterns in central Brazil.

Such full local ecological, evolutionary, and geoscientific analyses can be expected to help refine neotropical biogeography and to counteract the widespread abuse of the 'refuge model', leading thereby to a more balanced viewpoint about the relevance of ice-age restructuring of the landscape to evolutionary patterns.

Conflicting geoscientific evidence

It should be mentioned, in conclusion, that some geoscientists (e.g. Irion 1976) have questioned the model of fluctuating climate and landscape reorganization in the late Pleistocene, continuing to defend a traditional view of a highly stable 'Urwald' in the neotropics through much of the late Tertiary and Quaternary. Such a view is also favoured by many ecologists (see Colinvaux 1979; Benson 1982), but is amply belied by the relatively larger amount of palynological, geomorphological, and palaeoclimatic data available and reviewed in Chapter 1 which support the occurrence of large fluctuations in the vegetation. These data would have to be broadly reinterpreted if a model of stable forest were to be accepted. On the other hand, the data used by Irion to support forest continuity may also be directly incorporated into the model of late ice-age fragmentation of the forest, which emphasizes the importance of gallery forest along most major Amazonian rivers. Gifford's demonstration that gallery vegetation suitable to many humid-forest plants and animals can persist even on dystrophic soils in river-valleys in the Brazilian planalto today, under strongly seasonal and even semi-arid climates, supports the contentions of Prance (1973, 1978; Chapter 2) and Ab'Sáber (1977) that even during maximum forest contraction in the late Wisconsin–Würm glaciation, the Amazon Basin retained an extensive network of gallery forests which could have sheltered many humidity-dependent organisms. All Irion's riverine sampling points would be within such forests (often too narrow to be represented in Fig. 2.8). They reveal little evidence for vegetational fluctuations away from the rivers.

Furthermore, the present palaeoecological model no longer indicates that inter-refugial space conformed to the open Amazonian grassland savannas of today, unless it be near one of them and endowed with soils and vegetation indicating very open physiognomy in the past. Most of the blank areas in the model (Fig. 2.8) were probably clothed in various types of transition forest, interdigitated by gallery forest and types of cerrado as discussed in Chapter 2. Dispersal of most rain forest organisms across these drier forests or survival within them would be just as difficult as it is today where such formations predominate, due to low humidity, high insolation, scarcity of familiar resources, and unfavourable physical structure.

An alternative climatic model for the height of the Wisconsin–Würm glaciation (July–August, 18 000 years BP) has been published by Manabe and Hahn (1977). In contrast to the views of Damuth and Fairbridge (1970), Gates (1976), Brown and Ab'Sáber (1979), most Brazilian scientists, and this book (Chapter 1, Fig. 1.6), it suggests a minimal change in the Atlantic pressure cells and wind directions from those of the present. A very much reduced rainfall over continental South America is explained by the near extinction of the intertropical convergence zone (a facet of all models), and a strong surface outflow of air from the continent to the oceans. If the wind directions shown by Manabe and Hahn (1977, p. 3903) are substituted for those of Brown and Ab'Sáber (1979; Brown 1982*a*) in the determination of rainfall patterns in the southern winter (Fig. 1.6), some differences in precipitation patterns will occur, especially in central Brazil, eastern Pará, the Guiana Shield, the southwestern Amazon, and the Andean foothills. This will also modify, to a small

extent, the refuge model (Fig. 2.8). Surprisingly, these changes lead to few differences in the correlative overlap maps (Figs. 4.20 and 7.5). This is because, firstly, orographic rainfall is unchanged in many areas even by a 40° change in easterly wind direction, secondly, Fig. 1.6 contributes only one-fifth of the data used in the construction of the model (Fig. 2.8) and, thirdly, modern endemic areas are usually more extensive than the proposed refuges, overlapping the model in either case. It seems that the model in Fig. 2.8, derived from five different and independent sets of data on the physical environment (topography, palaeoclimate, geomorphology, soils, and vegetation structure), may be little affected by the continual minor changes which are inevitable in these data-sets.

Thus, as both the palaeoecological model and the biogeographical patterns become refined through the collection and analysis of more and better information, they should be progressively better able to incorporate and reinforce new data and be successfully used in predictions and interpretations, within the proper sphere of applicability of each.

THE ROAD AHEAD

It has been repeatedly emphasized throughout this book that neotropical biogeography still rests upon a very limited base of hard facts. In both the most fundamental systematic and geographical information and ecological and evolutionary studies of characters and divergence rates, most tropical American organisms are still virtually unknown. Further contributions to the biogeography of the region will be most useful if they include a broad data-base: biosystematic revision based on multiple characters and familiarity with natural populations, thorough and recent geographical sampling, evidence for evolutionary rates and tendencies, and even detailed genetic experiments and ecological observations related to adaptation and natural selection at different geographical scales. The interpretations should include experimental evidence that the patterns observed can be produced by the mechanisms invoked. The stimulus for the collection of such data is strong, and the rewards are great, for each complete new set of observations may be independently related to well-known present and past environmental parameters

and acquire its own significance in the light of historical and modern ecological and evolutionary processes.

More thorough ground coverage seems to be the most important course for future work on the geo-scientific evidence for palaeoecological conditions. It is especially important to determine the geological and radio carbon ages of Quaternary strata in many parts of the neotropics. Accurate dating, determination of the origins, and careful mapping of ice-age stone-lines and white sands under different forest soils perhaps have top priority. The potentially most definitive palaeoecological data are still the scarcest; massive emphasis on the collection, palynological analysis, and accurate dating of sediment cores obtained in the Amazon lowlands but away from rivers and modern savannas, and careful studies of the correlation between local vegetation and pollen rain to help in the evaluation of exogenous contributions to pollen profiles, will be necessary if a more accurate picture is to emerge about restructuring of the landscape in tropical America over the last 50 000 years. It is to be hoped that these will refine the geoscientific model (Fig. 2.8) to the point where it can be widely accepted as a plane of reference for biogeographical, evolutionary, and ecological correlation with the recent history of the region.

What about man?

The tenuous data-set on prehistoric man should be faced squarely in the near future by biogeographers and anthropologists alike, along with the traditional reluctance among the latter to apply evolutionary principles to cultural phenomena and to the interpretation of cultural adaptation. The correspondences between distributional patterns among groups of animals and plants and among archaeological, ethnographic, and linguistic phenomena may be coincidences or they may reflect similar processes of evolution and adaptation. Differences in time scale, rate of change, sensitivity to environmental alteration, niche breadth, and other aspects of the cultural and biological modes of adaptation must be identified and their significance assessed. More attention must be given to criteria for differentiating convergence, independent development, and diffusion in accounting for the origin and dispersal of cultural traits, as a basis for interpreting their distributional patterns.

A more definitive picture will only be produced when the field of ecological anthropology becomes a meeting place rather than a battleground for ecologists and social scientists. A more detailed geoscientific model of palaeoecological restructurings during the period of man's occupation of the neotropical forests will aid in establishing correlations (or the absence thereof) between changes in local environments and cultures in post-glacial times. The rapid deculturation of indigenous groups and the accelerating destruction of archaeological sites makes it urgent to collect relevant information before it is too late.

APPLICATIONS

Indicator organisms

The importance of indicator organisms in the study of tropical systems has been repeatedly emphasized. Modern tropical ecologists are learning to use a variety of better-known plant and animal species to characterize reliably some factors in the environment, including climatic parameters and their predictability, soil texture, fertility, and history, and long-range aptitude for different systems of agricultural or silvicultural management which are otherwise difficult to observe. When a sufficiently broad range of these well-studied organisms is available, it should become considerably simpler to analyse, evaluate, preserve, and manage tropical systems.

Conservation biology: planning and management

The primary practical application of the ideas and data presented in this book has been, and will be, for conservation and land-use planning in the neotropical forests. For optimal preservation of genetic resources, it is of central importance to know what they are, how they vary, and where they may be found. The observations analysed here have made a major contribution to this knowledge, but are as yet far from sufficient for effective conservation. They need to be extended to many other organisms from a variety of habitats. It also must be known how the organisms will respond to the restructuring of the surrounding landscape by man that may include significant modifications in regional rainfall, temperature, luminosity, air- and ground-water resources and hence opportunities for survival and evolution.

The prediction of biotic responses to environmental modifications, a principal goal of present ecosystem models, is still totally impractical in most tropical systems. Innumerable case histories exist which highlight the improper application of exogenous agricultural technology in the tropics, while others confirm the continuing value of local and empirical methods — especially those tested over centuries by the indigenous people, living at low density. These empirical approaches recognize the extremely complex biotic infrastructure of tropical communities, which is only beginning to be understood by non-resident ecologists. They also accept a large order of microheterogeneity and unpredictability in the physical environment, only recently confirmed by geoscientists.

In excessively complex systems such as these, the unpredictable response becomes an everyday ex-

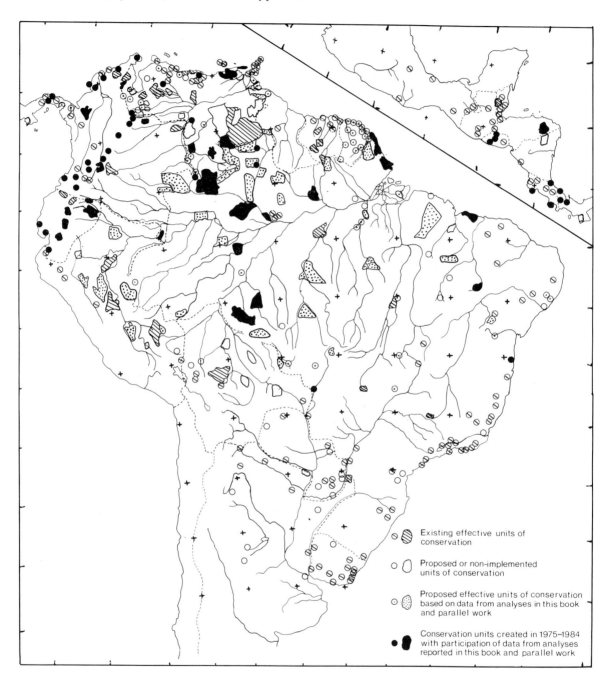

Fig. 7.7. Conservation areas, effective and proposed, in the forested regions of the American tropics (excluding the West Indies) (Wetterberg and Jorge Padua 1978; Wetterberg *et al.* 1978; de Granville 1975; Lamas 1979*b*; Brown and Ab'Sáber 1979). Areas conserved in 1975–1980 as a result of data, theory, and proposals related to endemism and refugia, totalling over 10^7 ha, are shown in black.

perience, tragically reflected in the failure of many agricultural experiments in the humid tropics. For this reason, conservation planning and land management must take into account both the modern failures and the discernible effects of past changes in local environments. A number of these effects have been perceived and discussed in the preceding chapters. It seems clear, for example, that heterogeneous areas may preserve a high genetic diversity, but the genes may be present in less accessible (more heterozygous) forms; therefore, a conservation programme should plan to cover both the nuclei and peripheries of recognized centres of endemism (Brown 1977a; Vanzolini 1980). Many deep-forest organisms depend also on successional species for their resources; their permanence may require the artificial maintenance of unpredictable secondary succession as part of the forest conservation unit (Gilbert 1980). Some tropical populations seem to be at high risk for unpredictable local extinction, and many supposedly 'stable' species have developed optimal colonization or readaptation strategies to help offset this. Management schemes must allow both the local extinctions and the re-establishments to occur. This requires, in turn, the establishment of very large contiguous areas of carefully managed permanent reserves, sufficient to guarantee numerous replicates of each microenvironment. All these recommendations and many others have emerged from the same recent studies of neotropical biology which have begun to provide hard facts about the present compositions and distributions of genetic resources. Further detailed work along these lines is urgently needed in order that the developing and visionary conservation programmes (Fig. 7.7; Wetterberg and Jorge Padua 1978; Wetterberg, Castro, Quintão, and Porto 1978) can rest on a firmer base.

The results of the pioneering studies of Steyermark (1976), de Granville (1975), Lamas (1979b), and Wetterberg and colleagues are already evident in the creation of many new conservation areas in the Amazon forests of Brazil, Colombia, Venezuela, Ecuador, and Peru. They must now be expanded to other countries and to other vegetation types and complemented by a programme of smaller local reserves designed to protect characteristic habitats with their accompanying soils, plants, animals, and water resources. It is worth mentioning, in conclusion, that the areas set aside in tropical America since 1975, for permanent preservation, largely as a result of the data incorporated in this book and parallel work (see Prance and Elias 1977), represent over 100 000 km^2 — more than half the area of all present conservation units in the neotropical forests (Fig. 7.7). They also cover 28 of the 44 endemic regions noted in Chapter 4 (Fig. 4.15) and 26 of the 62 refuges shown in Fig. 2.8. It is hoped that this conservation programme will continue, especially in implementation of further proposed areas (Fig. 7.7), including 8 endemic centres and 25 refuge areas still without effective conservation units, and consolidation of the protection and management plans for the areas already designated.

POSTSCRIPT

As this book is being printed (mid-1986), Neotropical refuge biogeography, with its accompanying concepts and applications, continues to be widely discussed and widely misunderstood, in university classrooms and graduate examinations, international congresses, conservation planning and management programmes, and the public sphere. Its successful application, contributing to the 1982 award of the Getty Conservation Prize to Maria Teresa Jorge Padua and Paulo Nogueira Neto in Brazil, is the subject of a forthcoming book by R. Foresta of the Resources for the Future Institute. Yet is is often still regarded as a highly controversial biological theory rather than a practical model, and is still often taught and discussed in terms which confuse the concepts of endemism, diversity, speciation processes, and paleoecology.

We live in a competitive time, when alternative scientific hypotheses and models tend to be presented in strong, antithetical terms, sometimes with a very limited data base or chance of experimental investigation. At the same time, clear synthetic thinking and voluminous hard data are increasingly necessary to permit the optimum application of scientific progress to the solution of major world problems. In the Neotropics, both a synthetic paleoecological model (developed from geoscientific studies, and understood as a recent chapter of vicariance bio-geography), and comprehensive biogeographical data-sets, can be of fundamental utility in the urgent task of preservation, sustainable utilisation, and manage-ment of the genetic heritage, be it in the form of endemism or diversity, or of forest, cerrado, grass-land, montane or even aquatic ecosystems, from the Tropic of Cancer in Mexico to the Chaco and Pampas of Argentina.

The concept of neotropical refuge biogeography has provided a frame of reference, impetus, and foundations for, and has given direction to, many systematic studies, biogeographical analyses, and palaeoecological investigations in tropical America over the past fifteen years. Previously, there were essentially no concepts available under which to study the biotas of areas like Amazonia.

We hope that the data presented here, with their synthesis in this final chapter, may help in the wider and clearer understanding of the foundation, scope, and application of neotropical refuge biogeography. Recent years have seen the formation of a continually more solid base for the theoretical model of evolution in the neotropics; see especially the recent books edited by Sioli (*The Amazon: Limology and Land-scape Ecology*) and Prance and Lovejoy (*Amazonia: Key Enviroments*). The field collection of broad biogeographical data, though aided by new roads into

formerly inaccessible areas, has become increasingly more hazardous with the increase of violence and environmental devastation in many parts of the neo-tropics. This is occasioned by heavy population pressure on the land unaccompanied by adequate technology for its sustained utilization, by continued exploitation of rare and thus highly valued natural resources (often for sale in luxury markets to the north), by intense trade in cocaine and equally intense repression, by ideological or neocolonial conflicts even involving palearctic superpowers, and by increased costs and difficulties in free movement of scientists within and between nations. It is already impossible to work in many key areas regarded as fundamental for testing the evolutionary models and in providing data on species relationships and adaptive potentials, and for preserving important segments of the genetic heritage. Some of these areas have little or no natural vegetation remaining. Time is running out for consolidating the foundations, for observing and testing the scope, and for fulfilling the application of neotropical refuge biogeography. We must resolutely confront and overcome the new difficulties in gath-ering data and the increasing pressure to over-exploit ecosystems for immediate material purposes. We hope that this book will provide incentives and courage towards the realization of these goals.

REFERENCES

Ab'Sáber, A. N. (1967). Problemas geomorfológicos da Amazonia brasileira. In *Atas Simpósio Biota Amazônica*, Vol. 1 (ed. H. Lent). Geociências.

—— (1977). Espaços ocupados pela expansão dos climas secas na América do Sul, por ocasião dos períodos glaciais quaternários. *Paleoclimas (Inst. Geogr., Univ. São Paulo)* 3, 1-19.

—— (1982). The paleoclimate and paleocology of Brazilian Amazonia. In *Biological diversification in the tropics* (ed. G. T. Prance). Colombia University Press, New York.

Absy, M. L. (1979). *A palynological study of Holocene sediments in the Amazon basin*. Academisch Proefschrift, University of Amsterdam.

—— and van der Hammen, T. (1976). Some palaeoecological data from Rondônia, southern part of the Amazon basin. *Acta Amazonica* 6, 293-9.

Ackery, P. R. (1975). A new pierine genus and species with notes on the genus *Tatochila* (Lepidoptera: Pieridae). *Bull. Allyn Mus., Sarasota* 30, 1-9.

Adams, M. J. (1973). Ecological zonation and the butterflies of the Sierra Nevada de Santa Marta, Colombia. *J. nat. Hist., Lond.* 7, 699-718.

—— (1977). Trapped in a Colombian Sierra. *Geogr. Mag.* 49, 250-4.

—— and Bernard, G. I. (1977). Pronophiline butterflies (Satyridae) of the Sierra Nevada de Santa Marta, Colombia. *Syst. Ent.* 2, 263-81.

—— —— (1979). Pronophiline butterflies (Satyridae) of the Serranía de Valledupar, Colombia–Venezuela border. *Syst. Ent.* 4, 95-118.

—— —— (1981). Pronophiline butterflies (Satyridae) of the Cordillera de Mérida, Venezuela. *Zool. J. Linn. Soc.* 71, 343-72.

Alexander, A. J. (1961a). A study of the biology and behavior of the caterpillars, pupae, and emerging butterflies of the subfamily Heliconiinae in Trinidad, West Indies. Part I. Some aspects of larval behavior. *Zoologica, NY* 46, 1-24.

—— (1961b). A study of the biology and behavior of the caterpillars, pupae and emerging butterflies of the subfamily Heliconiinae in Trinidad, West Indies. Part II. Molting, and the behavior of pupae and emerging adults. *Zoologica, NY* 46, 105-23.

Anderson, A. B. (1981). White-sand vegetation of Brazilian Amazonia. *Biotropica* 13, 199-210.

—— Prance, G. T., and de Albuquerque, B. W. P. (1975). Estudos sobre a vegetação das campinas Amazônicas III. A vegetação lenhosa da campina da Reserva Biológica – INPA – SUFRAMA. *Acta Amazonica* 5, 225-46.

Anderson, R. F. V. (1977). Ethological isolation and competition of allospecies in secondary contact. *Am. Nat.* 111, 939-49.

Andersson, L. (1979). Multi-layered distribution patterns and the hypothesis of rain forest refugia. *Bot. Notiser* 132, 185-90.

Andrade-Lima, D. de (1953). Notas sobre a dispersão de algumas espécies vegetais no Brasil. *Anais Sociedade Bio-logica Pernambuco* 11, 25-49.

—— (1982). Present day forest refuges in Northeastern Brazil. In *Biological diversification in the tropics* (ed. G. T. Prance). Columbia University Press, New York.

Ashton, P. S. (1969). Speciation among tropical forest trees: some deductions in the light of recent evidence. *Biol. J. Linn. Soc.* 1, 155-96.

Ávila-Pires, F. D. de (1974). Caracterização zoogeográfia de Província Amazônica. II. A familia Callithricidae e a zoogeografia Amazônica. *Anais Acad. bras. Ciênc.* 49, 159-81.

Ball, I. R. (1976). Nature and formulation of biogeographical hypotheses. *Syst. Zool.* 24, 407-30.

Basso, B. (1973). *The Kalapalo Indians of Central Brazil*. Holt, Rinehart and Winston, New York.

Bates, H. W. (1862). Contributions to an insect fauna of the Amazon Valley, Lepidoptera: Heliconidae. (Abstr.) Read Nov. 21, 1861. *J. Proc. Linn. Soc.* 6, 73-7.

—— (1864). *The naturalist on the River Amazons*, 2nd edn. John Murray, London.

Baur, G. N. (1968). *The ecological basis of rain forest management*. Sydney.

Beebe, W. (1955). Two little-known selective insect attractants. *Zoologica, NY* 40, 27-32.

—— Crane, J., and Fleming, H. (1960). A comparison of eggs, larvae and pupae in fourteen species of Heliconiine butterflies from Trinidad, W.I. *Zoologica, NY* 45, 111-54.

Beinroth, F. H. (1975). Relationship between U.S. Soil Taxonomy, the Brazilian soil classification and the FAO/UNESCO Soil Units. In *Soil management in tropical America* (ed. E. Bornemisza and A. Alvarado), Chapter 4, pp. 92-108. North Carolina State University Press, Raleigh.

Benson, W. W. (1978). Resource partitioning in passion vine butterflies. *Evolution* 32, 493-518.

—— (1982). Alternative models for infrageneric diversification in the humid tropics: tests with passion vine butterflies. In *Biological diversification in the tropics* (ed. G. T. Prance). Columbia University Press, New York.

—— Brown, K. S., Jr and Gilbert, L. E. (1976). Coevolution of plants and herbivores: passion flower butterflies. *Evolution* 29, 659-80.

Beurlen, K. (1970). *Geologie von Brasilien. Beitr. zur Regionalen Geologie der Erde*, 9. Borntraeger, Berlin.

Bibus, E. (1983). Die Klimamorphologische Bedeutung von stone-lines und Decksedimenten in mehrgliedrigen Bodenprofilen Brasiliens. *Z. Geomorph. N.F.*, Suppl.-Bd 48, 79-98.

Bigarella, J. J. (1971). Variações climáticas no Quaternário superior do Brasil e sua datação radiométrica pelo método do carbono 14. *Paleoclimas* 1. (Inst. Geogr., Univ. São Paulo 1, 1-22.

—— (1973). Geology of the Amazon and Parnaiba basins. In *The ocean basins and margins*, Vol. 1. *The South Atlantic* (ed. A. E. M. Nairn, and F. G. Stehli). Plenum Press, New York.

—— and de Andrade, G. O. (1965). Contribution to the study

of the Brazilian Quaternary. *Geol. Soc. Am., Spec. Paper* 84, 433–51.

—— and Becker, R. D. (eds.) (1975). International Symposium on the Quaternary. *Bol. Paranaense de Geociencias* 33, 379.

—— Mousinho, M. R., and da Silva, J. X. (1969). Processes and environments of the Brazilian Quaternary. In *The periglacial environment* (ed. T. L. Péwé). McGill University Press, Montreal.

Blake, E. R. (1977). *Manual of neotropical birds*, Vol. 1. University of Chicago Press.

Blandin, P. (1978). Contribution a la biogéographie néotropicale: espéces endémiques, espéces polytypiques et super-espéces chez les Brassolinae (Lepidoptera-Satyridae). *C. R. Soc. Biogéogr.* 471, 9–28.

Bonatti, E. and Gartner, S. (1973). Caribbean climate during Pleistocene ice ages. *Nature, Lond.* 244, 563–5.

Bourliere, F. (1973). The comparative ecology of rain forest mammals in Africa and South America: some introductory remarks. In *Tropical forest ecosystems in Africa and South America: a comparative review* (ed. B. J. Meggers, E. Ayensu, and W. D. Duckworth). Smithsonian Institution Press, Washington, DC.

Bradbury, J. P., Leyden, B., Salgado-Labouriau, M., Lewis Jr, W. M., Schubert, C., Binford, M. W., Frey, D. G., Whitehead, D. R., and Weibezahn, F. H. (1981). Late Quaternary environmental history of Lake Valencia, Venezuela. *Science, NY* 214, 1299–305.

Braun, E. H. and Ramos, J. R. de A. (1959). Estudo agrogeológico dos campos Puciari-Humaitá, Estado Amazonas e Território Federal de Rondônia. *Rev. Bras. Geogr., Rio de Janeiro* 21, 443–97.

Bristow, C. R. (1981). A revision of the brassoline genus *Catoblepia* (Lepidoptera: Rhopalocera). *Zool. J. Linn. Soc.* 72, 117–63.

Brower, L. P. and Brower, J. V. Z. (1964). Birds, butterflies and plant poisons: a study in ecological chemistry. *Zoologica, NY* 49, 137–59.

Brown, F. M. (1942a). Notes on Ecuadorian butterflies, V. *Cithaerias, Haetera* and *Pseudohaetera* N.G. (Satyridae). *J. NY ent. Soc.* 50, 309–33.

—— (1942b). Notes on Ecuadorian butterflies, III. The genus *Lymanopoda* Westwood (Satyridae). *Ann. ent. Soc. Am.* 36, 87–102.

—— (1948). Taxonomy and distribution of the genus *Pierella* (Lepidoptera). *Ann. Carn. Mus., Pittsburgh* 31, 49–87.

Brown, K. S. Jr (1973). The Heliconians of Brazil (Lepidoptera: Nymphalidae). Part V. Three new subspecies from Mato Grosso and Rondônia. *Bull. Allyn Mus., Sarasota* 13, 1–19.

—— (1976a). Geographical patterns of evolution in neotropical Lepidoptera. Systematics and derivation of known and new Heliconiini (Nymphalidae: Nymphalinae). *J. Ent. B, Lond.* 44, 201–42.

—— (1976b). An illustrated key to the silvaniform *Heliconius* (Lepidoptera: Nymphalidae), with description of new subspecies. *Trans. Am. ent. Soc.* 102, 373–484.

—— (1976c). In: 1975 Field Season Summary, Zone 10, South America (coord. T. C. Emmel). *News lepid. Soc.* 1976, no. 2 (May 15).

—— (1977a). Centros de evolução, refúgios quaternários e conservação de patrimônios genéticos na região neotropical: padrões de diferenciação em Ithomiinae (Lepidoptera: Nymphalidae). *Acta Amazonica* 7, 75–137.

—— (1977b). Geographical patterns of evolution in neotropical forest Lepidoptera (Nymphalidae: Ithomiinae and Nymphalinae-Heliconiini). In *Biogéographie et evolution en Amérique tropicale* (ed. H. Descimon), *Publ. Lab. zool. Ecole Norm. Super., Paris* 9, 118–60.

—— (1977c). Geographical patterns of evolution in neotropical Lepidoptera: differentiation of the species of *Melinaea* and *Mechanitis* (Nymphalidae, Ithomiinae), *Syst. Ent.* 2, 161–97.

—— (1978). Heterogeneidade: fator fundamental na teoria e prática de conservação de ambientes tropicais. *Encontro Nacional sobre Preservação de Fauna e Recursos Faunísticos, Brasília 1977* (Brasília, DF, IBDF) pp. 175–83.

—— (1979). *Ecologia Geográfica e Evolução nas Florestas Neotropicais.* Universidade Estadual de Campinas, São Paulo.

—— (1980). A review of the genus *Hypothyris* Hübner (Nymphalidae), with descriptions of three new subspecies and early stages of *H. daphnis. J. lepid. Soc.* 34, 152–72.

—— (1981). The biology of *Heliconius* and related genera. *Rev. Entom.* 26, 427–56.

—— (1982a). Paleoecology and regional patterns of evolution in neotropical forest butterflies. In *Biological diversification in the tropics* (ed. G. T. Prance), Columbia University Press, New York.

—— (1982b). Historical and ecological factors in the biogeography of aposematic neotropical butterflies. *Am. Zool.* 22, 453–71.

—— and Ab'Sáber, A. N. (1979). Ice-age refuges and evolution in the Neotropics: correlation of paleoclimatological, geomorphological and pedological data with modern biological endemism. *Paleoclimas (Inst. Geogr. Univ. São Paulo)* 5, 1–30.

—— and Benson, W. W. (1974). Adaptive polymorphism associated with multiple Müllerian mimicry in *Heliconius numata* (Lepid. Nymph.). *Biotropica* 6, 205–28.

—— —— (1977). Evolution in modern Amazonian non-forest islands: *Heliconius hermathena* [The Heliconians of Brazil (Lepidoptera: Nymphalidae). Part VII] *Biotropica* 9, 95–117.

—— and D'Almeida, R. F. (1970). The Ithomiinae of Brazil (Lepidoptera: Nymphalidae). II. A new genus and species of Ithomiinae with comments on the tribe Dircennini D'Almeida. *Trans. Am. ent. Soc.* 96, 1–17.

—— Damman, A. J., and Feeny, P. (1981). Troidine swallowtails (Lepidoptera: Papilionidae) in southeastern Brazil: natural history and foodplant relationships. *J. Res. Lepid.* 19, 199–226.

—— and Mielke, O. H. H. (1972). The Heliconians of Brazil (Lepidoptera: Nymphalidae). Part II. Introduction and general comments, with a supplementary revision of the tribe. *Zoologica, NY* 57, 1–40.

—— Sheppard, P. M., and Turner, J. R. G. (1974). Quaternary refugia in tropical America: evidence from race formation in *Heliconius* butterflies. *Proc. R. Soc., Lond.* B 187, 369–78.

—— and Vasconcellos-Neto, J. (1976). Predation on aposematic ithomiine butterflies by tanagers (*Pipraeidea melanonota*) [The Ithomiines of Brazil (Lepidoptera: Nymphalidae). Part IV]. *Biotropica* 8, 136–41.

Bruijning, C. F. A., Voorhoeve, J., and Cordijn, W. (eds.) (1977). Prehistorie. *Encyclopedie van Suriname*, pp. 506–15. Amsterdam/Brussel.

Bunting, B. T. (1965). *The geography of soils*. Aldine, Chicago.

Burtt Davy, J. (1938). The classification of tropical woody vegetation types. *Imperial Forestry Institute Paper* 13.

Bush, G. L. (1975). Modes of animal speciation. *A. Rev. Ecol. Syst.* 6, 339–64.

––, ––, and Arellano, L. J. (1985). The geology of the Río Beni: Further evidence for Holocene flooding in Amazonia. *Contributions in Science* (Nat. Hist. Mus. Los Angeles) no. 364, 18 pp.

The geology of the Río Beni: Further evidence for Holocene flooding in Amazonia. *Contributions in science* (Nat. Hist. Mus. Los Angeles) no. 364, 18 pp.

Camargo, J. M. F. de (1978). O grupo *Partamona* (*Partamona*) *testacea* (Klug): suas espécies, distribuição e diferenciação geográfica (Meliponinae, Apidae, Hymenoptera). MS thesis, Departamento de Zoologia, Universidade Federal do Paraná.

–– (1980). O grupo *Partamona* (*Partamona*) *testacea* (Klug): suas espécies, distribuição e diferenciação geográfica (Meliponinae, Apidae, Hymenoptera). *Acta Amazonica* 10 (Supl.) 1–175.

Carcasson, R. H. (1964). A preliminary survey of the zoogeography of African butterflies. *E. Afr. wildl. J.* 2, 122–57.

Carvalho Filho, R. and Leão, A. C. (1976). Solos do Projeto Ouro Preto, INCRA, Rondônia (Area de Expansão V-Jaru). *Boletim Técnico Centro de Pesquisas do Cacao, Itabuna, Bahia* 49.

Cei, J. M. (1979). The Patagonian herpetofauna. In *The South American Herpetofauna: its origin, evolution, and dispersal* (ed. W. E. Duellman). Mus. Nat. Hist., Univ. Kansas, Monograph 7.

Champion, H. G. (1936). A preliminary survey of the forest types of India and Burma. *Indian for. Rec.* n.s. 1, 1–286.

–– and Seth, S. K. (1968). *A revised survey of the forest types of India*. Manager of Publications, Delhi.

Chapman, F. M. (1917). The distribution of bird-life in Colombia: a contribution to a biological survey of South America. *Bull. Am. Mus. nat. Hist.* 36, 1–729.

–– (1921). The distribution of bird life in the Urubamba Valley of Peru. *Bull. US nat. Mus.* 117, 1–138.

–– (1926). The distribution of bird life in Ecuador, a contribution to the study of the origin of Andean bird life. *Bull. Am. Mus. nat. Hist.* 55, 1–784.

Clapperton, C. M. (1983). The glaciation of the Andes. *Quatern. Sci. Rev.* 2, 83–155.

Clench, H. K. (1964). A synopsis of the West Indian Lycaenidae, with remarks on their zoogeography. *J. Res. Lepid.* 2, 247–70.

–– (1972). A review of the genus *Lasaia* (Riodinidae). *J. Res. Lepid.* 10, 149–80.

Climap project members (1976). The surface of the ice-age Earth. *Science, NY* 191, 1131–7.

Colinvaux, P. (1979). The ice-age Amazon. *Nature, Lond.* 278, 399–400.

Comstock, W. P. (1961). *Butterflies of the American tropics. The genus* Anaea (*Lepidoptera Nymphalidae*). American Museum of Natural History, New York.

Connell, J. H. (1978). Diversity in tropical rain forests and coral reefs. *Science, NY* 199, 1302–10.

Cooke, R. E. (1974). Origin of the highland avifauna of southern Venezuela. *Systemat. Zool.* 23, 257–64.

Coope, G. R. (1970). Interpretations of Quaternary insect fossils. *A. Rev. Entom.* 15, 97–120.

–– (1978). Constancy of insect species versus inconstancy of Quaternary environments. In *Diversity of insect faunas* (ed. L. A. Mound and N. Waloff). Symposia of the Royal Entomological Society of London 9.

–– (1979). Late Cenozoic fossil Coleoptera: evolution, biogeography, and ecology. *A. Rev. Ecol. Syst.* 10, 247–67.

Correal, U. and Van der Hammen, T. (1977). *Investigaciones arqueológicas en los abrigos recosos del Tequendama*. Biblioteca Banco Popular, Premios de Arqueología 1, Bogotá.

Cracraft, J. (1973). Continental drift, paleoclimatology, and the evolution and biogeography of birds. *J. Zool., Lond.* 169, 455–545.

–– (1985a). Historical biogeography and patterns of differentiations within the South American avifauna: Areas of endemism. *Ornith. Monogr.* 36, 49–84.

Crane, J. (1955). Imaginal behavior of a Trinidad butterfly, *Heliconius erato hydara* Hewitson, with special reference to the social use of color. *Zoologica, NY* 40, 167–96.

–– (1957). Imaginal behavior in butterflies of the family Heliconiidae: changing social patterns and irrelevant actions. *Zoologica, NY* 42, 135–45.

–– and Fleming, H. (1953). Construction and operation of butterfly insectaries in the tropics. *Zoologica, NY* 38, 161–72.

Croizat, L. (1958). *Panbiogeography*, Vol. 1. *The New World*. Published by the author, Caracas.

–– (1976). *Biogeografía analítica y sintética ('Panbiogeografía') de las Américas*. Biblioteca de la Academia de Ciencias Fisicas, Matematicas y Naturales, Caracas, pp. 15–16.

Crowe, T. M. and Crowe Anna, A. (1982). Patterns of distribution, diversity and endemism in Afrotropical birds. *J. Zool. (London)* 198, 417–42.

Cruxent, J. M. and Rouse, I. (1958–9). *An archaeological chronology of Venezuela*. Social Science Monographs, Pan American Union, Washington, DC.

D'Almeida, R. F. (1935). Les *Actinote* de la partie orientale de l'Amérique du Sud. *Anais. Acad. bras. Ciênc.* 7, 69–88, 89–112.

–– (1951a). Ligeiras observações sôbre o gênero *Cithaerias* Hübner, 1819 (Lep. Satyridae). *Arq. Zool. São Paulo* 7, 493–505.

–– (1951b). Algumas considerações sôbre os gêneros *Mechanitis* Fabr. e *Melinaea* Huebn. (Lep. Ithomiidae). *Bol. Mus. Nac. (N.S.), Zoologia, Rio de Janeiro* 100, 1–27.

–– (1956). Notas sinonímicas sôbre Ithomiidae. (Lepidoptera, Rhopalocera). *Bol. Mus. Nac. (N.S.), Zoologia, Rio de Janeiro* 143, 1–18.

–– (1958). Espécies e subespécies novas de Ithomiidae. (Lepidoptera, Rhopalocera). *Bol. Mus. Nac. (N.S.), Zoologia, Rio de Janeiro* 173, 1–17.

–– (1960). Estudos sôbre algumas espécies da família Ithomiidae. (Lepidoptera, Rhopalocera). *Bol. Mus. Nac. (N.S.), Zoologia, Rio de Janeiro* 215, 1–31.

–– (1966). *Catálogo dos Papilionidae Americanos*. Sociedade Brasileira de Entomologia, São Paulo.

—— (1978). *Catálogo dos Ithomiidae Americanos* (*Lepidoptera*). Universidade Federal do Paraná, Curitiba.

Damuth, J. E. and Fairbridge, R. W. (1970). Equatorial Atlantic deep-sea arkosic sands and ice-age aridity in tropical South America. *Bull. Geol. Soc. Am.* **81**, 189–206.

Darlington, P. J. (1957) *Zoogeography: the geographical distribution of animals*. Wiley, New York.

Davis, I. (1968). Some Macro-Jê relationships. *Int. J. Am. Ling.* **34**, 42–7.

Dejonghe, J. F. and Mallet, B. (1978). Sur la redécouverte de la Pénélope à ailes blanches, *Penelope albipennis*. *Gerfaut* **68**, 204–9.

Delacour, J. and Amadon, D. (1973). *Curassows and related birds*. American Museum of Natural History, New York.

Descamps, M., Gasc, J. P., Lescure, J., and Sastre, C. 1978. Etude des écosystems guyanais II. Données biogéographiques sur la partie orientale des Guyanes. *C. R. Séanc. Soc. Biogéogr.* **467**, 55–82.

De Schauensee, R. M. (1966). *The species of birds of South America and their distribution*. Livingston, Narberth, Pennsylvania.

Descimon, H. (1977a). Biogéographie, mimétisme et spéciation dans le genre *Agrias* Doubleday (Lep. Nymphalidae Charaxinae). In *Biogéographie et evolution en Amérique tropicale* (ed. H. Descimon). *Publ. Lab. zool. Ecole Norm. Super., Paris* **9**, 307–44.

—— (ed.) (1977b). *Biogéographie et Evolution en Amérique Tropicale. Publ. Lab. zool. Ecole Norm. Super., Paris* **9**.

Diamond, A. W. and Hamilton, A. C. (1980). The distribution of forest passerine birds and Quaternary climatic change in tropical Africa. *J. Zool., Lond.* **191**, 379–402.

Diamond, J. (1972). Avifauna of the eastern highlands of New Guinea. *Publ. Nuttall. Ornith. Club* **11**.

—— (1973). Distributional ecology of New Guinea birds. *Science, NY* **179**, 759–69.

Diebold, A. Jr (1960). Determining the centers of dispersal of language groups. *Int. J. Am. Ling.* **26**, 1–10.

Dillon, L. S. (1948). The tribe Catagrammini (Lepidoptera: Nymphalidae). I. The genus *Catagramma* and allies. *Sci. Publ. Reading Publ. Mus. Art Gallery* **8**.

Dixon, J. F. (1979). Origin and distribution of reptiles in lowland tropical rainforests of South America. In *The South American Herpetofauna: its origin, evolution, and dispersal* (ed. W. E. Duellman). Mus. Nat. Hist., Univ. Kansas, Monograph 7.

Dorst, J. (1957). Contribution a l'étude écologique des oiseaux du haut Marañón (Pérou septentrional). *Oiseau* **27**, 235–69.

—— (1976). Historical factors influencing the richness and diversity of the South American avifauna. *Proc. 16th Int. Ornith. Congr., Canberra, Australia*, pp. 17–35.

Douglas, I. (1978). Tropical geomorphology. Present problems and future prospects. In *Geomorphology. Present problems and future prospects* (ed. C. Embleton *et al.*). Oxford.

Drummond, B. A. III (1976). Comparative ecology and mimetic relationships of Ithomiine butterflies in eastern Ecuador. Ph.D. thesis, University of Florida, Gainesville.

Ducke, A. and Black, G. (1953). Phytogeographical notes on the Brazilian Amazon. *An. Acad. bras. Ciênc.* **25**, 1–46.

—— —— (1954). Notas sobre a fitogeografia da Amazônia Brasileira. *Bol. Tecn. Inst. Agron. Norte* **29**, 1–62.

Duellman, W. E. (1972). South American frogs of the *Hyla rostrata* group (Amphibia, Anura, Hylidae). *Zool. Mededelingen* **47**, 177–92.

—— (ed.) (1979a). *The South American Herpetofauna: its origin, evolution and dispersal*. Mus. Nat. Hist., Univ. Kansas, Monograph 7.

—— (1979b). The herpetofauna of the Andes: patterns of distribution, origin, differentiation and present communities. In *The South American Herpetofauna: its origin, evolution and dispersal* (ed. W. E. Duellman). Mus. Nat. Hist. Univ. Kansas. Monograph 7.

—— (1982). Quaternary climatic-ecological fluctuations in the lowland tropics: frogs and forests. In *Biological diversification in the tropics* (ed. G. T. Prance). Columbia University Press, New York.

—— and Crump, M. L. (1974). Speciation in frogs of the *Hyla parviceps* group in the upper Amazon Basin. *Occ. Papers Mus. Nat. Hist., Kansas* **23**, 1–40.

Durbin, M. (1977). A survey of the Carib language family. In *Carib-speaking Indians* (ed. E. B. Basso). Anthropological Papers, University of Arizona 28.

Durden, C. J. and Rose, H. (1978). Butterflies from the middle Eocene: the earliest occurrence of fossil Papilionoidea (Lepidoptera). *Pearce-Sellards Series, Texas Memorial Museum* **29**, 1–25.

Eden, M. J. (1974). Paleoclimatic influences and the development of savanna in southern Venezuela. *J. Biogeog.* **1**, 95–109.

Ehrlich, P. R. and Gilbert, L. E. (1973). Population structure and dynamics of the tropical butterfly *Heliconius ethilla*. *Biotropica* **5**, 69–82.

—— and Raven, P. H. (1965). Butterflies and plants: a study in coevolution. *Evolution* **18**, 586–608.

Eidt, R. C. (1969). The climatology of South America. In *Biogeography and ecology in South America*, Vol. 1 (ed. E. J. Fittkau *et al.*). Junk, The Hague.

Eisenberg, J. E. (ed.) (1979). *Vertebrate ecology in the northern neotropics*. Smithsonian Institute, Washington, DC.

Eisenmann, E. (1955). The species of Middle American birds. *Trans. Linn. Soc. NY* **7**, 1–128.

Eiten, G. (1972). The cerrado vegetation of Brazil. *Bot. Rev.* **38**, 201–341.

—— (1975). The vegetation of the Serra do Roncador. *Biotropica* **7**, 112–35.

—— (1978). Delimitation of the cerrado concept. *Vegetatio* **36**, 169–78.

Eley, J. W. (1982). Systematic relationships and zoogeography of the White-winged Guan (*Penelope albipennis*) and related forms. *Wilson Bulletin* **94**, 241–59.

Emiliani, C. (1972). Quaternary paleotemperatures and the duration of the high-temperature intervals. *Science, NY* **178**, 398–401.

Emsley, M. G. (1963). A morphological study of imagine Heliconiinae (Lep.: Nymphalidae) with a consideration of the evolutionary relationships within the group. *Zoologica, NY* **48**, 85–130.

—— (1964). The geographical distribution of the color-pattern components of *Heliconius erato* and *Heliconius melpomene* with genetical evidence for the systematical relationship between the two species. *Zoologica, NY* **49**, 245–86.

—— (1965). Speciation in *Heliconius* (Lep. Nymphalidae): morphology and geographic distribution. *Zoologica, NY* **50**, 191-254.

Endler, J. A. (1977). *Geographic variation, speciation, and clines*. Monogr. Popul. Biol. 10. Princeton.

—— (1982). Pleistocene forest refuges: fact or fancy? In *Biological diversification in the tropics* (ed. G. T. Prance). Columbia University Press, New York.

Erwin, T. L. and Adis, J. (1982). Amazonian inundation forests: their role as short-term refuges and generators of species richness and taxon pulses. In *Biological diversification in the tropics* (ed. G. T. Prance). Columbia University Press, New York.

Escoto, J. A. V. (1964). In *Natural environments and early cultures. Handbook of middle American Indians*, Vol. 1 (ed. R. C. West). University of Texas, Austin.

Espinal, T. L. S. and Montenegro M. E. (1963). *Formaciones Vegetales de Colombia: Memoria Explicativa sobre el Mapa Ecológico*. Inst. Geogr. 'Agustin Codazzi', Bogotá, Colombia.

Evans, C. (1968). Archeological investigations on the Rio Napo, eastern Ecuador. *Smithson. Contrib. Anthropol.* 6.

—— and Meggers, B. J. (1960). Archeological investigations in British Guiana. *Bureau Am. Ethnol. Bull.* **177**.

—— —— and Cruxent, J. M. (1960). Preliminary results of archaeological investigations along the Orinoco and Ventuari Rivers, Venezuela. *Actas del 33° Congreso Internacional de Americanistas*, Vol. 2, pp. 359-69.

Evans, W. H. (1951). *A catalogue of the American Hesperiidae in the British Museum (Natural History)*. Part I. *Introduction and Group A, Pyrrhopyginae*. British Museum (Natural History), London.

—— (1952). *A catalogue of the American Hesperiidae in the British Museum (Natural History)*. Part II. *Groups B, C, D, Pyrginae, Section 1*. British Museum (Natural History), London.

—— (1953). *A catalogue of the American Hesperiidae in the British Museum (Natural History)*. Part III. *Groups E, F, G, Pyrginae, Section 2*. British Museum (Natural History), London.

—— (1955). *A catalogue of the American Hesperiidae in the British Museum (Natural History)*. Part IV. *Groups H to P, Hesperiinae and Megathyminae*. British Museum (Natural History), London.

Ewel, J. J., Madriz, A., and Tosi Jr, J. Jr. (1976). *Zonas de vida de Venezuela: Memoria explicativa sobre el mapa ecologica* (ed. 2). Ministerio de Agricultura y Cria, Fondo Nacional de Investigaciones Agropecuarias, Caracas.

Fairbridge, R. W. (1976). Shellfish-eating preceramic indians in coastal Brazil. *Science, NY* **191**, 353-9.

Falesi, I. C. (1972). Solos da rodovia Transamazônica. *Bol. Tech. IPEAN* 55.

FAO/UNESCO (1971). *World Soils Map, 4 South America*. UNESCO, Paris/Rome.

—— (1976). *World Soils Map, 3 Mexico and Central America*. UNESCO, Paris/Rome.

Fedorov, A. A. (1966). The structure of the tropical rain forest and speciation in the humid tropics. *J. Ecol.* **54**, 1-11.

Field, W. D. (1967a). Preliminary revision of butterflies of the genus *Calycopis* Scudder (Lycaenidae: Theclinae). *Proc. US natl. Mus.* **119**, 1-48.

—— (1967b). Butterflies of the new genus *Calystryma* (Lycaenidae: Theclinae, Strymonini). *Proc. US natl. Mus.* **123**, 1-31.

—— and Herrera, J. (1977). The pierid butterflies of the genera *Hypsochila* Ureta, *Phulia* Herrich-Schaffer, *Infraphulia* Field, *Pierphulia* Field, and *Piercolias* Staudinger. *Smithson. Contr. Zool.* 232.

Fittkau, E. J. (1974). Zur ökologischen Gliederung Amazoniens. I. Die erdgechichtliche Entwicklung Amazonlens. *Amazoniana* 5, 77-134.

Fitzpatrick, J. W. (1973). Speciation in the genus *Ochthoeca* (Aves: Tyrannidae). *Breviora* **402**, 1-13.

—— (1976). Systematics and biogeography of the tyrannid genus *Todirostrum* and related genera (Aves). *Bull. Mus. comp. Zool.* **147**, 435-63.

—— (1980). Some aspects of speciation in South American flycatchers. *Acta 17th. Congr. Int. Ornith., Berlin, 1978*, pp. 1273-9.

Fjeldså, J. (1985). Origin, evolution, and status of the avifauna of Andean wetlands. *Ornith. Monogr.* **36**, 85-112.

Flenley, J. R. (1979). *The equatorial rain forest. A geological history*. Butterworth, London.

Forbes, W. T. M. (1932). How old are the Lepidoptera? *Am. Nat.* **66**, 452-60.

—— (1948). A second review of *Melinaea* and *Mechanitis* (Lepidoptera, Ithomiinae). *J. NY ent. Soc.* **56**, 1-24.

Formas, J. R. (1979). La herpetofauna de los bosques temperados de Sudamérica. In *The South American Herpetofauna: its origin, evolution, and dispersal* (ed. W. E. Duellman). Mus. Nat. Hist., Univ. Kansas, Monograph 7.

Forster, W. (1964). Beiträge zur Kenntnis des Insektenfauna Boliviens XIX. Lepidoptera III. Satyridae. *Veröff. zool. Staatssamml. München* 8, 51-188.

Fox, J. F. (1979). Intermediate-disturbance hypothesis. *Science, NY* **204**, 1344-5.

Fox, R. M. (1940). A generic review of the Ithomiinae (Lepidoptera: Nymphalidae). *Trans. Am. ent. Soc.* **66**, 161-207.

—— (1949). The evolution and systematics of the Ithomiidae (Lepidoptera). *Univ. Pittsburgh Bull.* **45**, 36-47.

—— (1956). A monograph of the Ithomiidae (Lepidoptera). I. *Bull. Am. Mus. nat. Hist.* **111**, 1-76.

—— (1960a). A monograph of the Ithomiidae (Lepidoptera). II. The tribe Melinaeini Clark. *Trans. Am. ent. Soc.* **86**, 109-71.

—— (1960b). A postscript on the ithomine tribe Tithoreini. *J. NY ent. Soc.* **68**, 152-6.

—— (1965). Additional notes on *Melinaea* Hübner (Lepidoptera: Ithomiidae). *Proc. R. ent. Soc., Lond.* B **34**, 77-82.

—— (1967). A monograph of the Ithomiidae (Lepidoptera). III. The tribe Mechanitini Fox. *Mem. Am. ent. Soc.* **22**, 1-190.

—— (1968). Ithomiidae (Lepidoptera: Nymphaloidea) of Central America. *Trans. Am. ent. Soc.* **94**, 155-208.

—— and Real, H. G. (1971). A monograph of the Ithomiidae (Lepidoptera). IV. The tribe Napeogenini Fox. *Mem. Am. ent. Inst.* **15**, 1-368.

Futuyma, D. J. and Mayer, G. C. (1980). Non-allopatric speciation in animals. *Systemat. Zool.* **29**, 254-71.

Gallagher, P. (1976). La Pitía: an archaeological series in northwestern Venezuela. *Yale Univ. Publ. Anthropol.* 76.

Gallardo, J. M. (1979). Composición, distribución y origen de la herpetofauna chaqueña. In *The South American Herpetofauna: its origin, evolution, and dispersal* (ed. W. E. Duellman). Mus. Nat. Hist., Univ. Kansas, Monograph 7.

Galvão, E. (1969). Areas culturais indígenas do Brasil; 1900–1959. *Boletim do Museu Paraense Emílio Goeldi, Antropologia* 8.

Garner, H. F. (1966). Derangement of the Río Caroní, Venezuela. *Rev. Géomorph. Dynamique* 2, 53–80.

—— (1967). Rivers in the making. *Scient. Am.* 216, 84–94.

—— (1974). *The origin of landscapes*. Oxford.

—— (1975). Rainforests, deserts and evolution. *An. Acad. bras. Ciênc.* 47, 127–33, Supl.

Gates, W. L. (1976). Modeling the ice-age climate. *Science, NY* 191, 1138–44.

Gentry, A. H. (1978). Floristic needs in Pacific Tropical America. *Brittonia* 30, 134–53.

—— (1979). Distribution patterns of neotropical Bignoniaceae: some phytogeographic implications. In *Tropical botany* (ed. K. Larsen and L. B. Holm-Nielsen). Academic Press, London.

—— (1982). Phytogeographic patterns and evidence for a Chocó refuge. In *Biological diversification in the tropics* (ed. G. T. Prance). Columbia University Press, New York.

Gilbert, L. E. (1969). Some aspects of the ecology and community structure of ithomiid butterflies in Costa Rica. *Advanced Population Biology, Individual Research Reports, July–August*, Organization for Tropical Studies, Ciudad Universitaria, San José, Costa Rica.

—— (1972). Pollen feeding and reproductive biology of *Heliconius* butterflies. *Proc. natl. Acad. Sci.* 69, 1403–7.

—— (1975). Ecological consequences of a coevolved mutualism between butterflies and plants. In *Coevolution of animals and plants* (ed. L. E. Gilbert and P. H. Raven). University of Texas Press, Austin.

—— (1977). The role of insect–plant coevolution in the organization of ecosystems. *Coll. Internat. CNRS* (Comportement des Insectes et Milieu Trophique) 265, 399–413.

—— (1980). Food web organization and the conservation of neotropical diversity. In *Conservation biology* (ed. M. Soulé and B. Wilcox). Sinauer, Sunderland, Mass.

Gómez-Pompa, A. (1973). Ecology of the vegetation of Veracruz. In *Vegetation and vegetational history of northern Latin America* (ed. A. Graham) pp. 73–148. Elsevier, Amsterdam.

Grabert, H. (1983). Das Amazonas-Entwässerungssystem in Zeit und Raum. *Geol. Rundschau* 72, 671–83.

Graham, A. (ed.) (1973a). *Vegetation and vegetational history of Northern Latin America*. Elsevier, Amsterdam.

—— (1973b). Literature on vegetational history in Latin America. In *Vegetation and vegetational history of Northern Latin America* (ed. A. Graham). Elsevier, Amsterdam.

—— (1979). Literature on vegetational history in Latin America. Suppl. I. *Rev. Palaeobot. Palynol.* 27, 29–52.

—— (1982). Diversification beyond the Amazon Basin. In *Biological diversification in the tropics* (ed. G. T. Prance). Columbia University Press, New York.

Grant, V. (1981). *Plant speciation*, 2nd edn. Columbia University Press, New York.

Granville, J. J. de (1975). Projets de reserves botaniques et forestieres en Guyane. *Research Report, ORSTOM,* Cayenne, Cote B.7 (mimeo).

—— (1978). Recherches sur la Flore et la Vegetation Guyanaises. Thesis, Montpellier Université des Sciences et Techniques da Languedoc.

—— (1982). Rain forest flora and xeric flora refuges in French Guiana. In *Biological diversification in the tropics* (ed. G. T. Prance). Columbia University Press, New York.

Graves, G. R. (1982). Speciation in the Carbonated Flower-Piercer (*Diglossa carbonaria*) complex of the Andes. *Condor* 84, 1–14.

Greenberg, J. (1960). The general classification of Central and South American languages. *Selected Papers of the Fifth International Congress of Anthropological and Ethnological Sciences*, pp. 791–4.

Gyldenstolpe, N. (1945). A contribution to the ornithology of northern Bolivia. *Kungl. Svenska Vet.-Akad.*, Ser. 3, 23, 1–300.

Haber, W. A. (1978). Evolutionary ecology of tropical mimetic butterflies (Lepidoptera: Ithomiinae). Ph.D. thesis, University of Minnesota, Minneapolis.

Haffer, J. (1967a). Speciation in Colombian forest birds west of the Andes. *Am. Mus. Novit.* 2294.

—— (1967b). Zoogeographical notes on the 'nonforest' lowland bird faunas of northwestern South America. *Hornero* 10, 315–33.

—— (1969). Speciation in Amazonian forest birds. *Science, NY* 165, 131–7.

—— (1970a). Art-Entstehung bei einigen Waldvögeln Amazoniens. *J. Ornith.* 111, 285–331.

—— (1970b). Entstehung und Ausbreitung nord-Andiner Bergvögel. *Zool. Jahrb. Syst.* 97, 301–37.

—— (1974). Avian speciation in tropical South America. *Publ. Nuttall Ornith. Club* 14.

—— (1975). Avifauna of northwestern Colombia, South America. *Bonner Zool. Monogr.* 7.

—— (1977a). A systematic review of the Neotropical ground-cuckoos (Aves, *Neomorphus*). *Bonner Zool. Beitr.* 28, 48–76.

—— (1977b). Verbreitung und Hybridisation der *Pionites*-Papageien Amazoniens. *Bonner Zool. Beitr.* 28, 269–78.

—— (1978). Distribution of Amazon forest birds. *Bonner Zool. Beitr.* 29, 38–78.

—— (1979). Quaternary biogeography of tropical lowland South America. In *The South American Herpetofauna: its origin, evolution, and dispersal* (ed. W. E. Duellman). Mus. Nat. Hist., Univ. Kansas, Monograph 7.

—— (1981). Aspects of Neotropical bird speciation during the Cenozoic. In *Vicariance biogeography: a critique* (ed. G. Nelson and D. E. Rosen). Columbia University Press, New York.

—— (1982). General aspects of the refuge theory. In *Biological diversification in the tropics* (ed. G. T. Prance). Columbia University Press, New York.

—— (1983). Ergebnisse moderner ornithologischer Forschung im tropischen Amerika. *Spixiana* [München], Suppl. 9, 117–166.

—— (1985). Avian zoogeography of the neotropical lowlands. *Ornith. Monogr.* 36, 113–46.

—— (1986). Superspecies and species limits in vertebrates. *Z. zool. Syst. Evolut. forsch.*, in press.

Hall, A. (1928-30). A revision of the genus *Phyciodes* Hubner (Lepidoptera *Nymphalidae*). *Bull. Hill Mus. Witley*, Suppl. 2, 1–44; 3, 45–170; 4, 171–207.

—— (1938). On the types of *Adelpha* (Lep., Nymphalidae) in the collection of the British Museum. *Entomologist, Lond.* **71**, 184–7, 208–11, 232–5, 257–9, 284–5.

Hall, B. P. and Moreau, R. E. (1970). *An atlas of speciation in African passerine birds*. British Museum (Natural History), London.

Hamilton, A. C. (1976). The significance of patterns of distribution shown by forest plants and animals in tropical Africa for the reconstruction of upper Pleistocene palaeoenvironments: a review. In *Palaeoecology of Africa, the surrounding islands and Antarctica* (ed. E. M. Van Zinderen Bakker) Vol. 9, pp. 63–97.

Herrera, J. and Field, W. D. (1959). A revision of the butterfly genera *Theochila* and *Tatochila* (Lepidoptera: Pieridae). *Proc. US natl. Mus.* **108**, 467–514.

Hershkowitz, P. (1978). *Living New World monkeys (Platyrrhini)* Vol. I. University of Chicago Press.

Heyer, W. R. (1973). Systematics of the marmoratus group of the frog genus Leptodactylus (Amphibia, Leptodactylidae). *Nat. Hist. Mus. Los Angeles Cty., Contrib. Sci.* **251**, 1–50.

—— (1975). A preliminary analysis of the intergeneric relationships of the frog family Leptodactylidae. *Smithson. Contr. Zool.* **199**, 1–55.

—— and Maxson, L. R. (1982). Distribution, relationships and zoogeography of lowland frogs: the *Leptodactylus* complex in South America with special reference to Amazonia. In *Biological diversification in the tropics* (ed. G. T. Prance). Columbia University Press, New York.

Higgins, L. G. (1981). A revision of *Phyciodes* Hubner and related genera, with a review of the classification of the Melitaeinae (Lepidoptera: Nymphalidae). *Bull. Br. Mus. nat. Hist. Ent.* **43**, 77–243.

Hilbert, P. P. (1955). A cerâmica arqueológica da região de Oriximiná. *Instituto Antropológica e Etnológica do Pará* 9.

—— (1968). Archaologische Untersuchungen am Mittleren Amazonas. *Marburger Studien zur Volkerkunde* 1.

Hoogmoed, W. (1973). Notes on the Herpetofauna of Surinam. IV. The lizards and amphisbaenians of Surinam. *Biogeographica* **4**, 1–419.

—— (1979). The herpetofauna of the Guianan region. In *The South American Herpetofauna: its origin, evolution, and dispersal* (ed. W. E. Duellman). Mus. Nat. Hist., Univ. Kansas, Monograph 7.

Howard, G. D. (1947). Prehistoric ceramic styles of lowland South America, their distribution and history. *Yale Univ. Publ. Anthropol.* 37.

Howell, T. R. (1969). Avian distribution in Central America. *Auk* **86**, 293–326.

Huber, J. (1906). La vegetation de la vallee du Rio Purus (Amazone). *Bull. Herb. Boiss.* II. **6**, 249–76.

Huber, O. (1982). Significance of savanna vegetation in the Amazonian Territory of Venezuela. In *Biological diversification in the tropics* (ed. G. T. Prance). Columbia University Press, New York.

Hueck, K. and Seibert, P. (1972). *Vegetationskarte von Sudamerika*. Fischer, Stuttgart.

Irion, G. (1976). Die Entwicklung des zentral- und oberamazonischen Tieflands im Spät-Pleistozän und im Holozän. *Amazoniana* **6** 67–79.

—— and Absy, M. L. (1978). Paleoclimate in central Amazonnia as reflected by Quaternary sediments. *Abstracts, 10th Int. Congr. Sedimentology, Jerusalem*, pp. 331–2.

Izumi, S. and Sono, T. (1963). *Andes 2: Excavations at Kotosh, Peru, 1960*. Tokyo.

Jackson, J. F. (1978). Differentiation in the genera *Enyalius* and *Strobilurus* (Iguanidae): implications for Pleistocene climatic changes in eastern Brasil. *Arq. Zool., São Paulo* **30**, 1–79.

Journaux, A. (1975). Recherches géomorphologiques en Amazonie bresilienne. *Bull. Centre Géomorph. Caen (CNRS)* **20**, 1–67.

Kaye, W. J. (1907). Notes on the dominant Mullerian group of butterflies from the Potaro district of British Guiana. *Trans. R. ent. Soc., Lond.* **54**, 411–39.

Keast, A. (1961). Bird speciation on the Australian continent. *Bull. Mus. comp. Zool.* **123**, 305–495.

—— (1974). Avian speciation in Africa and Australia: some comparisons. *Emu* **74**, 261–9.

Keay, R. W. J. (1959). *Vegetation map of Africa. Explanatory notes*. Oxford.

Kingdon, J. (1971). *East African mammals. An atlas of evolution in Africa*, Vol. 1. Academic Press, London.

Kinzey, W. G. (1982). Distribution of primates and forest refuges. In *Biological diversification in the tropics* (ed. G. T. Prance). Columbia University Press, New York.

—— and Gentry, A. H. (1979). Habitat utilization in two species of Callicebus. In *Primate ecology: problem-oriented field studies* (ed. R. W. Sussman). Wiley, New York.

Klammer, G. (1971). Über plio-pleistozäne Terrassen und ihre Sedimente im unteren Amazonasgebiet. *Z. Geomorph. N.F.* **15**, 62–106.

—— (1981). Landforms, cyclic erosion and deposition, and Late Cenozoic changes in climate in southern Brazil. *Z. Geomorph. N.F.* **25**, 146–65.

—— (1982). Die Paläowüste des Pantanal von Mato Grosso und die pleistozäne Klimageschichte der brasilianischen Randtropen. *Z. Geomorph. N.F.* **26**, 393–416.

Kleinschmidt, O. (1926a). *Die Formenkreislehre und das Weltwerden des Lebens*. Halle a.s.

—— (1926b). Der weitere Ausbau der Formenkreislehre. *J. Ornith.* **74**, 405–8.

Klinge, H. (1973). Struktur und Artenreichtum des zentralamazonischen Regenwaldes. *Amazoniana, Kiel* **4**, 283–92.

—— and Medina, E. (1979). Rio Negro caatingas and campinas, Amazonas States of Venezuela and Brazil. In *Ecosystems of the World 9A: heathlands and related shrublands* (ed. R. L. Specht). Elsevier, Amsterdam.

Krüger, E. (1933). Verbreitung und Ableitung einiger Tagfalterfamilien des tropischen Amerikas. *Dt. Ent. Zeit., Frankfurt* 1932, 149–94.

Kubitzki, K. (1979). Ocorrência de *Kielmeyera* nos 'Campos de Humaitá' e a natureza dos 'campos' – Flora da Amazônia. *Acta Amazonica* **9**, 401–4.

—— (1983). Dissemination biology in the savanna vegetation of Amazonia. *Sonderb. Naturwiss. Ver. Hamburg* **7**, 353–7.

Lamas M., G. (1973). Taxonomia e evolução dos gêneros *Ituna* Doubleday (Danainae) e *Paititia*, gen. n., *Thyridia* Hubner e *Methona* Doubleday (Ithomiinae) (Lepidoptera, Nymphalidae). Doctor's thesis, Departamento de Zoologia, Instituto de Biociencias, Universidade de São Paulo.

—— (1976). Notes on Peruvian butterflies (Lepidoptera). II. New *Heliconius* (Nymphalidae) from Cusco and Madre de Diós. *Rev. peruana Ent.* **19**, 1–7.

—— (1979a). Los Dismorphiinae (Pieridae) de Mexico, America Central y las Antillas. *Rev. Soc. Mex. Lepid.* **5**, 3-37.

—— (1979b). Algunas reflexiones y sugerencias sobre la creación de parques nacionales en el Peru. *Rev. cienc. UNMSM, Lima* **71**, 101-14.

—— (1981). Los Ithomiinae (Lepidoptera, Nymphalidae) del Valle de Cosñipata, Cuzco, Peru. Estudio preliminar de un transecto altitudinal. *Resumens IV Congr. Latinoamericano Entom.*, **110**.

—— (1982). A preliminary zoogeographical division of Peru based on butterfly distributions (Lepidoptera, Papilionoidea). In *Biological diversification in the tropics* (ed. G. T. Prance). Columbia University Press, New York.

Langenheim, J. H., Lee, Y. T., and Martin, S. S. (1973). An evolutionary and ecological perspective of the Amazonian hylaean species of *Hymenaea* (Leguminosae: Caesalpinoideae). *Acta Amazonica* **3**, 5-37.

Lathrap, D. W. (1970). *The upper Amazon*. Praeger, New York.

Lattin, G. de (1957). Die Ausbreitungszentren der holarktischen Landtierwelt. *Verhandl. Dt. zool. Gesell. Hamburg* 1956, 380-410.

Lauer, W. (1952). Humide und aride Jahreszeiten in Afrika und Südamerika und ihre Beziehung zu den Vegetationsgürteln. *Bonner Geogr. Abhandl.* **9**, 15-98.

Lemoult, E. and Real, P. (1962). *Les Morpho d'Amérique du Sud et Centrale, Historique – Morphologie – Systematique*. Editions du Cabinet Entomol. E. LeMoult, Paris. (Suppl. a Novitates Entomologicae.)

Lesse, H. de (1967). Les nombres de chromosomes chez les Lépidoptères Rhopalocères néotropicaux. *Ann. Soc. ent. France, N.S* **3**, 67-136.

—— (1970a). Formules chromosomiques de quelques Lépidoptères Rhopalocères de Guyane. *Ann. Soc. ent. France, NS* **6**, 347-58.

—— (1970b). Les nombres de chromosomes chez les Lépidoptères Rhopalocères en Amérique Centrale et Colombie. *Ann. Soc. ent. France, NS* **6**, 347-58.

—— and Brown, K. S. Jr (1971). Formules chromosomiques de Lépidoptères Rhopalocères du Brésil. *Bull. Soc. ent. France* **76**, 131-7.

Lisboa, P. L. (1975). Observações gerais e revisão bibliográfica sobre as campinas amazonicas de areia branca. *Acta Amazonica* **5**, 211-23.

Livingstone, D. A. (1975). Late Quaternary climatic change in Africa. *A. Rev. Ecol. Syst.* **6**, 249-78.

—— and van der Hammen, T. (1978). Palaeogeography and palaeoclimatology. *Tropical forest ecosystems. A state-of-knowledge report*. UNESCO/UNEP/FAO, Paris.

Lleras, E. (1978). *Trigoniaceae*. Flora Neotropica. Monograph no. 19. N.Y. Botanical Garden, N.Y.

—— and Kirkbride, J. H. (1978). Alguns aspectos da vegetação da serra do Cachimbo. *Acta Amazonica* **8**, 51-65.

Loukotka, C. (1967). Ethno-linguistic distribution of South American Indians. *Ann. Ass. Am. Geog.* **57**, 2, map suppl. 8.

Lovejoy, T. E. (1975). Bird diversity and abundance in Amazon forest communities. *Living Bird* **13**, 127-91.

Lynch, J. D. (1979). The amphibians of the lowland tropical forest. In *The South American Herpetofauna: its origin, evolution, and dispersal* (ed. W. E. Duellman). Mus. Nat. Hist., Univ. Kansas, Monograph 7.

MacArthur, R. H. (1972). *Geographical ecology. Patterns in the distribution of species*. Harper and Row, New York.

Macedo, M. and Prance, G. T. (1978). Notes on the vegetation of Amazonia II. The dispersal of plants in Amazonian white sand campinas: the campinas as functional islands. *Brittonia* **30**, 203-15.

Manabe, S. and Hahn, D. G. (1977). Simulation of the tropical climate of an ice age. *J. geophys. Res.* **86**, 3889-911.

Mason, J. (1950). The languages of South American Indians. *Handbook of South American Indians*, vol. 6, pp. 157-317. Bureau of American Ethnology Bull. 143, Smithsonian Institution, Washington, DC.

Masters, J. H. (1968). Collecting Ithomiidae with Heliotrope. *J. lepid. Soc.* **22**, 108-10.

—— (1970). Bionomic notes on Haeterini and Biini in Venezuela (Satyridae). *J. lepid. Soc.* **24**, 15-18.

Maxson, L. R. and Heyer, W. R. (1982). Leptodactylid frogs and the Brazilian Shield: an old and continuing adaptive relationship. *Biotropica* **14**, 10-15.

Mayr, E. (1942). *Systematics and the origin of species*. Columbia University Press, New York.

—— (1963). *Animal species and evolution*. Harvard University Press, Cambridge, Massachusetts.

—— (1964). Neotropical Region. In *A new dictionary of birds* (ed. A. L. Thomson). McGraw-Hill, New York.

—— and Diamond, J. (1976). Birds on islands in the sky: Origin of the montane avifauna of northern Melanesia. *Proc. natl. Acad. Sci. USA* **73**, 1765-9.

—— and Phelps Jr, W. H. (1967). The origin of the bird fauna of the south Venezuelan highlands. *Bull. Am. Mus. nat. Hist.* **136**, 269-328.

—— and Short, L. L. (1970). Species taxa of North American birds. *Publ. Nuttall Ornith. Club* **9**.

Mees, G. F. (1977). Zur Verbreitung von *Phaethornis malaris* (Nordmann) (Aves, Trochilidae). *Zool. Mededelingen* **52**, 209-11.

Meggers, B. J. (1954). Environmental limitation on the development of culture. *Am. Anthropol.* **56**, 801-24.

—— (1971). *Amazonia; man and culture in a counterfeit paradise*. Aldine, Chicago.

—— (1974). Environment and culture in Amazonia. In *Man in the Amazon* (ed. C. Wagley). University of Florida Press, Gainesville.

—— (1975). Application of the biological model of diversification to cultural distributions in tropical lowland South America. *Biotropica* **7**, 141-61.

—— (1977). Vegetational fluctuation and prehistoric cultural adaptation in Amazonia: some tentative correlations. *Wld. Archaeol.* **8**, 287-303.

—— (1979). Climatic oscillation as a factor in the prehistory of Amazonia. *Am. Antiq.* **44**, 252-66.

—— (1982). Archeological and ethnographic evidence compatible with the model of forest fragmentation. In *Biological diversification in the tropics* (ed. G. T. Prance). Columbia University Press, New York.

—— and Estrada, E. (1965). Early formative period of coastal Ecuador. *Smithson. Contrib. Anthropol.* **1**.

—— and Evans, C. (1957). Archeological investigations at the mouth of the Amazon. *Bureau Am. Ethnol. Bull.* **167**, 57-73.

—— —— (1973). A reconstituição da pré-história amazônica. Algumas considerações teóricas. In *O Museu Goeldi no*

Ano do Sesquicentenário, Museu Paraense Emílio Goeldi, Publs. Avulsas, p. 20.

—— —— (1981). Un método cerámico para el reconocimiento de comunidades pre-históricas. *Boletin del Museo del Hombre Dominicano* 14.

Métraux, A. (1948). The hunting and gathering tribes of the Rio Negro Basin. *Handbook of South American Indians*, Vol. 3, pp. 861–7. Bureau of American Ethnology Bull. 143. Smithsonian Institution, Washington, DC.

Michael, O. (1911*a*). Lebensweise und Gewohnheiten der Morpho des Amazonasgebietes. *Fauna exot.* 1, 10–17.

—— (1911*b*). Beobachtungen über Vorkommen und Lebensweise der Agriasarten des Amazonasgebietes. *Fauna exot.* 1, 17–23.

—— (1912). Ueber die Lebensweise der Heliconiden. *Fauna exot.* 2, 8, 10–19, 21–2.

—— (1914–15). Die Papilio des Amazonasgebiets. *Ent. Z.* 27, 304–5; 28, 21–2, 31–3, 43–5, 52–4, 65–8, 79–80.

Mielke, O. H. H. (1971). Contribuição ao estudo faunístico dos Hesperiidae americanos – II. Distribição geográfica das espécies de *Aguna* Williams, 1927, com descrição de uma espécie nova e um novo sinónimo. (Lepidoptera: Hesperiidae). *Arq. Mus. Nac., Rio de Janeiro* 54, 203–9.

—— and Brown, K. S. Jr (1979). *Suplemento ao 'Catálogo dos Ithomiidae Americanos (Lepidoptera)' de R. Ferreira D'Almeida (Nymphalidae: Ithomiinae)*. Universidade Federal do Paraná, Curitiba.

Migliazza, E. C. (1982). Linguistic prehistory and the refugia model. In *Biological diversification in the tropics* (ed. G. T. Prance). Columbia University Press, New York.

Miller, D. B. and Feddes, R. G. (1971). *Global atlas of relative cloud cover 1967–1970 based on data from meteorological satellites*. US Department of Commerce, US Air Force, Washington.

Miller, L. D. (1965). Systematics and zoogeography of the genus *Phanus* (Hesperiidae) *J. Res. Lepid.* 4, 115–30.

—— (1968). The higher classification, phylogeny and zoogeography of the Satyridae (Lepidoptera). *Mem. Am. ent. Soc.* 24, 1–174.

Milliman, J. D., Summerhayes, C. P., and Barretto, H. T. (1975). Quaternary sedimentation on the Amazon continental margin: a model. *Bull. geol. Soc. Am.* 86, 610–14.

Mohr, E. C. J. and Van Baren, F. A. (1954). *Tropical soils*. W. Van Hoeve/Interscience, The Hague/London.

Money, D. C. (1965). *Climate, soils and vegetation*. University Tutorial, London.

Monroe, B. L. Jr (1968). A distributional survey of the birds of Honduras. *Ornith. Monogr.* 7.

Moore, H. E. Jr (1973). Palms in the tropical forest ecosystems of Africa and South America, 63–88. In *Tropical forest ecosystems in Africa and South America : a comparative review* (ed. B. J. Meggers, E. S. Ayensu, and W. D. Duckworth). Smithsonian Institution Press, Washington.

Moreau, R. E. (1963). Vicissitudes of the African biomes in the late Pleistocene. *Proc. zool. Soc., Lond.* 141, 395–421.

—— (1966). *The bird faunas of Africa and its islands*. Academic Press, New York.

Morley, T. (1975). The South American distribution of the Memecyleae (Melastomataceae) in relation to the Guiana area and to the question of forest refuges in Amazonia. *Phytologia* 31, 279–96.

Moss, A. M. (1919). The Papilios of Para. *Novitates Zool.* 26, 295–319.

—— (1947). Notes on the Syntomidae of Para, with special reference to wasp mimicry and fedegoso, *Heliotropium indicum* (Boraginaceae), as an attractant. *Entomologist, Lond.* 80, 30–5.

Moulton, J. C. (1909). On some of the principal mimetic (Müllerian) combinations of tropical American butterflies. *Trans. R. ent. Soc. Lond.* 56, 585–606.

Müller, P. (1972). Centres of dispersal and evolution in the Neotropical region. *Stud. Neotrop. Fauna* 7, 173–85.

—— (1973). The dispersal centres of terrestrial vertebrates in the Neotropical realm. *Biogeographica*, Vol. 2. Junk, The Hague.

Murdock, G. P. (1951). South American culture areas. *Southwest. J. Anthropol.* 7, 415–36.

Myers, T. P. (1970). *The late prehistoric period at Yarinacocha, Peru*. University Microfilms International, Ann Arbor.

Nelson, G. (1979). Refuges, humans and vicariance. *Syst. Zool.* 27, 484–7.

—— and Rosen, D. E. (eds.) (1981). *Vicariance biogeography: a critique*. Columbia University Press, New York.

Netto, L. (1885). Investigações sobre a archeologia brasileira. *Archivos Museu Nac., Rio de Janeiro* 6, 257–554.

Nicolay, S. S. (1971*a*). A review of the genus *Arcas* with descriptions of new species (Lycaenidae, Strymonini). *J. lepid. Soc.* 25, 87–108.

—— (1971*b*). A new genus of hairstreak from Central and South America. (Lycaenidae, Theclini). *J. lepid. Soc.* 25, Suppl. 1, 1–39.

—— (1976). A review of the Hübnerian genera *Panthiades* and *Cycnus* (Lycaenidae: Eumaeini). *Bull. Allyn Mus., Sarasota* 35, 1–30.

—— (1977). Studies in the genera of American hairstreaks. 4. A new genus of hairstreak from Central and South America (Lycaenidae: Eumaeini). *Bull. Allyn Mus., Sarasota* 44, 1–24.

—— (1979). Studies in the genera of American hairstreaks. 5. A review of the Hübnerian genus *Parrhasius* and description of a new genus *Michaelus* (Lycaenidae: Eumaeini). *Bull. Allyn Mus., Sarasota* 56, 1–51.

Noble, G. (1965). Proto-Arawakan and its descendents. *Int. J. Am. Linguist.* 31, 3.

Nordenskiold, E. (1913). Urnengräber und Mounds im Bolivianischen Flachländer. *Baessler Arch.* 3, 205–55.

—— (1924). The ethnography of South America seen from Mojos in Bolivia. *Comp. ethnograph. Stud.* 3.

Novaes, F. C. (1957). Notas de ornitologia Amazonica. I. Generos *Formicarius* e *Phlegopsis*. *Bol. Mus. Paraense E. Goeldi, nov. ser., Zool.* 8.

—— (1978). Sobre algumas aves pouco conhecidas da Amazônia brasileira. II. *Bol. Mus. Paraense E. Goeldi, nov. ser., Zool.* 9.

—— (1981). A estrutura da espécie nos periquitos do género *Pionites* Heine (Psittacidae, Aves). *Bol. Mus. Paraense E. Goeldi, nov. ser., Zool.* 106.

Oberg, K. (1953). Indian tribes of northern Mato Grosso, Brazil. *Institute of Social Anthropology Publication* 15.

O'Neill, J. P. (1974). The birds of Balta, a Peruvian dry tropical forest locality, with an analysis of their origins and ecological requirements. Ph.D. thesis, Louisiana State University, Baton Rouge.

Oren, D. C. and Willis, E. O. (1981). New Brazilian records ,for the golden parakeet (*Aratinga guarouba*). *Auk* **98**, 394-6.

Organização dos Estados Americanos (1974). *Marajó; um estudo para o seu desenvolvimento*. Secretaria Geral da Organizacão dos Estados Americanos, Washington, DC.

Otero, L. S. (1971). *Instruções para criação da borboleta 'capitão-do-mato'* (Morpho achillaena), *e outras espécies do gênero* Morpho *('azul-seda', 'bóia', 'azulão-branco', 'praia-grande')*. IBDF, Rio de Janeiro.

Paes de Camargo, A., Remo Alfonsi, A. R., Pinto, H. S., and Chiarini, J. V. (1977). Zoneamento da aptidão climática para culturas comerciais em áreas de cerrado. In *IV Simpósio sobre o Cerrado: bases para Utilização Agropecuária* (ed. M. G. Ferri). Itatiaia/EDUSP, Belo Horizonte.

Palmatary, H. C. (1939). Tapajo pottery. *Ethnographical Studies* 8, Gothenburg, Ethnographical Museum, Goteborg.

—— (1950). The pottery of Marajó Island, Brazil. *Trans. Am. Phil. Soc.* **39** (3).

Papageorgis, C. (1975). Mimicry in neotropical butterflies. *Am. Sci.* **63**, 522-32.

Parker, T. A. (1982). Observations of some unusual rainforest and marsh birds in southeastern Peru. *Wilson Bull.* **94**, 477-93.

Parmenter, C. and Folger, D. W. (1974). Eolian biogenic detritus in deep sea sediments: a possible index of equatorial ice age aridity. *Science, NY* **185**, 695-8.

Paynter, R. A. Jr (1972). Biology and evolution of the *Atlapetes schistaceus* species-group (Aves: Emberizinae). *Bull. Mus. comp. Zool.* **142**, 297-320.

—— (1978). Biology and evolution of the avian genus *Atlapetes* (Emberizinae). *Bull. Mus. comp. Zool.* **148**, 323-69.

Penna, D. S. F. (1879). Apontamentos sobre os cerâmicos do Pará. *Archivos Mus. Nac., Rio de Janeiro* 2, 47-67.

Peterson, G. M., Webb III, T., Kutzbach, J. E., van der Hammen, T., Wijmstra, T. A., and Street, F. A. (1979). The continental record of environmental conditions at 18,000 yr B.P.: an initial evaluation. *Quatern. Res.* **12**, 47-82.

Phelps, W. H. Jr (1968). Contribución al análisis de los elementos que componen la avifauna subtropical de las cordilleras de la costa norte de Venezuela. *Bol. Acad. Cienc. Físicas, Matemáticas y Naturales, Caracas* **26**, 7-43.

Pires, J. M. (1973). Tipos de vegetação de Amazônia. *Publ. Avulsas, Mus. Paraense E. Goeldi* **20**, 179-202.

Pliske, T. E. (1975). Attraction of Lepidoptera to plants containing pyrrolizidine alkaloids. *Environ. Ent.* **4**, 455-73.

Plowman, T. (1979). The genus *Brunfelsia*: a conspectus of the taxonomy and biogeography. In *The biology and taxonomy of the Solanaceae* (ed. J. G. Hawkes, R. N. Lester, and A. D. Skelding). *Linnaean Society Symposium Series*, 7.

—— (1981). Five new species of *Brunfelsia* from South America (Solanaceae). *Fieldiana, Botany, N.S.* **8**, i-vi. 1-16.

Porras, G. I. (1975). El formativo en el valle amazónico del Ecuador: Fase Pastaza. *Revista Univ. Católica, Quito año*

III 74-134.

Potts, R. W. L. (1943). Systematic notes concerning American Acraeinae (Lepidoptera: Nymphalidae). *Pan-Pac. Ent.* **19**, 31-2.

Prance, G. T. (1973). Phytogeographic support for the theory of Pleistocene forest refuges in the Amazon Basin, based on evidence from distribution pattern in Caryocaraceae, Chrysobalanaceae, Dichapetalaceae and Lecythidaceae. *Acta Amazonica* **3**, 5-28.

—— (1975). Flora and vegetation. In *Amazon jungle: green hell to red desert?* (ed. R. J. A. Goodland and H. S. Irwin). Elsevier, Amsterdam.

—— (1977). The phytogeographic subdivisions of Amazonia and their influence on the selection of biological reserves. In *Extinction is forever* (ed. G. T. Prance and T. S. Elias). New York Botanical Garden.

—— (1978). The origin and evolution of the Amazon Flora. *Interciencia* **3**, 297-322.

—— (1979). Notes on the vegetation of Amazonia III. The terminology of Amazonian forest types subject to inundation. *Brittonia* **31**, 26-38.

—— (1981). Discussion. In *Vicariance biogeography: a critique* (ed. G. Nelson and D. E. Rosen), pp. 395-405. Columbia University Press, New York.

—— (1982*a*). Forest refuges: evidences from woody angiosperms. In *Biological diversification in the tropics* (ed. G. T. Prance). Columbia University Press, New York.

—— (ed.) (1982*b*). *Biological diversification in the tropics*. Columbia University Press, New York.

—— (1982*c*). A review of the phytogeographic evidences for Pleistocene climate changes in the Neotropics. *Ann. Miss. Bot. Gard.* **69**, 594-624.

—— and Elias, T. S. (eds.). (197/). *Extinction is forever*. New York Botanical Garden.

—— Rodrigues, W. A., and da Silva, M. F. (1976). Inventario florestal de um hectare de mata de terra firme km 30 Estrada Manaus–Itacoatiara. *Acta Amazonica* **6**, 9-35.

—— and Schubart, H. O. R. (1978). Notes on the vegetation of Amazonia I. A preliminary note on the origin of the white sand campinas of the lower Rio Negro. *Brittonia* **30**, 60-3.

Prell, W. (1973). Evidence for Sargasso sea-like conditions in the Colombia basin, Caribbean Sea, during glacial periods. *Geol. Soc. Am., Abstr. Progr. 1973 Meetings*, Dallas.

Projeto RADAM (1973-75). *Levantamento de Recursos Naturais* 1-7. Ministério de Minas e Energia, Rio de Janeiro.

—— (1974). *Levantamento de Recursos Naturais* 4. Ministério de Minas e Energia, Rio de Janeiro.

Projeto RADAMBRASIL (1975-83). *Levantamento de Recursos Naturais* 8-32. Ministério de Minas e Energia, Rio de Janeiro.

—— (1976). *Levantamento de Recursos Naturais* 12. Ministério de Minas e Energia, Rio de Janeiro.

—— (1977). *Levantamento de Recursos Naturais* 13. Ministério de Minas e Energia, Rio de Janeiro.

Proyecto Radargrametrico del Amazonas (1979). *La Amazonia Colombiana y sus recursos*. Bogotá, Colombia. 4 Vols.

Ratisbona, L. R. (1976). The climate of Brazil. In *Climates of Central and South America. World survey of climatology* (ed. W. Schwerdtfeger) p. 12. Amsterdam, Elsevier.

Ratter, J. A., Richards, P. W., Argent, G., and Gifford, D. R. (1973). Observations on the vegetation of northern Mato Grosso. I. The woody vegetation types of the Xavantina–Cachimbo expedition area. *Phil. Trans. R. Soc.* **266B**, 449–92.

Reichholf, J. (1975). Biogeographie und Ökologie der Wasservögel im subtropisch–tropischen Südamerika. *Anz. Ornith. Ges. Bayern* **14**, 1–69.

Reinke, R. (1962). Das Klima Amazoniens. Ph.D. thesis, Universitat Tübingen.

Reissinger, E. (1972). Zur Taxonomie und Systematik der Gattung *Catasticta* Butler (Lepidoptera, Pieridae). *Ent. Z., Frankfurt* **82**, 97–112, 113–24.

Remington, C. L. (1968). Suture-zones of hybrid interaction between recently joined biotas. In *Evolutionary biology*, Vol. 2 (ed. T. Dobzhansky). Appleton-Century-Crofts, New York.

Remsen, J. V., Jr. and Parker, T.A., III. (1983). Contribution of river-created habitats to bird species richness in Amazonia. *Biotropica* **15**, 223–31.

Richards, P. W. (1952). *The tropical rain forest*. Cambridge University Press.

—— (1969). Speciation in the tropical rain forest and the concept of the niche. *Biol. J. Linn. Soc.* **1**, 149–53.

Robbins, R. K. and Small, G. B. Jr (1981). Wind dispersal of Panamanian hairstreak butterflies (Lepidoptera: Lycaenidae) and its evolutionary significance. *Biotropica* **13**, 308–15.

Rodrigues, A. D. (1955). As línguas 'impuras' da família Tupí-Guarani. *31ᵐᵒ Congresso Internacional de Americanistas, São Paulo. Anais*, Vol. 2, pp 1055–71.

—— (1958). Classification of Tupi-Guarani. *Int. J. Am. Ling.* **24**, 231–4.

—— (1974). Linguistic groups of Amazonia. In *Native South Americans* (ed. P. J. Lyon). Little, Brown, Boston.

Rodrigues, W. A. (1961a). Estudo preliminar de mata várzea alta de uma ilha do baixo Rio Negro de solo argiloso e úmido. *Publ. Bot. INPA* **10**, 1–50.

—— (1961b). Aspectos fitossociológicos das caatingas do Rio Negro. *Bol. Mus. Paraense Emílio Goeldi. Botanica* **15**, 1–41.

Roosevelt, A. C. (1980). *Parmana; prehistoric maize and manioc subsistence along the Amazon and Orinoco*. Academic Press, New York.

Rothschild, W. and Jordan, K. (1906). A revision of the American Papilios. *Novitates Zool.* **13**, 411–752.

Rouse, I. and Allaire, L. (1978). Caribbean. In *Chronologies in New World archaeology*, pp. 431–81. Academic Press, New York.

—— and Cruxent, J. M. (1963). *Venezuelan archaeology*. Yale University Press, New Haven.

Rydén, S. (1950). A study of South American Indian hunting traps. *Revista do Museu Paulista*, NS **4**, 247–352.

Salgado-Labouriau, M. L. (1979). Cambios climáticos durante el Cuaternario tardío paramero y su correlación con las tierras tropicales calientes. In *El Medio Ambiente Paramo* (ed. M. L. Salgado-Labouriau) pp. 67–78. Ediciones Centro de Estudios Avanzados, Mérida, Venezuela.

Sankoff, D. (1973). Parallels between genetics and lexico-statistics. In *Lexicostatistics in genetic linguistics* (ed. E. Dyen). Proceedings of the Yale conference, Yale University, 3–4 April 1971, Mouton, The Hague.

Sanchez, P. (1976). *Properties and management of soils in the tropics*. Wiley, New York.

Sarmiento, G. (1975). The dry plant formations of South America and their floristic connections. *J. Biogeog.* **2**, 233–51.

Sastre, C. (1977). Quelques aspects de la phytogéographie des milieux ouverts guyanais. In *Biogéographie et evolution en Amérique Tropicale* (ed. H. Descimon). *Publ. Lab. zool. Ecol. Norm. Super. Paris.* 9.

Scheuermann, R. G. (1977). Hallazgos del Paují *Crax mitu* (Aves, Cracidae) al norte del Río Amazonas y notas sobre su distribución. *Lozania (Acta zool. colomb.)* 22.

Schimper, A. F. W. (1898). *Pflanzengeographie auf physiologischer Grundlage*. Jena.

—— (1903). *Plant geography upon a physiological basis* (transl. W. R. Fischer, P. Groom, and I. B. Balfour). Oxford.

Schmitz, P. (1980). A evolução da cultura no sudoeste de Goiás. *Pesquisas, Antropologia* **31**, 185–225. Instituto Anchietano de Pesquisas, São Leopoldo.

Schreiber, H. (1978). Dispersal centres of Sphingidae (Lepidoptera) in the Neotropical region. *Biogeographica* **10**, 1–195.

Schubert, C. (1979). La zona del páramo: Morfología glacial y periglacial de los Andes de Venezuela. In *El Medio Ambiente Paramo* (ed. M. L. Salgado-Labouriau) pp. 11–28.

Shapiro, A. M. (1978a). Evidence for obligate monophenism in *Reliquia santamarta*, a neotropical alpine pierine butterfly (Lepidoptera: Pieridae). *Psyche* **84**, 183–90.

—— (1978b). The life history of *Reliquia santamarta*, a neotropical alpine pierine butterfly (Lepidoptera: Pieridae). *J. NY ent. Soc.* **86**, 45–50.

—— (1978c). The life history of an equatorial montane butterfly, *Tatochila xanthodice* (Lepidoptera: Pieridae). *J. NY ent. Soc.* **86**, 51–5.

—— (1978d). Development and phenotypic responses to photoperiod and temperature in an equatorial montane butterfly, *Tatochila xanthodice* (Lepidoptera: Pieridae). *Biotropica* **10**, 297–301.

—— (1979). Notes on the behavior and ecology of *Reliquia santamarta*, an alpine butterfly (Lepidoptera: Pieridae) from the Sierra Nevada de Santa Marta, Colombia, and comparisons with Nearctic alpine Pierini. *Stud. Neotrop. Fauna Env.* **14**, 161–70.

Sheppard, P. M. (1963). Some genetic studies on Müllerian mimics in butterflies of the genus *Heliconius*. *Zoologica, NY* **48**, 145–54.

Shields, O. (1976). Fossil butterflies and the evolution of Lepidoptera. *J. Res. Lepid.* **15**, 132–43.

—— and Dvorak, S. K. (1979). Butterfly distribution and continental drift between the Americas, the Caribbean and Africa. *J. nat. Hist., Lond.* **13**, 221–50.

Short, L. L. (1972). Relationships among the four species of the superspecies *Celeus elegans* (Aves, Picidae). *Am. Mus. Novitates* 2487.

—— (1974). Relationship of *Veniliornis 'cassini' chocoensis* and *V. 'cassini' caquetanus* with *V. affinis*. *Auk* **91**, 631–4.

—— (1975). A zoogeographic analysis of the South American Chaco avifauna. *Bull. Am. Mus. nat. Hist.* **154**, 163–352.

—— (1980). Speciation in South American Woodpeckers.

Acta 17th Congr. Intern. Ornith., Berlin, 1978, pp. 1268-72.

Sick, H. (1965). A fauna do cerrado. *Arq. Zool., São Paulo* **12**, 71–93.

—— (1966). As aves do cerrado como fauna arbóricola. *Anais Acad. bras. Ciênc.* **38**, 355–63.

—— (1980). Characteristics of the razor-billed curassow (*Mitu mitu mitu*). *Condor* **82**, 227–8.

—— (1985). Observations on the Andean-Patagonian component of southeastern Brazil's avifauna. *Ornith. Monogr.* **36**, 233–37.

Silverstone, P. A. (1975). A revision of the poison-arrow frogs of the genus *Dendrobates* Wagler. *Nat. Hist. Mus. Los Angeles Cty., Sci. Bull.* **21**, 1–55.

—— (1976). A revision of the poison-arrow frogs of the genus *Phyllobates* Bibron in Sagra (family Dendrobatidae). *Nat. Hist. Mus. Los Angeles Cty., Sci. Bull.* **27**, 1–53.

Simões, M. F. (1967). Considerações preliminares sobre a arqueologia do Alto Xingu (Mato Grosso). Programa Nacional de Pesquisas Arqueologicas, Resultados preliminares do primeiro ano, 1965–66. *Publ. Avulsas Mus. Paraense E. Goeldi* **6**, 129–44.

—— (1974). Contribuição à arqueologia dos arredores do baixo rio Negro. In Programa Nacional de Pesquisas Arqueológicas, Resultados Preliminares do Quinto Ano, 1969–1970. *Publ. Avulsas Mus. Paraense E. Goeldi* **26**, 165–88.

Simpson, B. B. (1975). Pleistocene changes in the flora of the high tropical Andes. *Palaeobiology* **1**, 273–94.

—— (1979). Quaternary biogeography of the high montane areas of South America. In *The South American Herpetofauna: its origin, evolution and dispersal* (ed. Duellman). Mus. Nat. Hist. Monograph 7.

—— and Haffer, J. (1978). Speciation patterns in the Amazon forest biota. *A. Rev. Ecol. Syst.* **9**, 497–518.

Simpson, D. R. (1972). Especiación en las plantas leñosas de la Amazonia peruana relacionada a las fluctuaciones climaticas durante el pleistoceno. *Resumenes do I Congresso Latinamericano de Botanica, México* 107.

Simpson-Vuilleumier, B. (1971). Pleistocene changes in the fauna and flora of South America. *Science, NY* **173**, 771–80.

Slud, P. (1960). The birds of Costa Rica: distribution and ecology. *Bull. Am. Mus. nat. Hist.* **128**, 1–430.

—— (1976). Geographic and climatic relationships of avifaunas with special reference to comparative distribution in the Neotropics. *Smithson. Contr. Zool.* 212.

Smiley, J. (1978). Plant chemistry and the evolution of host specificity: new evidence from *Heliconius* and *Passiflora*. *Science, NY* **201**, 745–7.

Snow, D. W. (1975). The classification of the manakins. *Bull. Br. ornith. Club* **95**, 20–7.

—— (1977). Duetting and other synchronized displays of the blue-backed manakins *Chiroxiphia* spp. In *Evolutionary Ecology* (ed. B. Stonehouse and C. Perrins). University Park Press, Baltimore.

—— (1982). *The cotingas, bellbirds, umbrellabirds and their allies*. British Museum (Natural History), London. Oxford University Press.

Soderstrom, T. R. and Calderón, C. E. (1974). Primitive forest grasses and evolution of the Bambusoideae. *Biotropica* **6**, 141–53.

Sombroek, P. (1966). *Amazon soils: a reconnaissance of the soils of the Brazilian Amazon region*. Centre Agr. Publ. & Document., Wageningen.

Sorensen, A. P. Jr (1967). Multilingualism in the Northwest Amazon. *Am. Anthropol.* **69**, 670–84.

—— (1973). South American Indian linguistics at the turn of the seventies. In *Peoples and cultures of native South America* (ed. D. R. Gross). Doubleday/The Natural History Press, New York.

Spassky, B., Richmond, R. C., Pérez-Salas, S., Pavlovsky, O., Mourão, C. A., Hunter, A. S., Hoenigsberg, H., Dobzhansky, T., and Aýala, F. J. (1971). Geography of the sibling species related to *Drosophila willistoni*, and of the semispecies of the *Drosophila paulistorum* complex. *Evolution* **25**, 129–43.

Stark, N. (1970). The nutrient content of plants and soils from Brazil and Suriname. *Biotropica* **2**, 51–60.

—— (1971*a*). Nutrient cycling. I. Elemental content of soils from South America. *Int. J. trop. Ecol.* **12**, 24–50.

—— (1971*b*). Nutrient cycling. II. Elemental content of plants from South America. *Int. J. trop. Ecol.* **12**, 117–21.

Stebbins, G. L. (1971). *Chromosomal evolution in higher plants*. Arnold, London.

Steward, J. H. (1948). Culture areas of the Tropical Forests. *Handbook of South American Indians*, Vol. 3, pp. 883–99. Bureau of American Ethnology Bull. 143. Smithsonian Institution, Washington, DC.

—— and Faron, L. C. (1959). *Native peoples of South America*. McGraw Hill, New York.

Steyermark, J. A. (1976). Areas de bosques humedos de Venezuela que requeiren proteccion. In *Conservación de los Bosques-Húmedos de Venezuela* (ed. L. S. Hamilton). Sierra Club/Consejo de Bienestar Rural, Caracas.

—— (1979). Plant refuge and dispersal centers in Venezuela: their relict and endemic element. In *Tropical botany* (ed. K. Larsen and L. B. Holm-Nielsen). Academic Press, London.

—— (1982). Relationships of some Venezuelan forest refuges with lowland tropical floras. In *Biological diversification in the tropics* (ed. G. T. Prance). Columbia University Press, New York.

Street, F. A. (1981). Tropical palaeoenvironments. *Progress in Physical Geography* **5**, 157–85.

Swadesh, M. (1959). Mapas de clasificación lingüística de México y las Américas. *Cuadernos del Instituto de Historia, Serie Antropológica* 8.

Takeuchi, M. (1960). A estrutura da vegetação na Amazonia. 1. A mata pluvial tropical. *Bol. Mus. Paraense Emílio Goeldi. Botanica* **6**, 1–17.

Templeton, A. (1981). Mechanisms of speciation – a population genetic approach. *A. Rev. Ecol. Syst.* **12**, 23–48.

Terborgh, J. (1971). Distribution on environmental gradients: theory and a preliminary interpretation of distributional patterns in the avifauna of the Cordillera Vilcabamba, Peru. *Ecology* **52**, 23–40.

——, Fitzpatrick, J. W., and Emmons, L. (1984). Annotated checklist of bird and mammal species of Cocha Cashu Biological Station, Manu National Park, Peru. *Fieldiana, Zool. N.S.* No. 21, 29p.

—— and Weske, J. S. (1975). The role of competition in the distribution of Andean birds. *Ecology* **56**, 562–76.

—— and Winter, B. (1982). ˙Evolutionary circumstances of

species with small ranges. In *Biological diversification in the tropics* (ed. G. T. Prance). Columbia University Press, New York.

Tindale, N. B. (1981). Origin of the Lepidoptera, with description of a new mid-Triassic species and notes on the origin of the butterfly stem. *J. lepid. Soc.* **34**, 263–85.

Toledo, V. M. (1976). Los cambios climaticos del Pleistoceno y sus efectos sobre la vegetación tropical calida y humeda de México. M.S. thesis, Univ. Nacional Antonoma de Mexico.

—— (1982). Pleistocene changes of vegetation in tropical Mexico. In *Biological diversification in the tropics* (ed. G. T. Prance). Columbia University Press, New York.

Tosi, J. Jr (1960). Zonas de vida natural en el Perú. OEA, Zona Andina, Boletín Técnico 5. Lima.

Traylor, M. A., Jr. (1985). Species limits in the *Ochthoeca diadema* species-group (Tyrannidae). *Ornith. Monogr.* **36**, 431–42.

Trewartha, G. T. (1954). *An introduction to climate.* McGraw Hill, New York.

—— (1961). *The earth's problem climates.* University of Wisconsin Press.

Tricart, J. (1974). Existence de périodes sèches au Quaternaire en Amazonie et dans les régions voisines. *Rev. Géomorph. dynam.* **4**, 145–58.

Turner, J. R. G. (1965). Evolution of complex polymorphism and mimicry in distasteful South American butterflies. *Proc. XII Int. Congr. Entomol., London* 1964, p. 267.

—— (1966). A rare mimetic *Heliconius* (Lepidoptera: Nymphalidae). *Proc. R. ent. Soc., Lond. B* **35**, 128–32.

—— (1967). A little-recognized species of *Heliconius* butterfly (Nymphalidae). *J. Res. Lepid.* **5**, 97–112.

—— (1968). Natural selection for and against a polymorphism which interacts with sex. *Evolution* **22**, 481–95.

—— (1971). Two thousand generations of hybridisation in a *Heliconius* butterfly. *Evolution* **25**, 471–82.

—— (1972). The genetics of some polymorphic forms of the butterflies *Heliconius melpomene* (Linnaeus) and *H. erato* (Linnaeus). II. The hybridization of subspecies of *H. melpomene* from Suriname and Trinidad. *Zoologica, NY* **56**, 125–57.

—— (1974). Breeding *Heliconius* in a temperate climate. *J. lepid. Soc.* **28**, 26–33.

—— (1976). Muellerian mimicry: classical 'beanbag' evolution and the role of ecological islands in adaptive race formation. In *Population genetics and ecology* (ed. S. Karlin and E. Nevo). Academic Press, London.

—— (1977). Forest refuges as ecological islands: disorderly extinction and the adaptive radiation of muellerian mimics. In *Biogéographie et evolution en Amérique tropicale* (ed. H. Descimon). *Publ. Lab. zool. École Norm. Super., Paris* **9**, 98–117.

—— (1978). Butterfly mimicry: the genetical evolution of an adaptation. In *Evolutionary biology* Vol. 10 (ed. M. K. Hecht, W. C. Steere, and B. Wallace). Plenum Press, New York.

—— (1982). How do refuges produce biological diversity? Allopatry and parapatry, extinction and gene flow in mimetic butterflies. In *Biological diversification in the tropics* (ed. G. T. Prance). Columbia University Press, New York.

—— and Crane, J. (1963). The genetics of some polymorphic forms of the butterflies *Heliconius melpomene* Linnaeus and *H. erato* Linnaeus. I. Major genes. *Zoologica, NY* **47**, 141–52.

Udvardy, M. D. F. (1969). *Dynamic zoogeography.* Reinhold, New York.

Uhle, M. (1920). Los principios de la civilización en la sierra peruana. *Bol. Acad. Nac. Historia* **1**, 44–56.

Van der Hammen, T. (1974). The Pleistocene changes of vegetation and climate in tropical South America. *J. Biogeogr.* **1**, 3–26.

—— (1979). Changes in life conditions on earth during the past one million years. *Biol. Skrifter (Danske Videnskabernes Selskab)* **22** (6), 1–22.

—— (1983). The paleoecology and paleogeography of savannas. In *Tropical savannas* (ed. F. Bourlière). *Ecosystems of the world* vol. 13. Elsevier, The Hague.

—— and Gonzales, E. (1960). Upper Pleistocene and Holocene climate and vegetation of the 'Sabana de Bogotá' (Colombia, South America). *Leidse Geol. Meded.* **25**, 261–315.

—— and Gonzales, T. (1964). A pollen diagram from the Quaternary of the Sabana de Bogotá (Colombia) and its significance for the geology of the northern Andes. *Geol. Mijnbouw* **43**, 113–17.

—— Werner, J. H., and van Dommelen, H. (1973). Palynological record of the upheaval of the northern Andes: a study of the Pliocene and Lower Quaternary of the Colombian Eastern Cordillera and the early evolution of its high-Andean biota. *Rev. Palaeobot. Palynol.* **16**, 1–122.

Vane-Wright, R. I. (1979). The coloration, identification and phylogeny of Nessaea butterflies (Lepidoptera: Nymphalidae). *Bull. Br. Mus. nat. Hist. Ent.* **38**, 27–56.

Van Geel, B. and van der Hammen, T. (1973). Upper Quaternary vegetational and climatic sequence of the Fúquene area (Eastern Cordillera, Colombia). *Palaeogeogr. Palaeoclimatol. Palaeoecol.* **14**, 9–92.

Van Wambeke, A. (1978). Properties and potentials of soils in the Amazon Basin. *Interciencia* **3**, 233–42.

Vanzolini, P. E. (1970). Zoologia sistemática, geografia e a origem das espécies. *Inst. Geografico São Paulo. Serie téses e monografias* **3**, 1–56.

—— (1973). Paleoclimates, relief, and species multiplication in tropical forests. In *Tropical forest ecosystems in Africa and South America: a comparative review* (ed. B. J. Meggers, E. S. Ayensu, and W. D. Duckworth). Smithsonian Institution Press, Washington.

—— (1974). Ecological and geographical distribution of lizards in Pernambuco, northeastern Brasil (Sauria). *Pap. Avulsos Zool., São Paulo* **28**, 61–90.

—— (1976). On the lizards of a cerrado-caatinga contact: evolutionary and zoogeographical implications. *Pap. Avulsos Zool., São Paulo* **29**, 111–19.

—— (1980). Algumas questões ecológicas ligadas à conservação da natureza no Brazil. *Interfacies, São José do Rio Preto, São Paulo* **21**, 1–23.

—— (1981). A quasi-historical approach to the natural history of the differentiation of reptiles in tropical geographical isolates. *Pap. Avulsos Zool., São Paulo* **34**, 189–204.

—— and Ab'Sáber, A. N. (1968). Divergence rate in South American lizards of the genus *Liolaemus* (Sauria, Iguanidae). *Pap. Avulsos Zool., São Paulo* **21**, 205–8.

—— and Williams, E. E. (1970). South American anoles: geographic differentiation and evolution of the *Anolis chrysolepis* species group (Sauria, Iguanidae). *Arq. Zool., São Paulo* **19**, 1-298.

—— —— (1981). The vanishing refuge: a mechanism for eco-geographic speciation. *Papeis Avulsos Zool., São Paulo* **34**, 251-5.

Vargas Arenas, I. (1979). La tradición saladoide del oriente de Venezuela; la Fase Cuartel. *Biblioteca de la Academia Nacional de la Historia, Serie: Estudios, Monografías y Ensayos* 5.

Vaurie, C. (1966*a*). Systematic notes on the bird family Cracidae, no. 5: *Penelope purpurascens, Penelope jacquaçu*, and *Penelope obscura. Am. Mus. Novit.* 2250.

—— (1966*b*). Systematic notes on the bird family Cracidae, no. 6: Reviews of nine species of *Penelope. Am. Mus. Novit.* 2251.

—— (1967). Systematic notes on the bird family Cracidae, no. 10: the genera *Mitu* and *Pauxi* and the generic relationships of the Cracini. *Am. Mus. Novit.* 2307.

—— (1980). Taxonomy and geographical distribution of the Furnariidae (Aves, Passeriformes). *Bull. Am. Mus. nat. Hist.* **166**, 1-357.

Velloso, H. P. (1966). *Atlas Florestal do Brasil.* Rio de Janeiro.

—— Japiassu, A. M. S., Goes Filho, L. and Leite, P. F. (1974). As regiões fitoecológicas, sua natureza e seus recursos econômicas. Estudo fitogeográfico. In *Projeto RAĎAM. Folha SB22, Araguaia e parte da folha SC22 Tocantins.* Rio de Janeiro, Levantamento de Recursos Naturais. Vol. 4.

Verneau, R. (1920). Sur la répartition en Amérique des poteries décorées au 'champlevé'. *J. Soc. Am. Paris* **12**, 1-10.

Verstappen, H. T. (1975). On paleo-climates and landform development in Malesia. In *Modern Quaternary research in Southeast Asia* (ed. G. J. Bartstra and W. A Caspari). Balkema, Amsterdam.

Vuilleumier, F. (1965). Relationships and evolution within the Cracidae (Aves, Galliformes). *Bull. Mus. comp. Zool.* **134**, 1-27.

—— (1969). Pleistocene speciation in birds living in the high Andes. *Nature, Lond.* **223**, 1179-80.

—— (1970). Insular biogeography in continental regions. I. The northern Andes of South America. *Am. Nat.* **104**, 373-88.

—— (1971). Generic relationships and speciation patterns in *Ochthoeca, Myiotheretes, Xolmis, Neoxolmis, Agriornis*, and *Muscisaxicola.* In *Evolutionary relationships of some South American ground tyrants* (ed. W. J. Smith and F. Vuilleumier). *Bull. Mus. comp, Zool.* 141.

—— (1972). Speciation in South American birds: a progress report. *Acta IV Congr. Latin. Zoologica 1 ('1970')* 239-55.

—— (1975). Zoogeography. In *Avian biology*, Vol. 5 (ed. D. S. Farner and J. R. King). Academic Press, New York.

—— (1980). Speciation in birds of the high Andes. *Acta 17th Congr. Intern. Ornith., Berlin 1978*, pp. 1256-61.

—— (1985). Forest birds of Patagonia: Ecological geography, speciation, endemism, and faunal history. *Ornith. Monogr.* **36**, 255-304.

—— and Simberloff, D. (1980). Ecology versus history as determinants of patchy and insular distributions in high Andean birds. In *Evolutionary biology*, Vol. 12 (ed.

M. K. Hecht, W. C. Steere, and B. Wallace). Plenum Press, New York.

Walker, D. (ed.) (1972). *Bridge and barrier: the natural and cultural history of Torres Strait.* Australian National University Research School of Pacific Studies, Dept. of Biogeography and Geomorphology, Pubn. B9/3, Canberra.

—— (1982). Speculations on the origin and evolution of Sunda-Sahul rain forests. In *Biological diversification in the tropics* (ed. G. T. Prance). Columbia University Press, New York.

Webb, S. D. (1978). A history of savanna vertebrates in the New World. Part II: South America and the great interchange. *A. Rev. Ecol. Syst.* **9**, 393-426.

Webster, P. J. and Streten, N. A. (1978). Late Quaternary ice age climates of tropical Australasia: interpretations and reconstructions. *Quatern. Res.* **10**, 279-309.

Welty, J. C. (1975). *The life of birds.* Philadelphia.

Wesley, D. J. and Emmel, T. C. (1975). The chromosomes of neotropical butterflies from Trinidad and Tobago. *Biotropica* **7**, 24-31.

Wetterberg, G. B., Castro, C. S. de, Quintão, A. T. B., and Porto, E. R. (1978). Estado atual dos parques nacionais e reservas equivalentes na América do Sul – 1978. *Brasil Florestal* **36**, 11-36. [Revised English translation; 1980. The 1978 status of national parks and equivalent reserves in South America. Int. Park Affairs Div., Natl. Park Service, USDI, Washington, DC.]

—— and Jorge Padua, M. T. (1978). Preservação da natureza na Amazônia brasileira. Situação em 1978. PRODEPEF Série Técnica (PNUD/FAO/IBDF/BRA-545) 13. Brasília, IBDF.

Weyl, R. (1956). Eiszeitliche Gletscherspuren in Costa Rica (Mittelamerika). *Z. Gletscherkunde Glazialgeologie* **3**, 317-25.

White, F. (1962). Geographic variation and speciation in Africa with particular reference to Diospyros. *Syst. Assoc. Publ.* 4, pp. 71-103.

White, M. J. D. (1978). *Modes of speciation.* Freeman, San Francisco.

Whitehead, D. J. (1976). Classification and evolution of *Rhinochenus* Lucas (Coleoptera: Curculionidae: Cryptorhynchiinae), and Quaternary Middle American zoogeography. *Quaest. Entom.* **12**, 118-201.

Whitmore, T. C. (1975). *Tropical rain forests of the Far East.* Clarendon Press, Oxford.

—— (1981). Palaeoclimate and vegetation history. In *Wallace's line and plate tectonics* (ed. T. C. Whitmore). Clarendon Press, Oxford.

Wijmstra, T. A. (1967). A pollen diagram from the Upper Holocene of the lower Magdalena Valley, Colombia. *Leidse Geol. Meded.* **39**, 261-7.

—— (1971). The palynology of the Guiana coastal basin. *Adademisch Proefschrift, Univ. Amsterdam.*

—— and van der Hammen, T. (1966). Palynological data on the history of tropical savannas in northern South America. *Leidse Geol. Meded.* **38**, 71-83.

Willey, G. R. (1971). *An introduction to American archaeology. 2. South America.* Prentice Hall, Englewood Cliffs.

Williams, E. E. and Vanzolini, P. E. (1980). Notes and biogeographic comments on anoles from Brasil. *Pap. Avulsos Zool., São Paulo* **34**, 99-108.

Williams, M. D. (1980). First description of the eggs of the white-winged guan, *Penelope albipennis*, with notes on its nest. *Auk* **97**, 889–92.

Willis, E. O. (1968). Studies of the behavior of lunulated and Salvin's antbirds. *Condor* **70**, 128–48.

–– (1969). On the behavior of five species of *Rhegmatorhina*, ant-following antbirds of the Amazon basin. *Wilson Bull.* **81**, 363–95.

–– (1979). Comportamento e ecologia da mãe-de-taoca, *Phlegopsis nigromaculata* (D'Orbigny & Lafresnaye (Aves, Formicariidae). *Rev. Brasil. Biol.* **39**, 117–59.

Winge, H. (1973). Races of *Drosophila willistoni* sibling species: probable origin in Quaternary forest refuges of South America. *Genetics* **74**, Suppl., 297–8.

Young, A. M. (1973). Notes on the comparative ethology and ecology of several species of *Morpho* butterflies in Costa Rica. *Stud. neotrop. Fauna* **8**, 17–50.

–– and Muyshondt, A. (1972). Geographical and ecological expansion in tropical butterflies of the genus *Morpho* in evolutionary time. *Rev. Biol. Trop.* **20**, 231–63.

Zeuner, F. E. (1962). Notes on the evolution of the Rhopalocera (Lep.) *Proc. XI Congr. Entomol. Wien*, Vol. 1, pp. 310–13.

Zimmer, J. T. (1936). Studies of Peruvian birds no. 22: Notes on the Pipridae. *Am. Mus. Novit.* 889.

SUPPLEMENTARY REFERENCES
added in proof

Absy, M. L. (1985). Palynology of Amazonia: The history of the forests as revealed by the palynological record: In *Amazonia* (ed. G. T. Prance and T. E. Lovejoy). Pergamon Press, Oxford and New York.

Beven, S., Connor, E. F., and Beven, K. (1984). Avian biogeography in the Amazon basin and the biological model of diversification. *J. Biogeog.* **11**, 383–99.

Bigarella, J. J. and Ferreira, A. M. M. (1985). Amazonian geology and the Pleistocene and the Cenozoic environments and paleoclimates. In *Amazonia*, (ed. G. T. Prance and T. E. Lovejoy). Pergamon Press, Oxford and New York.

Bigarella, J. J., de Andrade-Lime, D., and Riehs, P. J. (1981). Consideraçoes a respeito das mudancas paleoambientais na distribução de algumas especies vegetais e animais no Brasil. *Anais Acad. Brasil. Cienç.* (1975) **47**, Supl., 411–64.

Binford, M. W. (1982). Ecological history of Lake Valencia, Venezuela: Interpretation of animal microfossils and some chemical, physical, and geological features. *Ecol. Monogr.* **52**, 307–33.

Cerqueira, R. (1982). South American landscapes and their mammals. In *Mammalian biology in South America* (ed. M. A. Mares and H. H. Genoways). Spec. Pub. 6, Pymatuning Laboratory of Ecology, University of Pittsburgh.

Connor, E. F. and McKenney, M. S. (1985). The biological model of diversification and the biogeography of endemism. *Geol. Soc. Am., Ann. Meeting*, abstr.

Cracraft, J. (1982). Geographic differentiation, cladistics, and vicariance biogeography: reconstructing the tempo and mode of evolution. *Am. Zool.* **22**, 411–24.

–– (1983). Cladistic analysis and vicariance biogeography. *Amr. Sci.* **71**, 273–81.

–– (1985*b*). Biological diversification and its causes. *Ann. Missouri Bot. Garden* **72**, 794–822.

Graves, G. R. (1985). Elevational correlates of speciation and intraspecific geographic variation in plumage in Andean forest birds. *Auk* **102**, 556–79.

Hooghiemstra, H. (1984). *Vegetational and climatic history of the high plain of Bogotá, Colombia: A continuous record of the last 3.5 million years*. Cramer, Liechtenstein.

Klammer, G. (1984). The relief of the extra-Andean Amazon basin. In *The Amazon. Limnology and landscape ecology of a mighty tropical river and its basin* (ed. H. Sioli). Monogr. Biol. 56. Junk, Dordrecht. Boston, Lancaster.

Lewin, R. (1984). Fragile forests implied by Pleistocene pollen. *Science* **226**, 36–7.

Liu, Kam-biu and Colinvaux, P. A. (1985). Forest changes in the Amazon basin during the last glacial maximum. *Nature, Lond.* **318**, 556–7.

Mayr, E. and O'Hara, R. J. (1986). The biogeographic evidence supporting the Pleistocene forest refuge hypothesis. *Evolution* **40**, 55–67.

Ochsenius, C. 1983. Aridity and biogeography in northernmost South America during the late Pleistocene (peri-Caribbean arid belt, 62°–74° W). *Zbl. Geol. Paläontol.*, Teil 1: 264–78.

Prance, G. T. 1985. The changing forests. In *Amazonia* (ed. G. T. Prance and T. E. Lovejoy), pp. 146–65. Pergamon Press, Oxford and New York.

Prance, G. T. and Lovejoy, T. E. (eds), (1985). *Amazonia, Key Environments*. Pergamon Press, Oxford and New York.

Remsen, J. V., Jr. (1984). High incidence of "leapfrog" pattern of geographic variation in Andean birds: implications for the speciation process. *Science* **224**, 171–3.

Roth, P. (1982). Habitat-Aufteilung bei sumpatrischen Papageien des südlichen Amazonasgebietes. Dissertation Universität Zürich. (Portug. transl., *Acta Amazonica*, 1985).

Salati, E. and Marques, J. (1984). Climatology of the Amazon region, pp. 85–126. In *The Amazon, Limnology and lanscape ecology of a mighty tropical river and its basin* (ed. H. Sioli). Mongr. Biol. 56. Junk, Dordrecht. Boston, Lancaster.

Simpson, B. B. (1983). An historical phytogeography of the high Andean flora. *Revista Chilena de Historia Natural* **56**, 109–22.

Sioli, H. (ed.) (1984). *The Amazon. Limnology and landscape ecology of a mighty tropical river and its basin.* Mongr. Biol. 56. Junk, Dordrecht. Boston, Lancaster.

Smith, R. T. 1982. Quaternary environmental change in equatorial regions with particular reference to vegetation history: a bibliography. *Palaeogeogr., Palaeoclimatol., Palaeocol.* 39, 331–45.

Stehli, F. G. and Webb, S. D. (eds.) 1985. *The Great American Biotic interchange.* Plenum Press, New York and London.

Vuilleumier, F. 1985b. Fossil and Recent avifaunas and the Interamerican Interchange. In *The Great American Biotic Interchange* (ed. F. G. Stehli and S. D. Webb). Plenum Press, New York and London.

INDEX